Lecture Notes in Computer Science 2149

Edited by G. Goos, J. Hartmanis, and J. van Leeuwen

W0091211

Springer
Berlin
Heidelberg
New York
Barcelona
Hong Kong
London
Milan
Paris
Tokyo

Olivier Gascuel Bernard M.E. Moret (Eds.)

Algorithms in Bioinformatics

First International Workshop, WABI 2001
Århus Denmark, August 28-31, 2001
Proceedings

 Springer

Series Editors

Gerhard Goos, Karlsruhe University, Germany
Juris Hartmanis, Cornell University, NY, USA
Jan van Leeuwen, Utrecht University, The Netherlands

Volume Editors

Olivier Gascuel
LIRMM, 161 rue Ada
34392 Montpellier, France
E-mail: gascuel@lirmm.fr

Bernard M.E. Moret
University of New Mexico
Department of Computer Science
Albuquerque, NM 87131, USA
E-mail: moret@cs.unm.edu

Cataloging-in-Publication Data applied for

Die Deutsche Bibliothek - CIP-Einheitsaufnahme

Algorithms in bioinformatics : first international workshop ; proceedings /
WABI 2001, Aarhus, Denmark, August 28 - 31, 2001. Olivier Gascuel ; Bernard
M. E. Moret (ed.). - Berlin ; Heidelberg ; New York ; Barcelona ; Hong Kong ;
London ; Milan ; Paris ; Singapore ; Tokyo : Springer, 2001
  (Lecture notes in computer science ; Vol. 2149)
  ISBN 3-540-42516-0

CR Subject Classification (1998): F.1, E.1, F.2.2, G.1, G.2.1, G.3, J.3

ISSN 0302-9743
ISBN 3-540-42516-0 Springer-Verlag Berlin Heidelberg New York

Springer-Verlag Berlin Heidelberg New York
a member of BertelsmannSpringer Science+Business Media GmbH

http://www.springer.de

© Springer-Verlag Berlin Heidelberg 2001
Printed in Germany

Typesetting: Camera-ready by author, data conversion by Christian Grosche, Hamburg
Printed on acid-free paper   SPIN 10958735   06/3111   5 4 3 2 1

Preface

We are very pleased to present the proceedings of the *First Workshop on Bioinformatics (WABI 2001)*, which took place in Aarhus on August 28–31, 2001, under the auspices of the *European Association for Theoretical Computer Science (EATCS)* and the Danish Center for *Basic Research in Computer Science (BRICS)*.

The *Workshop on Algorithms in Bioinformatics* covers research on all aspects of algorithmic work in bioinformatics. The emphasis is on discrete algorithms that address important problems in molecular biology. These are founded on sound models, are computationally efficient, and have been implemented and tested in simulations and on real datasets. The goal is to present recent research results, including significant work-in-progress, and to identify and explore directions of future research. Specific topics of interest include, but are not limited to:

- Exact and approximate algorithms for genomics, sequence analysis, gene and signal recognition, alignment, molecular evolution, structure determination or prediction, gene expression and gene networks, proteomics, functional genomics, and drug design.
- Methods, software and dataset repositories for development and testing of such algorithms and their underlying models.
- High-performance approaches to computationally hard problems in bioinformatics, particularly optimization problems.

A major goal of the workshop is to bring together researchers spanning the range from abstract algorithm design to biological dataset analysis, to encourage dialogue between application specialists and algorithm designers, mediated by algorithm engineers and high-performance computing specialists. We believe that such a dialogue is necessary for the progress of computational biology, inasmuch as application specialists cannot analyze their datasets without fast and robust algorithms and, conversely, algorithm designers cannot produce useful algorithms without being aware of the problems faced by biologists. Part of this mix was achieved automatically this year by colocating into a single large conference, *ALGO 2001*, three workshops: *WABI 2001*, the *5th Workshop on Algorithm Engineering (WAE 2001)*, and the *9th European Symposium on Algorithms (ESA 2001)*, and sharing keynote addresses among the three workshops. *ESA* attracts algorithm designers, mostly with a theoretical leaning, while *WAE* is explicitly targeted at algorithm engineers and algorithm experimentalists.

These proceedings reflect such a mix. We received over 50 submissions in response to our call and were able to accept 23 of them, ranging from mathematical tools through to experimental studies of approximation algorithms and reports on significant computational analyses. Numerous biological problems are dealt with, including genetic mapping, sequence alignment and sequence analysis, phylogeny, comparative genomics, and protein structure.

We were also fortunate to attract Dr. Gene Myers, Vice-President for Informatics Research at Celera Genomics, and Prof. Jotun Hein, Aarhus University, to address the joint workshops, joining five other distinguished speakers (Profs. Herbert Edelsbrunner and Lars Arge from Duke University, Prof. Susanne Albers from Dortmund University, Prof. Uri Zwick from Tel Aviv University, and Dr. Andrei Broder from Alta Vista). The quality of the submissions and the interest expressed in the workshop is promising – plans for next year's workshop are under way.

We would like to thank all the authors for submitting their work to the workshop and all the presenters and attendees for their participation. We were particularly fortunate in enlisting the help of a very distinguished panel of researchers for our program committee, which undoubtedly accounts for the large number of submissions and the high quality of the presentations. Our heartfelt thanks go to all:

Craig Benham (Mt Sinai School of Medicine, New York, USA)
Mikhail Gelfand (Integrated Genomics, Moscow, Russia)
Raffaele Giancarlo (U. di Palermo, Italy)
Michael Hallett (McGill U., Canada)
Jotun Hein (Aarhus U., Denmark)
Michael Hendy (Massey U., New Zealand)
Inge Jonassen (Bergen U., Norway)
Junhyong Kim (Yale U., New Haven, USA)
Jens Lagergren (KTH Stockholm, Sweden)
Edward Marcotte (U. Texas Austin, USA)
Satoru Miyano (Tokyo U., Japan)
Gene Myers (Celera Genomics, USA)
Marie-France Sagot (Institut Pasteur, France)
David Sankoff (U. Montreal, Canada)
Thomas Schiex (INRA Toulouse, France)
Joao Setubal (U. Campinas, Sao Paolo, Brazil)
Ron Shamir (Tel Aviv U., Israel)
Lisa Vawter (GlaxoSmithKline, USA)
Martin Vingron (Max Planck Inst. Berlin, Germany)
Tandy Warnow (U. Texas Austin, USA)

In addition, the opinion of several other researchers was solicited. These subreferees include Tim Beissbarth, Vincent Berry, Benny Chor, Eivind Coward, Ingvar Eidhammer, Thomas Faraut, Nicolas Galtier, Michel Goulard, Jacques van Helden, Anja von Heydebreck, Ina Koch, Chaim Linhart, Hannes Luz, Vsevolod Yu, Michal Ozery, Itsik Pe'er, Sven Rahmann, Katja Rateitschak, Eric Rivals, Mikhail A. Roytberg, Roded Sharan, Jens Stoye, Dekel Tsur, and Jian Zhang. We thank them all.

Lastly, we thank Prof. Erik Meineche-Schmidt, BRICS codirector, who started the entire enterprise by calling on one of us (Bernard Moret) to set up the workshop and who led the team of committee chairs and organizers through the setup, development, and actual events of the three combined workshops, with the assistance of Prof. Gerth Brødal.

We hope that you will consider contributing to *WABI 2002*, through a submission or by participating in the workshop.

June 2001   Olivier Gascuel and Bernard M.E. Moret

Table of Contents

An Improved Model for Statistical Alignment

István Miklós[1] and Zoltán Toroczkai[2]

[1] Department of Plant Taxonomy and Ecology
Eötvös University, Ludovika tér 2, H-1083 Budapest, Hungary
miklosi@ludens.elte.hu
[2] Theoretical Division and Center for Nonlinear Studies
Los Alamos National Laboratory, Los Alamos, NM87545, USA
toro@lanl.gov

Abstract. The statistical approach to molecular sequence evolution involves the stochastic modeling of the substitution, insertion and deletion processes. Substitution has been modeled in a reliable way for more than three decades by using finite Markov-processes. Insertion and deletion, however, seem to be more difficult to model, and the recent approaches cannot acceptably deal with multiple insertions and deletions. A new method based on a generating function approach is introduced to describe the multiple insertion process. The presented algorithm computes the approximate joint probability of two sequences in $O(l^3)$ running time where l is the geometric mean of the sequence lengths.

1 Introduction

The traditional sequence analysis [1] needs proper evolutionary parameters. These parameters depend on the actual divergence time, which is usually unknown as well. Another major problem is that the evolutionary parameters cannot be estimated from a single alignment. Incorrectly determined parameters might cause unrecognizable bias in the sequence alignment.

One way to break this vicious circle is the maximum likelihood parameter estimation. In the pioneering work of Bishop and Thompson [2], an approximate likelihood calculation was introduced. Several years later, Thorne, Kishino, and Felsenstein wrote a landmark paper [3], in which they presented an improved maximum likelihood algorithm, which estimates the evolutionary distance between two sequences involving all possible alignments in the likelihood calculation. Their 1991 model (frequently referred to as the TKF91 model) considers only single insertions and deletions, but this consideration is rather unrealistic [4,5]. Later it was further improved by allowing longer insertions and deletions [4] in the model, which is usually coined as the TKF92 model. However, this model assumes that sequences contain unbreakable fragments, and only whole fragments are inserted and deleted. As it was shown [4], the fragment model has a flaw: considering unbreakable fragments, there is no possible explanation for overlapping deletions with a scenario of just two events. This problem is solvable by assuming that the ancestral sequence was fragmented independently on both branches immediately after the split, and sequences evolved since then according to the fragment model [6]. However, this assumption does not solve the problem completely: fragments do not

O. Gascuel and B.M.E. Moret (Eds.): WABI 2001, LNCS 2149, pp. 1–10, 2001.

have biological realism. The lack of the biological realism is revealed when we want to generalize this split model for multiple sequence comparison. For example, consider that we have proteins from humans, gorillas and chimps. When we want to analyze the three sequences simultaneously, two pairs of fragmentation are needed: one pair at the gorilla-(human and chimp) split and one at the human-chimp split. When only sequences from gorillas and humans are compared, the fragmentation at the human-chimp split is omitted. Thus, the description of the evolution of two sequences depends on the number of the introduced splits, and there is no sensible interpretation to this dependence.

1.1 The Thorne–Kishino–Felsenstein Model

Since our model is related to the TKF91 model we describe it briefly. Most of the definitions and notations are introduced in here.

The TKF model is the fusion of two independent time-continuous Markov processes, the substitution and the insertion-deletion process.

The Substitution Process: Each character can be substituted independently for another character dictated by one of the well-known substitution processes [7],[8]. The substitution process is described by a system of linear differential equations

$$\frac{d\mathbf{x}(t)}{dt} = \mathbf{Q} \cdot \mathbf{x}(t) \tag{1}$$

where $\mathbf{Q}$ is the rate matrix. Since $\mathbf{Q}$ contains too many parameters, it is usually separated into two components, $\mathbf{Q}_0 s$, where $\mathbf{Q}_0$ is kept constant and is estimated with a less rigorous method than maximum likelihood [4]. The solution of (1) is

$$\mathbf{x}(t) = e^{\mathbf{Q}_0 st}\mathbf{x}(0) \tag{2}$$

The Insertion-Deletion Process: The insertion-deletion process is traditionally described not in terms of amino acids or nucleotides but in terms of imaginary links. A mortal link is associated to the right of each character, and additionally, there is an immortal link at the left end of the sequence. Each link can give birth to a mortal link with birth rate λ. The newborn link always appears at the right side of its parent. Accompanying the birth of a mortal link, is the birth of a character drawn from the equilibrium distribution. Only mortal links can die out with death rate μ, taking their character to the left with them. Assuming independence between links, it is sufficient to describe the fate of single mortal link and the immortal one. According to the possible histories of links (Figure 1), three types of functions are considered. Let $p_k^{(1)}(t)$ denote the probability that after time t, a mortal link has survived, and has exactly k descendants including itself. Let $p_k^{(2)}(t)$ denote the probability that after time t, a mortal link died, but it left exactly k descendants. Let $p_k(t)$ denote the probability that after time t, the immortal link has exactly k descendants, including itself.

Time	Immortal	Mortal	Mortal
0	o	*	*
t	o*...*	**...*	–*...*
k	$p_k(t)$	$p_k^{(1)}(t)$	$p_k^{(2)}(t)$

Fig. 1. The Possible Fates of Links. The second column shows the fate of the immortal link (o). After a time period t it has k descendants including itself. The third column describes the fate of a survived mortal link (*). It has k descendants including itself after time t. The fourth column depicts the fate of a mortal link that died, but left k descendants after time t.

Calculating the Joint Probability of Two Sequences: The joint probability of two sequences A and B is calculated as the equilibrium probability of sequence A times the probability that sequence B evolved from A under time $2t$, where t is the divergence time.

$$P(A, B) = P_\infty(A)P_{2t}(B \mid A) \tag{3}$$

A possible transition is described as an alignment. The upper sequence is the ancestor; the lower sequence is the descendant. For example the following alignment describes that the immortal link o has one descendant, the first mortal link * died out, and the second mortal link has two descendants including itself.

```
o  –  A*  U*  –
o  G*  –  C*  A*
```

The probability of an alignment is the probability of the ancestor, times the probability of the transition. For example, the probability of the above alignment is

$$\gamma_2 \pi(A)\pi(U)p_2(t)\pi(G)p_0^{(2)}(t)p_2^{(1)}(t)f_{UC}(2t)\pi(A) \tag{4}$$

where γ_n is the probability that a sequence contains n mortal links, $\pi(X)$ is the frequency of the character X, and $f_{ij}(2t)$ is the probability that a character i is of j at time $2t$. The joint probability of two sequences is the summation of the alignment probabilities.

2   The Model

Our model differs from the TKF models in the insertion-deletion process. The TKF91 model assumes only single insertions and deletions, as illustrated in Figure 2. Long insertions and deletions are allowed in the TKF92 model, as illustrated in Figure 3. However, these long indels are considered as unbreakable fragments as they have only one common mortal link. The death of the mortal link causes the deletion of every character in the long insertion. The distinction from the previous model is that in our model every character has its own mortal link in the long insertions, as illustrated in Figure 4.

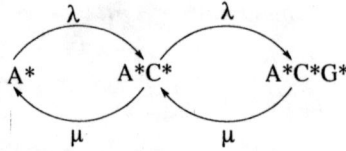

Fig. 2. The Flow-chart of the TKF91 Model. Each link can give birth to a mortal link with birth rate $\lambda > 0$. Mortal links die with death rate $\mu > 0$.

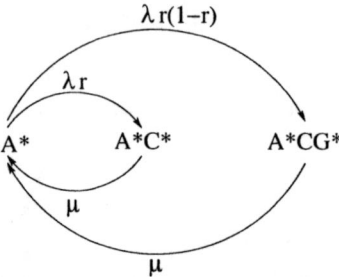

Fig. 3. The Flowchart of the Thorne–Kishino–Felsenstein Fragment Model. A link can give birth to a fragment of length k with birth rate $\lambda r(1 - r)^{k-1}$, with $\lambda > 0$ and $0 < r < 1$. Fragments are unbreakable so that only whole fragments can die with death rate $\mu > 0$.

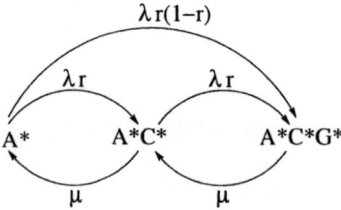

Fig. 4. The Flowchart of Our Model. Each link can give birth to k mortal links with birth rate $\lambda r(1 - r)^{k-1}$, with $\lambda > 0$ and $0 < r < 1$. Each newborn link can die independently with death rate $\mu > 0$.

Thus, this model allows long insertions without considering unbreakable fragments. It is possible that a long fragment is inserted into the sequence first and some of the inserted links die and some of them survive after then. A link gives birth to a block of k mortal links with rate λ_k, where

$$\lambda_k = \lambda r(1 - r)^{k-1}, \; k = 1, 2, \ldots, \; \lambda > 0, \; 0 < r < 1 \tag{5}$$

Only mortal links can die with rate $\mu > 0$.

2.1 Calculating the Generating Functions

The Master Equation: First, the probabilities of the possible fates of the immortal link is computed. Collecting the gain and loss terms for this birth-death process, the following Master equation is obtained:

$$\frac{dp_n}{dt} = \sum_{j=1}^{n-1}(n-j)\lambda_j p_{n-j} + n\mu p_{n+1} - \left(n\sum_{j=1}^{\infty}\lambda_j + (n-1)\mu\right)p_n \tag{6}$$

Using $\sum_{j=1}^{\infty}\lambda_j = \lambda$ and $\sum_{j=1}^{n-1}(n-j)\lambda_j p_{n-j} = \sum_{k=1}^{n-1}k\lambda_{n-k}p_k$, we have:

$$\frac{dp_n}{dt} = \lambda r \sum_{k=1}^{n-1} k(1-r)^{n-k-1}p_k + n\mu p_{n+1} - (n\lambda + (n-1)\mu)\,p_n \tag{7}$$

Due to the immortal link, we have $\forall t,\ p_0(t) = 0$. For $n = 1$, the sum in (7) is void. The initial conditions are given by:

$$p_n(0) = \delta_{n,1} \tag{8}$$

Next, we introduce the generating function [9]:

$$P(\xi;t) = \sum_{n=0}^{\infty}\xi^n p_n(t) \tag{9}$$

Multiplying (7) by ξ^n, then summing over n, we obtain a linear PDE for the generating function:

$$\frac{\partial P}{\partial t} - (1-\xi)\left(\mu - \frac{\lambda\xi}{1-\xi(1-r)}\right)\frac{\partial P}{\partial \xi} = -(1-\xi)\frac{\mu}{\xi}P \tag{10}$$

with initial condition $P(\xi;0) = \xi$.

Solution to the PDE for the Generating Function: We use the method of Lagrange:

$$\frac{dt}{1} = \frac{d\xi}{-(1-\xi)\left(\mu - \frac{\lambda\xi}{1-\xi(1-r)}\right)} = \frac{dP}{-(1-\xi)\frac{\mu}{\xi}P} \tag{11}$$

The two equalities define two, one-parameter families of surfaces, namely $v(t;\xi;P)$ and $w(t;\xi;P)$. After integrating the first and the second equalities in (11) the following families of surfaces are obtained:

$$v(\xi;t) = \frac{(1-\xi)^r}{(\mu - a\xi)^{\lambda/a}}e^{-t(\mu r - \lambda)} = c_1 \tag{12}$$

$$w(\xi;t;P) = P\frac{(\mu - a\xi)^r}{\xi} = c_2 \tag{13}$$

with $a \equiv \lambda + \mu(1 - r) > 0$. The general form of the solution is an arbitrary function of $w = g(v)$. This means:

$$P(\xi;t) = \xi(\mu - a\xi)^{-\lambda/a}g\left(\frac{(1-\xi)^r}{(\mu - a\xi)^{\lambda/a}}e^{-t(\mu r - \lambda)}\right) \tag{14}$$

The function g is fixed from the initial condition $P(\xi, 0) = \xi$:

$$g(z) = \left(\mu - af^{-1}(z)\right)^{\lambda/a} \tag{15}$$

where

$$f(x) = \frac{(1-x)^r}{(\mu - ax)^{\lambda/a}} \tag{16}$$

Thus the exact form for the generating function becomes:

$$P(\xi; t) = \xi \left(\frac{\mu - af^{-1}(f(\xi)e^{-(\mu r - \lambda)t})}{\mu - a\xi} \right)^{\lambda/a} \tag{17}$$

The Probabilities for the Fate of the Mortal Links: The Master Equations for the probabilities $p_n^{(1)}(t)$ and $p_n^{(2)}(t)$ are given by

$$\frac{dp_n^{(1)}}{dt} = \sum_{j=1}^{n-1}(n-j)\lambda_j p_{n-j}^{(1)} + n\mu p_{n+1}^{(1)} - \left(n\sum_{j=1}^{\infty}\lambda_j + n\mu \right)p_n^{(1)} \tag{18}$$

$$\frac{dp_n^{(2)}}{dt} = \sum_{j=1}^{n-1}(n-j)\lambda_j p_{n-j}^{(2)} + (n+1)\mu p_{n+1}^{(2)} + \mu p_{n+1}^{(1)} - \left(n\sum_{j=1}^{\infty}\lambda_j + n\mu \right)p_n^{(2)} \tag{19}$$

We have the following conditions to be fulfilled:

$$\forall t \geq 0, \ p_0^{(1)}(t) = 0 \tag{20}$$

and the initial conditions:

$$\forall n \geq 0, \ p_n^{(1)}(0) = \delta_{n,1}, \ p_n^{(2)}(0) = 0 \tag{21}$$

The corresponding partial differential equations for the generating functions, $P^{(i)}(\xi, t) = \sum_{n=0}^{\infty} \xi^n p_n^{(i)}(t)$, for $i = 1, 2$, are given by

$$\frac{\partial P^{(1)}}{\partial t} - (1-\xi)\left(\mu - \frac{\lambda\xi}{1 - \xi(1-r)} \right)\frac{\partial P^{(1)}}{\partial \xi} = -\frac{\mu}{\xi}P^{(1)} \tag{22}$$

$$\frac{\partial P^{(2)}}{\partial t} - (1-\xi)\left(\mu - \frac{\lambda\xi}{1 - \xi(1-r)} \right)\frac{\partial P^{(2)}}{\partial \xi} = \frac{\mu}{\xi}P^{(1)} \tag{23}$$

Solution to the PDEs for the Generating Functions of the Mortal Links: First, we solve (22) using the method of Lagrange

$$\frac{dt}{1} = \frac{d\xi}{-(1-\xi)\left(\mu - \frac{\lambda\xi}{1-\xi(1-r)}\right)} = \frac{dP^{(1)}}{-\frac{\mu}{\xi}P^{(1)}} \tag{24}$$

The two, one-parameter families of surfaces are $v(t; \xi; P^{(1)})$ and $w(t; \xi; P^{(1)})$. Since v comes from the integration of the first equality in (22), it is the same as (12). Integrating the second equality yields:

$$w(\xi; t; P^{(1)}) = P^{(1)} \frac{(1 - \xi)^{\mu r / (\mu r - \lambda)}}{\xi(\mu - a\xi)^{\lambda / (\mu r - \lambda)}} = c_2 \tag{25}$$

Proceeding as in the previous section, we have:

$$P^{(1)}(\xi; t) =$$
$$\xi \left(\frac{\mu - a\xi}{\mu - af^{-1}(f(\xi)e^{-(\mu r - \lambda)t})} \right)^{\frac{\lambda}{\mu r - \lambda}} \left(\frac{1 - f^{-1}(f(\xi)e^{-(\mu r - \lambda)t})}{1 - \xi} \right)^{\frac{\mu r}{\mu r - \lambda}} \tag{26}$$

with f given by (16). To calculate $P^{(2)}(\xi; t)$, we first define $Q(\xi; t) = P^{(1)}(\xi; t) + P^{(2)}(\xi; t)$. Summing (22) and (23) the following equation is obtained for Q:

$$\frac{\partial Q}{\partial t} - (1 - \xi) \left(\mu - \frac{\lambda \xi}{1 - \xi(1 - r)} \right) \frac{\partial Q}{\partial \xi} = 0 \tag{27}$$

This is again easily solved with the method of characteristics. First, we integrate the characteristic equation, which is the first equation in (24), to obtain the family of characteristic curves, given by $v(\xi; t) = c_1$ as in (12). Thus, $Q(\xi; t) = g(v)$ is the general solution, where $g(x)$ is an arbitrary, differentiable function, to be set by the initial conditions. Using (20) and (21), we have $Q(\xi; 0) = \xi$. This leads to:

$$Q(\xi; t) = f^{-1}(f(\xi)e^{-(\mu r - \lambda)t}) \tag{28}$$

with f given by (16), and therefore:

$$P^{(2)}(\xi; t) = f^{-1}(f(\xi)e^{-(\mu r - \lambda)t}) - P^{(1)}(\xi; t) \tag{29}$$

with $P^{(1)}(\xi; t)$ given by (26).

2.2 The Equilibrium Length Distribution

The generating function of the equilibrium length distribution can be obtained from (17) by considering the limit $t \to \infty$. Since $f^{-1}(0) = 1$ and due to the immortal link, the generating function becomes

$$\Gamma(\xi) = \left(\frac{\mu - a}{\mu - a\xi} \right)^{\frac{\lambda}{a}} \tag{30}$$

Calculating the Taylor-series of $\Gamma(\xi)$ around 0, we get for the equilibrium probabilities:

$$\gamma_n = (\mu - a)^{\frac{\lambda}{a}} \frac{\Pi_{i=0}^{n-1}(\lambda + ia)}{n! \mu^{n-1+\lambda/a}} \tag{31}$$

From $\frac{d\Gamma(\xi)}{d\xi}$ in the limit of $\xi \to 1^-$, the expected value of the sequence length is obtained as:

$$E(\gamma) = \frac{\lambda}{\mu r - \lambda} \tag{32}$$

3 The Algorithm

3.1 Calculating the Transition Probabilities

Unfortunately, the inverse of f given by (16) does not have a closed form. Thus a numerical approach is needed for calculating the transition probability functions $p_n(t)$, $p_n^{(1)}(t)$, and $p_n^{(2)}(t)$. We calculate the generating functions $P(\xi;t)$, $P^{(1)}(\xi;t)$ and $P^{(2)}(\xi;t)$ in $l_1 + 1$ points around $\xi = 0$, where l_1 is the length of the shorter sequence. For doing this, the following equation must be solved for x numerically where ξ, μ, λ, r, t, and a are given.

$$f(\xi)e^{-(\mu r - \lambda)t} = \frac{(1-x)^r}{(\mu - ax)^{\frac{\lambda}{a}}} \tag{33}$$

Given $l_1 = 1$ points, the functions are partially derived l_1 times. After this

$$p_n(t) = \frac{\partial^n P(\xi, t)}{\partial \xi^n} \frac{1}{n!} \tag{34}$$

and similarly for $p_n^{(1)}(t)$ and for $p_n^{(2)}(t)$. Thus, the transition probability functions can be calculated in $O(l^2)$ time.

3.2 Dynamic Programming for the Joint Probability

Without loss of generality we can suppose that the shorter sequence is sequence B. The equilibrium probability of sequence A is

$$P_\infty(A) = (\mu - a)^{\frac{\lambda}{a}} \frac{\Pi_{i=0}^{l(A)-1)}(\lambda + ia)}{l(A)\mu^{l(A)-1+\lambda/a}\Pi_{i=1}^{l(A)}\pi(a_i)} \tag{35}$$

where a_i is the ith character in A and $l(A)$ is the length of the sequence.

Let A_i denote the i-long prefix of A and let B_j denote the j-long prefix of B. There is a dynamic programming algorithm for calculating the transition probabilities $P_t(A_i \mid B_j)$. The initial conditions are given by:

$$P_t(A_o \mid B_j) = p_{n+1}(t)\Pi_{k=1}^j\pi(b_k) \tag{36}$$

To save computation time, we calculate $\Pi_{k=l}^k\pi(b_k)$ for every $l < j$ before the recursion. Then the recursion follows

$$P_t(A_i \mid B_j) = \sum_{l=0}^j P_t(A_{i-1} \mid B_l)p_{j-l}^{(2)}(t)\Pi_{k=l+1}^j\pi(b_k)$$

$$+ \sum_{l=0}^{j-1} P_t(A_{i-1} \mid B_l)p_{j-l}^{(1)}(f)f_{a_ib_{l+1}}\Pi_{k=l+2}^j\pi(b_k) \tag{37}$$

The dynamic programming is the most time-consuming part of the algorithm, it takes $O(l^3)$ running time.

3.3   Finding the Maximum Likelihood Parameters

As mentioned earlier, the substitution process is described with only one parameter, st. (A general phenomenon is that the time and rate parameters can not be estimated individually, only their product.) The insertion-deletion model is described with three parameters, λt, μt, and r, which however, can be reduced to two, if the following equation is taken under consideration

$$\frac{\lambda}{\mu r - \lambda} = \frac{l(A) + l(B)}{2} \tag{38}$$

namely, the mean of the sequence lengths is the maximum likelihood estimator for the expected value of the length distribution.

The maximum likelihood values of the three remaining parameters can be obtained using one of the well-known numerical methods (gradient method, etc.).

4   Discussion and Conclusions

There is an increasing desire for statistical methods of sequence analysis in the bioinformatics community. The statistical alignment provides a sensitive homology testing [5], which is better than the traditional, similarity-based methods [10]. The summation over the possible alignments leads to a good evolutionary parameter estimation [3], while the parameter estimation from a single alignment is doubtful [3,11].

Methods based on evolutionary models integrate the multiple alignment and the evolutionary tree reconstruction. The generalization of the Thorne–Kishino–Felsenstein model to arbitrary number of sequences is straightforward [12,13]. A novel approach is to treat the evolutionary models as HMM. The TKF model fits into the concept of pair-HMM [14]. Similarly, the generalization to n sequences can be handled as multiple-HMM. Following this approach, one can sample alignments related to a tree providing an objective approximation to the multiple alignment problem [15]. Sampling pairwise alignments and evolutionary parameters allows further investigations of the evolutionary process [16].

The weak point of the statistical approach is the lack of an appropriate evolutionary model. A new model and an associated algorithm for computing the joint probability were introduced. This new model is superior to the Thorne–Kishino–Felsenstein model: it allows long insertions without considering unbreakable fragments. However, it is only a small inch to the reality, as it contains at least two unrealistic properties. It cannot deal with long deletions, and the rates for the long insertions form a geometric series. The elimination of both these problems seems to be rather difficult but not impossible. Other rate functions for long insertions lead to more difficult PDE-s whose characteristic equations may not be integrated without a rather involved computational overhead. The same situation appears when long deletions are allowed. Moreover, in this case calculating only the fates of the individual links is not sufficient. Thus, for achieving more appropriate models, numerical calculations are needed in an earlier state of the procedure. Nevertheless, we hope that the generating function approach will open some novel avenues for further research.

Acknowledgments

We thank Carsten Wiuf and the anonymous referees for useful discussions and suggestions. Z.T. was supported by the DOE under contract W-7405-ENG-36.

References

1. Needleman, S.B., Wunsch, C.D.: A general method applicable to the search for similarites in the amino acid sequences of two proteins. *J. Mol. Biol.* **48** (1970), 443–453.
2. Bishop, M. J., Thompson, E.A.: Maximum likelihood alignment of DNA sequences. *J. Mol. Biol.* **190** (1986), 159–165.
3. Thorne, J.L., Kishino, H., Felsenstein, J.: An evolutionary model for maximum likelihood alignment of DNA sequences. *J. Mol. Evol.* **33** (1991), 114–124.
4. Thorne, J.L., Kishino, H., Felsenstein, J.: Inching toward reality: an improved likelihood model of sequence evolution. *J. Mol. Evol.* **34** (1992), 3–16.
5. Hein, J., Wiuf, C., Knudsen, B., Moller, M.B., Wiblig, G.: Statistical alignment: computational properties, homology testing and goodness-of-fit. *J. Mol. Biol.* **302** (2000), 265–279.
6. Miklós, I.: Irreversible likelihood models, European Mathematical Genetics Meeting, 20–21. April, 2001, Lille, France.
7. Dayhoff, M.O., Schwartz, R.M., Orcutt, B.C.: A model for evolutionary change in proteins, matrices for detecting distant relationships. In: Dayhoff, M.O. (ed.): *Atlas of Protein Sequence and Structure, Vol. 5.* Cambridge University Press, Washingtown DC. (1978), 343–352.
8. Tavare, S.: Some probabilistic and statistical problems in the analysis of DNA sequences. *Lec. Math. Life Sci.* **17** (1986), 57–86.
9. Feller, W.: *An introduction to the probability theory and its applications, Vol. 1.* McGraw-Hill, New York (1968), 264–269.
10. Altschul, S.F.: A protein alignment scoring system sensitive at all evolutionary distances. *J. Mol. Evol.* **36** (1993), 290–300.
11. Fleissner, R., Metzler, D., von Haeseler, A.: Can one estimate distances from pairwise sequence alignments? In: Bornberg-Bauer, E., Rost, U., Stoye, J., Vingron, M. (eds) *GCB 2000, Proceedings of the German Conference on Bioinformatics*, Heidelberg (2000), Logos Verlag, Berlin, 89–95.
12. Hein, J.: Algorithm for statistical alignment of sequences related by a binary tree. In: Altman, R.B., Dunker, A.K., Hunter, L., Lauderdale, K., Klein, T.E. (eds), *Pacific Symposium on Biocomputing*, World Scientific, Singapore (2001), 179–190.
13. Hein, J., Jensen, J.L., Pedersen, C.S.N.: Algorithm for statistical multiple alignment. *Bioinformatics 2001*, Skövde, Sweden.
14. Durbin, R., Eddy, S., Krogh, A, Mitchison, G.: *Biological Sequence Analysis: Probabilistic Models of Proteins and Nucleic Acids.* Cambridge University Press, Cambridge (1998).
15. Holmes, I., Bruno, W.J.: Evolutionary HMMs: A Bayesian Approach to Multiple Alignment, *Bioinformatics* (2001), accepted.
16. http://www.math.uni-frankfurt.de/~stoch/software/mcmcalgn/

Improving Profile-Profile Alignments via Log Average Scoring

Niklas von Öhsen and Ralf Zimmer

SCAI—Institute for Algorithms and Scientific Computing
GMD—German National Research Center for Information Technology
Schloss Birlinghoven, Sankt Augustin, 53754, Germany
niklas.von-oehsen@gmd.de

Abstract. Alignments of frequency profiles against frequency profiles have a wide scope of applications in currently used bioinformatic analysis tools ranging from multiple alignment methods based on the progressive alignment approach to detecting of structural similarities based on remote sequence homology. We present the new *log average scoring* approach to calculating the score to be used with alignment algorithms like dynamic programming and show that it significantly outperforms the commonly used average scoring and dot product approach on a fold recognition benchmark. The score is also applicable to the problem of aligning two multiple alignments since every multiple alignment induces a frequency profile.

1 Introduction

The use of alignment algorithms for the establishing of protein homology relationships has a long tradition in the field of bioinformatics. When first developed, these algorithms aimed at assessing the homology of two protein sequences and at constructing their best mapping onto each other in terms of homology. By extending these algorithms to align sequences of amino acids not only to their counterparts but to frequency profiles, which was first proposed by Gribskov [10], it became feasible to analyse the relationship of a single protein with a whole family of proteins described by the frequency profile. Based on this idea the PSI-Blast program [2] was developed which belongs to the most well known and heavily used tools in computational biology. Recently, a further abstraction has proven to be of considerable use in protein structure prediction. In the CAFASP2 contest of fully automated protein structure prediction the group of Rychlewski *et al.* reached the second rank using a profile-profile alignment method called FFAS [18]. The notion of alignment is thus extended to provide a mapping between two protein families represented by their frequency profiles. Rychlewski *et al.* used the dot product to calculate the alignment score for a pair of profile vectors. In this paper we present a new approach which allows to choose an amino acid substitution model like the BLO-SUM model [12] and leads to a score that not only increases the ability to judge the relatedness of two proteins by the alignment score but also has a meaning in terms of the underlying substitution model.

We start by introducing the definition of profiles and subsequently discuss the three candidate methods for scoring profile vectors against each other. In the second part

O. Gascuel and B.M.E. Moret (Eds.): WABI 2001, LNCS 2149, pp. 11–26, 2001.

the fold recognition experiments we performed are described and discussed. In the appendix further technical information on the benchmarks can be found.

2   Theory

The use of profiles is to represent a set of related proteins by a statistical model that does not increase in size even when the set of proteins gets large. This is done by making a multiple alignment of all the sequences in the set and then counting the relative frequencies of occurrence of each amino acid in each position of the multiple alignment. Usually it is assumed that the underlying set of proteins is not known completely, but that we have a small subset of representatives, for instance from a homology search over a database. Extensive work has been done on the issue of estimating the "real" frequencies of the full set from the sample retrieved by the homology search. Any of these methods like pseudo counts [20], Dirichlet mixture models [6], minimal-risk estimation [24], sequence weighting methods [13], [16], [19] may be used to preprocess the sample to get the best estimation of the frequencies before one of the following scoring methods is applied. In any case will the construction yield a vector of probability vectors which are in our setting of dimension 20 (one for each amino acid). These probabilities are positive real numbers that sum up to one and stand for the probability of seeing a certain amino acid in this position of a multiple alignment of all family members. This sequence of vectors will be called frequency profile or profile throughout the paper. The gaps occurring in the multiple alignments are not accounted for in our models, therefore the frequency vectors must eventually be scaled up to reach a total probability sum of one. All of the profile-to-profile scoring methods introduced will be defined by a formula which gives the corresponding score depending on two probability vectors (or profile positions) named α and β.

2.1   Dot Product Scoring

The simplest and fastest method is the dot product method as used by Rychlewski *et al.* [18]. This is a rather heuristic approach since a possible interpretation of the sum of these scoring terms over all aligned positions remains unclear. The score is calculated as

$$\text{score}_{\text{dot product}}(\alpha, \beta) = \sum_{i=1}^{20} \alpha_i \beta_i$$

which is in fact the probability that identical amino acids are produced by drawing from the distribution α and β independently. The log of this score might therefore serve as a meaningful measure of profile similarity but this is not discussed here. As can be seen this scoring approach does not incorporate any knowledge about the similarities between the amino acids and is therefore independent of any substitution matrix.

2.2   Sequence-Sequence Alignment

When aligning two amino acid sequences, the score is calculated as a likelihood ratio between the likelihood that the alignment occurs between "related" sequences and the likelihood that the alignment occurs between "unrelated" sequences. The notion of relatedness is defined here by the employed substitution model, which incorporates two probability distributions describing each case. The first distribution, called null model, describes the average case in which the two positions are each distributed like the amino acid background and are unrelated, yielding $P(X = i, Y = j) = p_i p_j$. Here p_k stands for the probability of seeing an amino acid k when randomly picking one amino acid from an amino acid sequence database. The probability of seeing a pair of amino acids in a "related" pair of sequences in corresponding positions has been estimated by some authors using different methods. M. Dayhoff derived the distribution from observations of single point mutations resulting in the series of PAM Matrices [8]. In the case of the BLOSUM matrix series [12] the distribution is derived from blocks of multiply aligned sequences, which are clustered up to a certain amount of sequence identity. We introduce an event called "related" for the case that the values of X and Y are related amino acids and call the probability distribution $P(X = i, Y = j|\text{related}) = p_{\text{rel}}(i, j)$. Using this, we receive the formula for the log odds score ("log" always standing for the natural logarithm)

$$M(i, j) = \log\left(\frac{p_{\text{rel}}(i, j)}{p_i p_j}\right) \tag{1}$$

which are the values stored in the substitution matrices except for a constant factor which is $\frac{10}{\log 10}$ in the Dayhoff models and $\frac{2}{\log 2}$ in the BLOSUM matrices.

Using Bayes' formula we get an interpretation of the likelihood ratio term defining the log odds alignment score:

$$P(\text{related}|X = i, Y = j) = P(\text{related})\frac{P(X = i, Y = j|\text{related})}{P(X = i, Y = j)} \tag{2}$$

$$= P(\text{related})\frac{p_{\text{rel}}(i, j)}{p_i p_j} \tag{3}$$

This means that except for the prior probability $P(\text{related})$, which is a constant, the usual sequence-sequence alignment score is the log of the probability that the two amino acids come from related positions, given the data.

If different positions are assumed to be independent of each other, the log odds score summed up over all aligned amino acid pairs is the log of the probability that the alignment occurs in case the sequences are related divided by the probability that the alignment occurs between unrelated sequences. It is therefore in a certain statistical sense the best means to decide whether the alignment maps related or unrelated sequences onto each other (Neyman-Pearson lemma, e.g. [23]). This quantity will be maximised by the dynamic programming approach yielding the alignment that maximises the likelihood ratio in favour of the "related" hypothesis. The gap parameters add penalties to this log-likelihood-ratio score, which indicate that the more gaps an alignment has, the more likely it is to occur between unrelated sequences rather than between related sequences.

2.3 Average Scoring

The average scoring method has been the very first approach to scoring frequency profiles against amino acid sequences [10]. The basic idea is that the score for a distribution of amino acids is calculated by taking the expected value of the sequence-sequence score under the profile distribution while keeping the amino acid from the sequence fixed. This can be extended to to profile-profile alignments in a straightforward fashion and has been used in ClustalW [22]. There, two multiple alignments are aligned using as score an average over all pairwise scores between residues, which is equivalent to the average scoring approach as used here. The formula which we obtain this way is the following:

$$\text{score}_{\text{average}}(\alpha, \beta) = \sum_{i=1}^{20} \sum_{j=1}^{20} \alpha_i \beta_j \log \frac{p_{\text{rel}}(i,j)}{p_i p_j} \tag{4}$$

It can easily be shown that this score has an interpretation: Let N be a large integer and lets take a sample of size N from the two profile positions (each sample being a pair of amino acids, the distribution being $(\alpha_i \beta_j)_{i,j \in 1,\ldots,20}$). Then this score divides the likelihood that the related distribution produced the sample by the likelihood that the unrelated distribution produced the sample, takes the log and divides this by N. The average score summed up over all aligned profile positions thus has the following meaning: If we draw for each aligned pair of profile positions a sample of size N which happens to show $N \alpha_i \beta_j$ times the amino acid pair (i, j), then the summed up average score is the best means to decide whether this happens rather under the "related" or the "unrelated" model.

The problem with the approach is, that this is not the question we are asking. The two distributions ("related" and "unrelated") that are suggested as only options are both known to be inappropriate since their marginal distributions (the distributions that are obtained by fixing the first letter and allowing the second to take any value and vice versa) are the background distribution of amino acids by the definition of the substitution model. The appropriate setting for this model describes a situation in which each profile position would in fact be occupied by a completely random amino acid (determined by the background probability distribution) meaning that, if we drew more and more amino acids from a position, then the observed distribution would have to converge to the background amino acid distribution. This is not compatible with the meaning usually associated with a profile vector which is thought of being itself the limiting distribution to which the distribution of such a sample should converge.

Another drawback to this method is the fact that the special case of this formula, when one of the profiles degenerates to a single sequence (at each position a probability distribution which has probability one for a single amino acid), has not the expected behaviour of a good scoring system. This will be shown in the following section, where we will extend the commonly used sequence-sequence score in a first step to the profile-sequence setting such that a strict statistical interpretation of the score is at hand and then further to the profile-profile setting which will be evaluated further on.

2.4 Profile-Sequence Scoring

The sequence-sequence scoring (1) can be extended to the profile-sequence case in a straightforward manner. It has been noted several times (e.g. [7]) that for the case that the target distribution of amino acids in a profile position α is known, the score given by

$$score(\alpha, j) = \log \frac{\alpha_j}{p_j} \tag{5}$$

yields an optimal test statistics by which to decide whether the amino acid j is a sample from the distribution α or rather from the background distribution p. These values summed up over all aligned positions therefore give a direct measure of how likely it is that the amino acid sequence is a sample from the profile rather than being random. If for a protein family only the corresponding profile is known, calculating this score is an optimal way to decide whether an amino acid sequence is from this family or not. This is a rather limited question to ask if we want to explore distant relationships. Therefore, in our setting it is of interest whether the sequence is somehow evolutionary related to the family characterised by the profile or not.

Evolutionary Profile-Sequence Scoring. One method for evaluating this in the profile-sequence case is the evolutionary profile method [11,7] which only makes use of the evolutionary model underlying the amino acid similarity matrices. The values $P(i \to j) = \frac{p_{rel}(i,j)}{p_i}$ can, due to the construction of the similarity matrix, be interpreted as the transition probabilities for a probabilistic transition (mutation) of the amino acid i to j. From this point of view the value $M(i,j)$ from (1) can be written as $M(i,j) = \log \frac{P(i \to j)}{p_j}$ which can be read as the likelihood ratio of amino acid j having occurred by transition from amino acid i against j occurring just by chance. This can be extended to the profile-sequence case where i is replaced with the profile vector α and letting the same probabilistic transition take place on a random amino acid with distribution α instead of on the fixed amino acid i. The resulting probability of j occurring by transition from an amino acid distributed like α is given by

$$\sum_{i=1}^{20} \alpha_i P(i \to j) = \sum_{i=1}^{20} \alpha_i \frac{p_{rel}(i,j)}{p_i} \tag{6}$$

which leads to the score

$$Score(\alpha, j) = \log \frac{\sum_{i=1}^{20} \alpha_i P(i \to j)}{p_j} = \log \sum_{i=1}^{20} \alpha_i \frac{p_{rel}(i,j)}{p_i p_j} \tag{7}$$

This score summed up over all aligned positions in an alignment of a profile against a sequence is therefore an optimal means by which to decide whether the sequence is more likely the result of sampling from the profile which has undergone the probabilistic evolutionary transition or whether the sequence occurs just by chance (optimality in a statistical sense).

It is apparent that the formula 7 is not a special case of the earlier introduced average scoring (4). This is a drawback for the average scoring approach since it fails to yield an intuitively correct result in a simple example: If the profile position is distributed like the amino acid background distribution, i. e. $\alpha_i = p_i$ for all i, we would expect that we have no information available on which to decide whether an amino acid j is related with the profile position or not. Thus it is a desirable property of a scoring system that any amino acid j should yield zero when scored against the background distribution. This is the case for the evolutionary profile-sequence score but is not the case for the average score where we receive (with p being the background distribution and e_j being the j-th unit vector)

$$\text{score}(\alpha, j) = \text{score}_{\text{average}}(p, e_j) = \sum_{i=1,\ldots,20} p_i M(i, j)$$

which is never positive due to Jensen's inequality (see e.g. [4]) and will always be negative for the amino acid background distribution commonly observed. Thus the average score would propose that we have evidence against the hypothesis that the profile position and the amino acid are related, which seems questionable. This is the motivation to look for a generalisation of the evolutionary sequence-profile scoring scheme to the profile-profile case. The results are explained in the following section which introduces the new scoring function proposed in this paper.

2.5 Log Average Scoring

Let again (X, Y) be a pair of random variables with values in $\{1, \ldots, 20\}$ which represent positions in profiles for which the question whether they are related is to be answered. Since the goal here is to score profile positions against profile positions we have to incorporate into our model the fact that the special X and Y we are observing have the amino acid distribution $(\alpha_i)_{i=1,\ldots,20}$ and $(\beta_j)_{j=1,\ldots,20}$, respectively. This is done by introducing an event E which has the following property:

$$P(X = i, Y = j | E) = \alpha_i \beta_j \tag{8}$$

This leads to the equations

$$P(\text{related}|E) = \sum_{i=1}^{20} \sum_{j=1}^{20} P(X = i, Y = j, \text{ related}|E) \tag{9}$$

$$= \sum_{i=1}^{20} \sum_{j=1}^{20} P(X = i, Y = j | E) P(\text{related}|X = i, Y = j, E) \tag{10}$$

Since a substitution model that directly addresses the case E with its special distributions of α and β is not available for the calculation of the last factor, we use the standard model (see equation (3)) as an approximation instead and exploit the knowledge on the amino acid distributions (see (8)) at the current profile positions for the first factor:

$$\approx \sum_{i=1}^{20} \sum_{j=1}^{20} P(X = i, Y = j | E) P(\text{related}|X = i, Y = j) \tag{11}$$

$$= P(\text{related}) \sum_{i=1}^{20} \sum_{j=1}^{20} \alpha_i \beta_j \frac{p_{\text{rel}}(i, j)}{p_i p_j} \tag{12}$$

If the prior probability is set to 1 and the log is taken like in the usual sequence-sequence score we receive the following formula for the *log average score*

$$\text{score}_{\text{logaverage}}(\alpha, \beta) = \log \sum_{i=1}^{20} \sum_{j=1}^{20} \alpha_i \beta_j \frac{p_{\text{rel}}(i,j)}{p_i p_j} \tag{13}$$

It is interesting to note that the only difference between this formula and the average score is the exchanged order of the log and the sums. As can be seen this formula is an extension of the evolutionary profile score for the profile-sequence case with the advantages discussed above. If these scoring terms are summed up over all aligned positions in a profile-profile alignment the resulting alignment score is thus the log of the probability that the profiles are related under the substitution model given the data they provide (except for the prior).

3 Evaluation

In order to evaluate whether the different scores are a good measure of the relatedness of two profiles, we performed fold recognition and related pair recognition benchmarks. Additionally, we investigated how a confidence measure for the protein fold prediction depending on the underlying scoring system performed on the benchmark set of proteins.

3.1 Data Set

The experiments were carried out using a protein sequence set which consists of 1511 chains from a subset of the PDB with a maximum of 40% pairwise sequence identity (see [5]). The composition of the test set in terms of relationships on different SCOP levels is shown in figure 1. Throughout the experiments the SCOP version 1.50 is used [17].

Note that there are 34 proteins in the set which are the only representatives of their SCOP fold in the test set. They were deliberately left in the test set even though it is not possible to recognise their correct fold class because this way the results resemble the numbers in the application case of a query with unknown structure.

For all sequences a structure of the same SCOP class can be found in the benchmark set, there are 34 chains in the set without a corresponding fold representative (i.e. single members of their fold class in the benchmark), SCOP superfamily and SCOP family representatives can be found for 1360 and 1113 sequences of the test benchmark set, respectively.

Only chains contributing to a single domain according to the SCOP database were used in order to allow for a one-to-one mapping of the chains to their SCOP classification. For each chain a frequency profile representing a set of possibly homologous sequences was constructed based on PSI-Blast searches on a non redundant sequence database following a procedure described in the appendix.

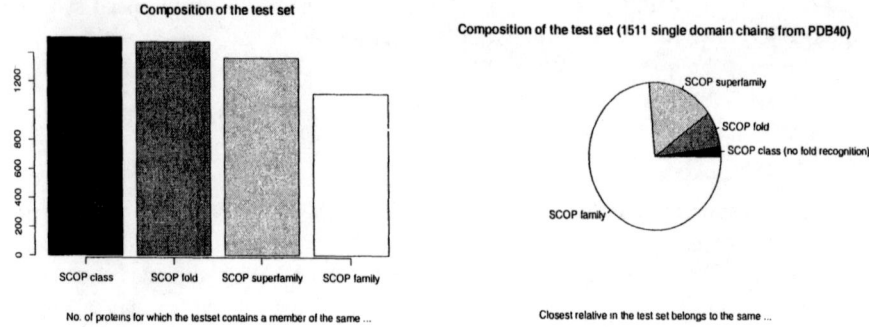

Fig. 1. Composition of the Test Set. Left: Number of proteins for which the test set contains a member of the indicated SCOP level. Right: Number of proteins whose closest relative (in terms of SCOP level) in the test set belongs to the indicated SCOP level. This is a partition of the test set in terms of fold recognition difficulty; ranging from SCOP family being the easiest to SCOP class being impossible.

3.2   Implementation Details

For each examined scoring approach we then used a JAVA implementation of the Gotoh global alignment algorithm [9] to align a query profile against each of the remaining 1510 profiles in the test set. For a query sequence of length 150 about 6 alignments per second can be computed on a $400 MHz$ Ultra Sparc 10 workstation.

It should be noted that for the case of fold recognition where one profile is subsequently being aligned against a whole database of profiles a significant speedup can be achieved by preprocessing the query profile α and calculating

$$\widetilde{\alpha} := \left(\sum_{i=1}^{20} \alpha_i \frac{p_{\text{rel}}(i,j)}{p_i p_j} \right)_{j=1,\ldots,20}$$

thus reducing the score calculation

$$\text{score}_{\text{logaverage}}(\alpha, \beta) = \log \sum_{j=1}^{20} \widetilde{\alpha}_j \beta_j$$

to one scalar product and one logarithm. This can be done in a similar manner with the average scoring approach where the complexity reduces to only the scalar product. The running time of the algorithm could be reduced by a factor of more than 6 using this technique.

3.3   Alignment Parameters

The appropriate gap penalties were determined separately for every scoring method using a machine learning approach (see appendix, [25]) and are shown in table 1. Throughout the experiments shown here we used the BLOSUM 62 substitution model

Table 1. Gap Penalties Used for the Experiments.

scoring	gap open	gap extension
dot product	3.12	0.68
average	5.60	1.22
log average	10.35	0.16

[12]. The average scoring alignments were calculated using the values from the BLO-SUM 62 scoring matrix and, thus, contain the above mentioned scaling factor of $f = \frac{2}{\log 2}$. To keep the results comparable we also applied the factor to the log average score. Therefore, the gap penalties for the log average score in table 1 must be divided by f if the score is calculated exactly as in formula (13).

3.4   Results

For each of the three profile scoring system discussed in section 2 the following test were performed using the constructed frequency profiles. In order to assess the superiority of the profile methods over simple sequence methods we also performed the tests for plain sequence-sequence alignment on the original chains using the BLOSUM 62 substitution matrix and the same gap penalties as for the log average scoring.

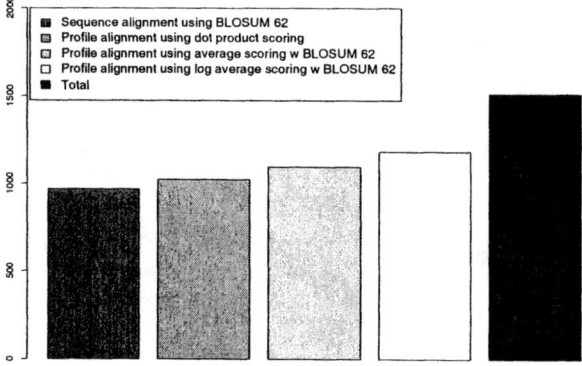

Fig. 2. Total Fold Recognition Performance.

Fold Recognition. The goal here is to identify the SCOP fold to which the query protein belongs by examining the alignment scores of all 1510 alignments of the query profile against the other profiles. The scores are sorted in a list together with the name of the protein which produced the score and the fold prediction is the SCOP fold of the

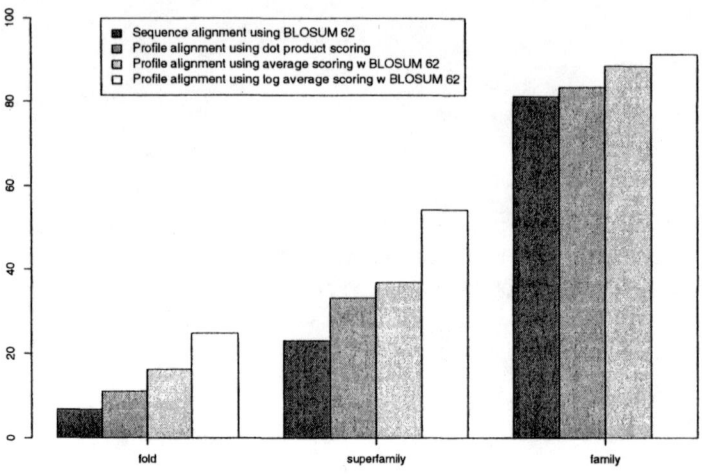

Fig. 3. Fold Recognition Performance for Each of the Difficulty Classes.

highest scoring protein in the list. Since all the proteins in the list are aligned against the same query and the scores are compared, a possible systematic bias of the score by special features of the query sequence is not relevant for this test (e. g. length dependence). The test was performed following a leave-one-out procedure, e. g. for each of the 1511 proteins the fold was predicted using the list of alignments against the 1510 other profiles. The fold recognition rate is then defined as the percentage of all proteins for which the fold prediction yielded a correct result.

Out of the 1511 test sequences log average scoring is able to assign correct folds for 1181 cases or 78.1%, whereas the usual average scoring correctly predicts 1097 (72.6%) and dot product scoring 1024 (67.7%) sequences, both improving on simple sequence-sequence alignment with 969 (64.1%) correct assignments. This improvement becomes more distinctive for more difficult cases towards the twilight of detectable sequence similarity. Figure 3 shows the fold recognition rates for family, superfamily, fold predictions separately. Here, all four methods perform well for the easiest case, family recognition, with 81.2% for sequence alignment performing worst and log average profile scoring with 91.5% performing best. For the hardest case of fold detection, log average scoring (24.8%) significantly outperforms (at least 50% improvement) both other profile methods (11.1% and 16.2%), whereas sequence alignment hardly is able to make correct predictions (6.8%). However, the effect of performance improvement is most marked for the superfamily level, where some remote evolutionary relationships should, by definition, be detectable via sensitive sequence methods. Here, the new scoring scheme again achieves a 50% improvement over the second best (average profile scoring) methods, thereby increasing the recognition rate from 36.8% to 54.3%. This almost doubles the recognition rate of simple sequence alignment (23.0%).

A more detailed look on the fold recognition results can be achieved by using confidence measures which measure the quality of the fold prediction a priori. Here we use the z-score gap which is defined as follows. First the mean and standard deviation for the scores in the list are calculated and the raw scores are transformed into z-scores with respect to the determined normal distribution, i. e. the following formula is applied:

$$z - \text{score} = \frac{\text{score} - \text{mean}}{\text{standard deviation}}$$

Then the difference of the z-score between the top scoring protein and the next best belonging to a SCOP fold different from the predicted one is calculated yielding the z-score gap. A list L which contains all 1511 fold predictions together with their z-score gap is set up and sorted with respect to the z-score gap. Entries $l \in L$ which represent correct fold predictions are termed *positives*, others *negatives*. If i is an index in this list, figure 4 shows the percentage of correct fold predictions if only the top i entries of the list are predicted. It also demonstrates a clear improvement of fold prediction sensi-

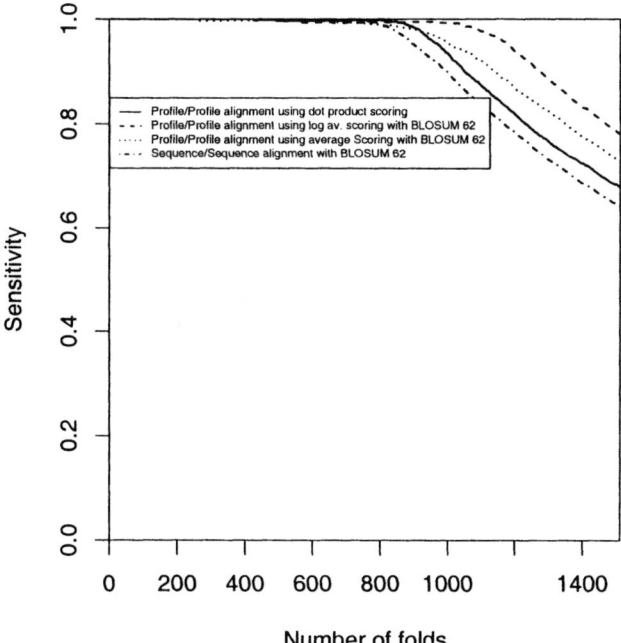

Fig. 4. Fold Recognition Ranked with Respect to the z-score Gap (See Text).

tivity and specificity for the log average scoring as compared to the competing scoring schemes. Again, all profile methods perform better than pure sequence alignment, but dot product only shows a slight improvement.

Related Pair Recognition. This protocol aims at a slightly different question. The goal is to decide whether two proteins have the same SCOP fold by only looking at the score of their profile alignment. Therefore, a good performance in this test means that the scoring system is a good absolute measure of similarity between the sequences. Length dependency and other systematic biases will decrease the performance of a scoring system here.

The calculations done here also rely on the 1511 lists calculated in the fold recognition setting. These are merged into one large list following two different procedures:

- z-scores: Before merging, the mean and standard deviation for each of the lists are calculated and the raw scores are transformed into z-scores as in (3.4). This setting is related with the fold recognition setting since biases introduced by the query profile should be removed by the rescaling.
- raw scores: No transformation is applied.

The resulting list L contains in each entry $l \in L$ a score score(l) and the two proteins whose alignment produced the score. An entry $l \in L$ will be called *positive* if the two proteins have the same SCOP fold and *negative* if not. The list of $1\,511 * 1\,510 = 2\,281\,610$ entries is then sorted with respect to the alignment score and for all scores s in the list specificity and sensitivity are calculated from the following formulas:

$$\text{spec}(s) = \frac{\#\{l \in L \mid l \text{ positive, score}(l) > s\}}{\#\{l \in L \mid \text{score}(l) > s\}} \tag{14}$$

$$\text{sens}(s) = \frac{\#\{l \in L \mid l \text{ positive, score}(l) > s\}}{\#\{l \in L \mid l \text{ positive}\}} \tag{15}$$

The plots of these quantities for the whole range of score values are shown in figure 5, which clearly exhibits the recognition performance of the new scoring scheme over the whole range of specificities. The ranking of the respective methods is again sequence alignment, dot product, average scoring, and log average scoring best, almost doubling the performance of average scoring. Using z-scores, sequence alignment and dot product scoring improve somewhat, but still, log average scoring consistently shows doubled performance over the second best method.

4   Discussion

All experiments we performed show a clear improvement of recognition performance when using the introduced log average score over average scoring as well as over dot product scoring. The results of the fold recognition test are most interesting for the protein targets that fall into the superfamily difficulty class since the SCOP hierarchy suggests here a "probable common evolutionary origin" which would make this problem tractable to sequence homology methods as the ones discussed here. The increase in performance over the best previously known profile method (average scoring) achieved by using log average scoring becomes as large as 48 % (from 36.8% to 54.3%) and is still greater on the fold level.

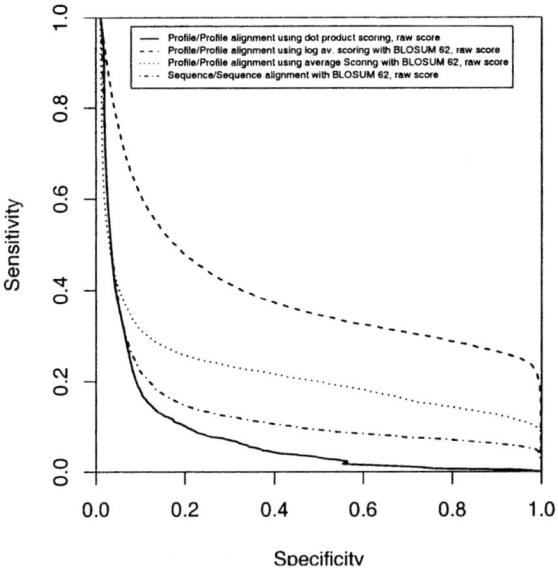

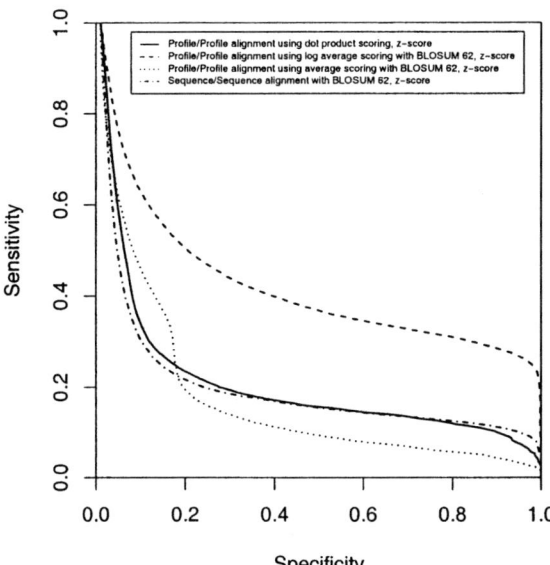

Fig. 5. Related Pair Recognition. Top: Specificity-sensitivity plots for the raw scores. Bottom: Specificity-sensitivity plots for the z-scores (see text).

The pair recognition test for the raw score provides a good measure of how well the alignment score represents a quantitative measure for the relationship between two proteins. The log average score outperforms all other methods here and the plain sequence-sequence alignment score even outperforms the dot product method which indicates that the latter approach is heavily dependent on some re-weighting procedure like the z-score rescaling. When performing this z-score rescaling the average scoring becomes significantly worse which is an unexpected effect since the objective is to make the scores comparable independent of the scoring method used. It is interesting that the log average score shows only a slight improvement here over the raw score performance suggesting that the raw score alone is already a good measure of similarity for the two profiles.

In conclusion, we see that the proposed log average score leads to a superior performance of profile-profile alignment methods in the disciplines fold recognition and related pair recognition suggesting that it is a better measure for the similarity of two profiles than the previously described other methods tested here. This is the effect of simply exchanging the log and the weighted average in the definition of the average score. A more general fact might also be learned from this: When a scoring function that maps a state to a score is to be extended to a more general setting where a score is assigned to a distribution of states, it is not always the best way to simply take the expected value (i. e. average scoring). Following this, future developments might include an incorporation of the log average scoring into a new scoring approach for protein threading as well as an application of the technique in the context of progressive multiple alignment tools.

Acknowledgements

This work was supported by the DFG priority programme "Informatikmethoden zur Analyse und Interpretation großer genomischer Datenmengen", grant ZI616/1-1. We thank Alexander Zien and Ingolf Sommer for the construction of the profiles and many helpful discussions.

A   Appendix

Two distinct sets of proteins from the PDB [3] are used in the described experiments. The first one is a set introduced by [1, 21] of 251 single domain proteins with known structure. It is derived from a non-redundant subset of the PDB introduced by [14] where the sequences have no more than 25 % pairwise sequence identity. From this set all single-domain proteins with all atom coordinates available are selected yielding the training set S_{train} of 251 proteins (see also [25]).

A.1   Adjusting Gap Costs

To provide each scoring approach with appropriate gap penalties we use the iterative approach *VALP* (for Violated Inequality Minimization Approximation Linear Programming) introduced in [25] which is based on a machine learning approach. We use a training set TR of 81 proteins from the data set mentioned above belonging to 11 fold classes each of which contain at least five of the sequences from TR. In every iteration each of the members of TR is used as a query and aligned against all 251 protein profiles. If we call the alignments of the best scoring fold class member for each of the 81 proteins the 81 *good* alignments and all the alignments of each of the 81 proteins against a member of a different fold class a *bad* alignment then the iteration tries to maximise the difference of the alignment scores between the good and the bad alignments. The iterations were stopped when a convergence could be observed which always happened before 16 iterations were completed.

A.2 Construction of Frequency Profiles

For each amino acid sequence in the two sets a homology search is performed using PSI-Blast [2] with 10 iterations against the KIND [15] database of non redundant amino acid sequences. The resulting multiple alignment from the last iteration is restricted to the query sequence. A frequency profile is calculated via a sequence weighting procedure that minimises the relative entropies of the frequency vectors regarding the background amino acid distribution [16]. Finally, a constant number of pseudo counts is added to account for amino acids that may occur by chance at this position. This is necessary since the goal is to end up with an estimation of the "true" amino acid distribution in a certain position of a protein family and it is not advisable to conclude from a finite number of observations which failed to show a certain amino acid that it is impossible (zero probability) to observe this amino acid in this position. Finally, all the profiles are scaled such that the total probability for all amino acids in each position yields one.

References

1. Nick Alexandrov, Ruth Nussinov, and Ralf Zimmer. Fast protein fold recognition via sequence to structure alignment and contact capacity potentials. In Lawrence Hunter and Teri E. Klein, editors, *Pacific Symposium on Biocomputing'96*, pages 53–72. World Scientific Publishing Co., 1996.
2. Stephen F. Altschul, Thomas L. Madden, Alejandro A. Schäffer, Jinghui Zhang, Zheng Zhang, Webb Miller, and David J. Lipman. Gapped BLAST and PSI-BLAST: a new generation of protein database search programs. *Nucleic Acids Research*, 25(17):3389–3402, September 1997.
3. F.C. Bernstein, T.F. Koetzle, G.J.B. Williams, E.F. Jr. Meyer, M.D. Brice, J.R. Rodgers, O. Kennard, T. Shimanouchi, and M. Tasumi. The protein data bank: a computer based archival file for macromolecular structures. *J.Mol.Biol.*, 112:535–542, 1977.
4. Patrick Billingsley. *Probability and Measure*. Wiley, 1995.
5. S. E. Brenner, P. Koehl, and M. Levitt. The ASTRAL compendium for protein structure and sequence analysis. *Nucleic Acids Res*, 28(1):254–6., 2000.
6. Michael Brown, Richard Hughey, Anders Krogh, I. Saira Mian, Kimmen Sjölander, and David Haussler. Using dirichlet mixture priors to derive hidden markov models for protein families. In *Proceedings of the Second Conference on Intelligent Systems for Molecular Biology*, volume 2, Washington, DC, July 1993. AAAI Press. preprint.
7. Jean-Michel Claverie. Some useful statistical properties of position-weight matrices. *Computers Chem.*, 18(3):287–294, 1994.

8. Margaret O. Dayhoff, R.M. Schwartz, and B.C. Orcutt. A model of evolutionary change in proteins. In *Atlas of Protein Sequence and Structure*, volume 5, Supplement 3, chapter 22, pages 345–352. National Biochemical Research Foundation, Washington DC, 1978.
9. Osamu Gotoh. An improved algorithm for matching biological sequences. *Journal of Molecular Biology*, 162:705–708, 1982.
10. Michael Gribskov, A. D. McLachlan, and David Eisenberg. Profile analysis: Detection of distantly related proteins. *Proceedings of the National Academy of Sciences of the United States of America*, 84(13):4355–4358, 1987.
11. Michael Gribskov and Stella Veretnik. Identification of sequence patterns with profile analysis. In *Methods in Enzymology*, volume 266, chapter 13, pages 198–212. Academic Press, Inc., 1996.
12. Steven Henikoff and Jorja G. Henikoff. Amino acid substitution matrices from protein blocks. *Proceedings of the National Academy of Sciences of the United States of America*, 89(22):10915–10919, 1992.
13. Steven Henikoff and Jorja G. Henikoff. Position–based sequence weights. *Journal of Molecular Biology*, 243(4):574–578, 1994. 4. November.
14. Uwe Hobohm and Chris Sander. Enlarged representative set of protein structures. *Protein Science*, 3:522–524, 1994.
15. Yvonne Kallberg and Bengt Persson. KIND – A non-redundant protein database. *Bioinformatics*, 15(3):260–261, March 1999.
16. Anders Krogh and Graeme Mitchison. Maximum entropy weighting of aligned sequences of protein or DNA. In C. Rawlings, D. Clark, R. Altman, L. Hunter, T. Lengauer, and S. Wodak, editors, *Proceedings of ISMB 95*, pages 215–221, Menlo Park, California 94025, 1995. AAAI Press.
17. L. Lo Conte, B. Ailey, T. J. Hubbard, S. E. Brenner, A. G. Murzin, and C. Chothia. SCOP: a structural classification of proteins database. *Nucleic Acids Res*, 28(1):257–9., 2000.
18. Leszek Rychlewski, Lukasz Jaroszewski, Weizhong Li, and Adam Godzik. Comparison of sequence profiles. Strategies for structural predictions using sequence information. *Protein Science*, 9:232–241, 2000.
19. Shamil R. Sunyaev, Frank Eisenhaber, Igor V. Rodchenkov, Birgit Eisenhaber, Vladimir G. Tumanyan, and Eugene N. Kuznetsov. PSIC: profile extraction from sequence alignments with position-specific counts of independent observations. *Protein Engineering*, 12(5):387–394, 1999.
20. Roman L Tatusov, Stephen F. Altschul, and Eugene V. Koonin. Detection of conserved segments in proteins: Iterative scanning of sequence databases with alignment blocks. *Proceedings of the National Academy of Sciences of the United States of America*, 91:12091–12095, December 1994.
21. Ralf Thiele, Ralf Zimmer, and Thomas Lengauer. Protein threading by recursive dynamic programming. *Journal of Molecular Biology*, 290(3):757–779, 1999.
22. Julie D. Thompson, Desmond G. Higgins, and Toby J.Gibson. CLUSTAL W: Improving the sensitivity of progressive multiple sequence alignment through sequence weighting, position-specific gap penalties and weight matrix choice. *Nucleic Acids Research*, 22(22):4673–4680, Nov 1994.
23. Hermann Witting. *Mathematische Statistik*. Teubner, 1966.
24. Thomas D. Wu, Craig G. Nevill-Manning, and Douglas L. Brutlag. Minimal-risk scoring matrices for sequence analysis. *Journal of Computational Biology*, 6(2):219–235, 1999.
25. Alexander Zien, Ralf Zimmer, and Thomas Lengauer. A simple iterative approach to parameter optimization. *Journal of Computational Biology*, 7(3):483–501, 2000.

False Positives in Genomic Map Assembly and Sequence Validation

Thomas Anantharaman[1] and Bud Mishra[2]

[1] Department of Biostatistics and Medical Informatics
University of Wisconsin, Madison, WI
tsa@carbon.biotech.wisc.edu
[2] Courant Institute, New York, NY

Abstract. This paper outlines an algorithm for whole genome order restriction optical map assembly. The algorithm can run very reliably in polynomial time by exploiting a strict limit on the probability that two maps that appear to overlap are in fact unrelated (false positives). The main result of this paper is a tight bound on the false positive probability based on a careful model of the experimental errors in the maps found in practice. Using this false positive probability bound, we show that the probability of failure to compute the correct map can be limited to acceptable levels if the input map error rates satisfy certain sharply delineated conditions. Thus careful experimental design must be used to ensure that whole genome map assembly can be done quickly and reliably.

1 Introduction

In the recent years, genome-wide shot-gun restriction mapping of several microorganisms using optical mapping [8, 7] have led to high-resolution restriction maps that directly facilitated sequence assembly avoiding gaps and compressions or validated shot-gun sequence assembly [4]. The simplicity and scalability of shot-gun optical mapping suggests obvious extensions to bigger and more complex genomes, and in fact, its applications to human and rice are underway. Furthermore, a good-quality human map is likely to play a critical role in validating several currently available but unverified sequences.

The key computational component of this process involves the assembly of large numbers of partial restriction maps with errors into an accurate restriction map of the complete genome. The general solution has been shown to be NP-complete, but a polynomial time solution is possible if a small fraction of false negatives (wasted data) is permitted. The critical component of this algorithm is an accurate bound for the false positive probability that two maps that appear to match are in fact unrelated.

The map assembly and alignment problems are related to the much more widely studied sequence assembly and alignment problems. The primary difference in the problem domains is that the sequence alignment problem involves only discrete data in which errors can be modeled as discrete probabilities, whereas map alignment involves fragment sizing errors and hence requires continuous error models. However, even in the case of sequence alignment, statistical significance tests play a key role in eliminating false positive matches and are included in many sequence alignment tools such as BLAST (see for example chapter 2 in [5]).

A simple bound using Brun's sieve can be easily derived [2], but such a bound often fails to exploit the full power of optical mapping. Here, we derive a much tighter but more complex bound that characterizes the sharp transition from infeasible experiments (requiring exponential computation time) to feasible experiments (polynomial computation time) much more accurately. Based on these bounds, a newer implementation of the Gentig algorithm for assembling genome-wide shot-gun maps [2] has improved its performance in practice.

A close examination shows that the false positive probability bound exhibits a computational phase-transition: that is, for poor choice of experimental parameters the probability of obtaining a solution map is close to zero, but improves suddenly to probability one as the experimental parameters are improved continuously. Thus careful optimized choice of the experimental parameters analytically has strong implication to experiment design in solving the problem accurately without incurring unnecessary laboratory or computational cost. In this paper, we explicitly delineate the interdependencies among these parameters and explore the trade-offs in parameter space: e.g., sizing error vs. digestion rate vs. total coverage. There are many direct applications of these bounds apart from the alignment and assembly of maps in Gentig: Comparing two related maps (e.g. chromosomal aberrations), Validating a sequence (e.g. shot-gun assembly-sequence) or a map (e.g., a clone map) against a map, etc. Specific usage of our bounds in these applications will appear elsewhere [3].

1.1   A Sub-quadratic Time Map Assembly Algorithm: Gentig

For the sake of completeness we give a brief but general description of the basic Gentig (GENomic conTIG) map assembly algorithm previously described elsewhere in details [2]. Roughly, Gentig can be thought of as a greedy algorithm that in any step considers two islands (individual maps or map contigs) and postulates the best possible way these two maps can be aligned. Next, it examines the overlapped region between these two islands and weighs the evidence in favor of the hypothesis that "these two islands are unrelated and the overlap is simply a chance occurrence." If enough evidence favors this "false positive" hypothesis, Gentig rejects the postulated overlap. In the absence of such evidence, the overlap is accepted and the islands are fused into a bigger/deeper island. What complicates these simple ideas is that one needs a very quantitative approach to calculate the probabilities, the most likely alignment and the criteria for rejecting a false positive overlap—all of these steps depending on the models of the error processes governing the observations of individual single molecule maps. Ultimately, the Gentig algorithm can be seen to be solving a constrained optimization problem with a Bayesian inference algorithm to find the most likely overlaps among the maps subject to the constraints imposed by the acceptable false positive probability. False Positive constraints limit the search space, thus obviating full-scale back-tracking and avoiding an exponential time complexity. As a result, the Gentig algorithm is able to achieve a sub-quadratic time complexity.

The Bayesian probability density estimate for a proposed placement is an approximation of the probability density that the two distinct component maps could have been derived from that placement while allowing for various modeled data errors: *sizing errors*, *missing restriction cut sites*, and *false optical cuts sites*.

The posterior conditional probability density for a hypothesized placement $\mathcal{H}$, given the maps, consists of the product of a prior probability density for the hypothesized placement and a conditional density of the errors in the component maps relative to the hypothesized placement. Let the M input maps to be contiged be denoted by data vectors D_j ($1 \leq j \leq M$) specifying the restriction site locations and enzymes. Then the Bayesian probability density for $\mathcal{H}$, given the data can be written using Bayes rule as in [1]:

$$f(\mathcal{H}|D_1\ldots D_M) \;=\; f(\mathcal{H})\prod_{j=1}^{M}f(D_j|\mathcal{H})\Big/\prod_{j=1}^{M}f(D_j) \;\propto\; f(\mathcal{H})\prod_{j=1}^{M}f(D_j|\mathcal{H}).$$

The conditional probability density function $f(D_j|\mathcal{H})$ depends on the error model used. We model the following errors in the input data:

1. Each orientation is equally likely to be correct.
2. Each fragment size in data D_j is assumed to have an independent error distributed as a Gaussian with standard deviation σ. (It is also possible to model the standard deviation as some polynomial of the true fragment size.)
3. Missing restriction sites in input maps D_j are modeled by a probability p_c of an actual restriction site being present in the data.
4. False restriction sites in the input maps D_j are modeled by a rate parameter p_f, which specifies the expected false cut density in the input maps, and is assumed to be uniformly and randomly distributed over the input maps.

The Bayesian probability density components $f(\mathcal{H})$ and $f(D_j|\mathcal{H})$ are computed separately for each contig (island) of the proposed placement and the overall probability density is equal to their products. For computational convenience, we actually compute a *penalty function*, Λ, proportional to the logarithm of the probability density as follows:

$$f(\mathcal{H}) = \left(\prod_{j=1}^{M}\frac{1}{(\sqrt{2\pi}\sigma)^{m_j}}\right)\exp(-\Lambda/(2\sigma^2)).$$

Here m_j is the number of cuts in input map D_j.

For fragment sizing errors, consider each fragment of the proposed contig, and let the contig fragment be composed of overlaps from several map fragments of length $x_1, \ldots, x_N$. If $p_c = 1$ and $p_f = 0$ (the ideal situation), it is easy to show that the hypothesized fragment size μ and the penalty Λ are:

$$\mu = \frac{\sum_i^N x_i}{N}, \quad \text{and} \quad \Lambda = \sum_{i=1}^{N}(x_i - \mu)^2.$$

Now consider the presence of missing cuts (restriction sites) with $p_c < 1$. To model the multiplicative error of p_c for each cut present in the contig we add a penalty $\Lambda_c = 2\sigma^2\log[1/p_c]$ and to model the multiplicative error of $(1 - p_c)$ for each missing cut in the contig we add a penalty $\Lambda_n = 2\sigma^2 \log[1/(1 - p_c)]$. The alignment computed by a Dynamic Programming algorithm determines which cuts are missing.

The computation of μ is modified in the case of missing cuts by assuming that the missing cuts are located in the same relative location (as a fraction of length) as in overlapping maps that do not have the corresponding cut missing. Finally, consider the presence of false optical cuts when $p_f > 0$. For each false cut, we add a penalty $\Lambda_f = 2\sigma^2 \log[1/(p_f\sqrt{2\pi}\sigma)]$ in order to model a "scaled" multiplicative penalty of p_f. A modified penalty term is required for the end fragments of each map which might be partial fragments, as described in [2]. When combining contigs of maps rather than input maps, the Dynamic programming structure is the same, except that the exact penalty values are slightly different and computed as the increase in penalty of the new contig over the penalty of the two shallower contigs being combined.

The resulting alignment algorithm has a time complexity of $O(m_i^2 m_j^2)$ in the worst case, but an average case complexity of $O(m_i + m_j)$, achieved with several simple heuristics. The basic dynamic programming is combined with a global search that tries all possible pairs of the M input maps for possible overlaps. A sophisticated implementation in Gentig achieves an average case time complexity of $O([mM]^{1+\epsilon})$ ($\epsilon = 0.40$ is typical for the errors we encounter), where m is the average value of m_j. It relies on several heuristics based on "geometric hashing" while avoiding any backtracking.

1.2 Summary of the New Results

Before proceeding further with the technical details of our probabilistic analysis, we summarize the two main formulae that can be used directly in estimating the false positive probability for a particular map alignment, or in designing a biochemical experiment with the goal of bounding the false positive probability below some acceptable small value (typically $\leq 10^{-3}$).

The Formula for False Positive Probability. Consider a population of M ordered restriction maps with errors of the kind described earlier. Assume that the best matching pair of maps (under a Bayesian formulation) has n aligned cuts and r misaligned cuts, and R is some average of the relative sizing error of aligned fragments in the overlap. Then FPT_r denotes the probability that the two maps are unrelated and the detected overlap is purely by chance.

$$FPT_r \leq 4\binom{M}{2}\binom{2n + r + 2}{r} P_n e^{\frac{rR}{\sqrt{2\pi}}}$$

$$\text{where } P_n = \frac{(R\sqrt{\frac{\pi\epsilon}{8}})^n}{\sqrt{n\pi}}.$$

Note that if $r = 0$ (implying that the best match has all the cuts aligned and the only error source is sizing error), then $FPT_0 = 4\binom{M}{2}P_n$. If $R \ll 1$ then as n gets larger FPT_0 exhibits an exponential decay to 0, and this property remains true for non-zero values of r.

The Formula for Feasible Genome-Wide Shotgun Optical Mapping. Consider an optical mapping experiment for genome-wide shotgun mapping for a genome of size G and involving M molecules each of length L_d. Thus the coverage is ML_d/G. Let the a fragment of true size X have a measured size $\sim \mathcal{N}(X, \sigma^2 X)$. Let the average true fragment size be L, and the digestion rate of the restriction enzyme be P_d. Thus the average relative sizing error $R = \sigma\sqrt{P_d/L}$ and the average size of aligned fragments will be $L/P_d{}^2$. As usual, let θ represent the minimum "overlap threshold." Hence the expected number of aligned fragments in a valid overlap is at least $n = \theta L_d P_d{}^2/L$. Let $d = 1/P_d$, the inverse of the digest rate. Feasible experimental parameters are those that result in an acceptable (e.g., $\leq 10^{-3}$) False Positive rate FPT:

$$FPT \approx 2M^2 \left(\frac{\lceil 2nd + 2 \rceil}{\lfloor 2n(d-1) \rfloor} \right) \frac{(R\sqrt{\frac{\pi e}{8}})^n}{\sqrt{n\pi}} e^{\frac{2(d-1)nR}{\sqrt{2\pi}}}$$

To achieve acceptable false positive rate, one needs to choose an acceptable value for the experimental parameters: P_d, σ, L_d and coverage. FPT exhibits a sharp phase transition in the space of experimental parameters. Thus the success of a mapping project depends extremely critically on a prudent combination of experimental errors (digestion rate, sizing), sample size (molecule length and number of molecules) and problem size (genome length). Relative sizing error can be lowered simply by increasing L with a choice of rarer-cutting enzyme and digestion rate can be improved by better chemistry [6].

As an example, for a human genome of size $G = 3,300Mb$ and a desired coverage of 6×, consider the following experiment. Assume a typical value of molecule length $L_d = 2Mb$. If the enzyme of choice is PAC I, the average true fragment length is about $25Kb$. Assume a minimum overlap[1] of $\theta = 30\%$. Assume that the sizing error for a fragment of $30kb$ is about $3.0kb$, and hence $\sigma^2 = 0.3kb$. With a digest rate of $P_d = 82\%$ we get an unacceptable $FPT \approx 0.0362$. However just increasing P_d to 86% results in an acceptable $FPT \approx 0.0009$. Alternately, reducing average sizing error from $3.0kb$ to $2.4kb$ while keeping $P_d = 82\%$ also produces an acceptable $FPT \approx 0.0007$.

Obviously one should allow some margin in choosing experimental parameters so that the actual experimental parameters will be a reasonable distance from the phase transition boundary. This is needed both to allow some slippage in experimental errors as well as the possibility that there may be additional small errors not modeled by the error model.

2 A Technical Probabilistic Lemma

The key to understanding the false positive bound is the following technical lemma that forms the basis of further computation. Let $\mathbf{X} = \langle x_1, \ldots, x_n \rangle$ and $\mathbf{Y} = \langle y_1, \ldots, y_n \rangle$ be a pair of sequences of positive real numbers, each sequence representing sizes of an ordered sequence of restriction fragments. We rely on a "matching rule" to decide whether $\mathbf{X}$ and $\mathbf{Y}$ represent the same restriction fragments in a genome, by comparing

---

[1] This value should be selected to minimize FPT.

the individual component fragments. We proceed by computing a "weighted squared relative sizing error" that is then compared to a specific threshold Θ. The "weighted squared relative sizing error" is simply

$$\sum_{i=1}^{n} w_i \left(\frac{X_i - Y_i}{X_i + Y_i} \right)^2,$$

where w_i's are chosen to match the error model. For example, if the sizing error variance for a fragment with true size X is $\sigma^2 X^p$, where $p \in [0, 2]$, we can use $w_i \equiv \frac{x_i + y_i}{\sigma^2}^{2-p}$.

Lemma 1. *Let* $\mathbf{X} = \langle X_1, \ldots, X_n \rangle$ *and* $\mathbf{Y} = \langle Y_1, \ldots, Y_n \rangle$ *be a pair of sequences of IID random variables* X_i's *and* Y_i's *with exponential distributions and pdf's* $f(x) = \frac{1}{L} e^{-x/L}$. *Then*

1. $\Pr(|X_i - Y_i|/(X_i + Y_i) \leq \Theta) \leq \Theta$, *for all* $0 \leq \Theta$ *and with equality holding, if* $\Theta \leq 1$.
2. $\Pr(\sum_{i=1}^{n} w_i (\frac{X_i - Y_i}{X_i + Y_i})^2 \leq \Theta) \leq \frac{(\frac{\pi}{4}\Theta)^{n/2}}{(\frac{n}{2})! \prod_{i=1}^{n} \sqrt{w_i}}$, *for all* $0 \leq \Theta$ *and with equality holding, if* $\Theta \leq \min_{1 \leq i \leq n} w_i$.

Proof. The first identity can be shown by integrating the relevant portion of the joint distribution of X_i and Y_i:

$$\Pr(|X_i - Y_i|/(X_i + Y_i) \leq \Theta)$$
$$= \int_{X_i=0}^{\infty} \int_{Y_i=X_i \frac{1-\Theta}{1+\Theta}}^{X_i \frac{1+\Theta}{1-\Theta}} \frac{1}{L^2} e^{-\frac{X_i+Y_i}{L}} \, dY_i \, dX_i = \Theta.$$

Note that this means that for each pair of random fragment sizes X_i, Y_i the statistic $U_i \equiv |X_i - Y_i|/(X_i + Y_i)$ is uniformly distributed between 0 and 1.

We can now compute the overall probability P_n for all n fragment pairs:

$$P_n \equiv \Pr(\sum_{i=1}^{n} w_i (\frac{X_i - Y_i}{X_i + Y_i})^2 \leq \Theta)$$
$$= \Pr(\sum_{i=1}^{n} w_i U_i^2 \leq \Theta)$$

Note that $U_1, \ldots, U_n$ are IID uniform distributions over $[0,1]$, hence this probability is just that part of the volume of the n-dimensional unit cube that satisfies the condition $\sum_1^n w_i U_i^2 \leq \Theta$. For small sizing errors such that $\Theta \leq \min(w_1, \ldots, w_n)$, this region is one orthant of an n dimensional ellipsoid with radius values of $\sqrt{\Theta/w_i}$ in the ith dimension. In general this volume is an upper bound and hence:

$$P_n \leq \frac{(\frac{\pi}{4}\Theta)^{n/2}}{(\frac{n}{2})! \prod_{i=1}^{n} \sqrt{w_i}}$$

Here $n!$ is defined in terms of the Gamma function for fractional n: $n! \equiv \Gamma(n + 1)$.
QED

Lemma 2. *Let* $\mathbf{X} = \langle X_1, \ldots, X_n \rangle$ *and* $\mathbf{Y} = \langle Y_1, \ldots, Y_n \rangle$ *be a pair of sequences such that variables* X_i's *and* Y_i's *are given in terms of IID random variables* Z_j's *with exponential distributions and pdf's* $f(z) = \frac{1}{L}e^{-z/L}$. *In particular, for* $i = 1, \ldots, n$, *if we can express* X_i *and* Y_i *in terms of exponential IID random variables* $Z_1, \ldots, Z_{r_i}$, $Z_{r_i+1}, \ldots, Z_{r_i+s_i}$ *as follows:*

$$\min(X_i, Y_i) = \frac{1}{2} \sum_{k=1}^{s_i} Z_{r_i+k}$$

$$\max(X_i, Y_i) = \sum_{k=1}^{r_i} Z_k + \frac{1}{2} \sum_{k=1}^{s_i} Z_{r_i+k}.$$

Then

1. $\Pr(|X_i - Y_i|/(X_i + Y_i) \le \Theta) \le \binom{s_i+r_i-1}{r_i}\Theta^{r_i}$, *for all* $\Theta \ge 0$.
2. $\Pr(\sum_{i=1}^n w_i(\frac{X_i-Y_i}{X_i+Y_i})^2 \le \Theta) \le \frac{\prod_{i=1}^n (r_i/2)!\binom{s_i+r_i-1}{r_i}(\Theta/w_i)^{r_i/2}}{(\sum_{i=1}^n r_i/2)!}$, *for all* $\Theta \ge 0$.

Proof. Similar to the previous lemma. **QED**

3 Model of Random Maps

Our model of random maps is that cut sites are randomly and uniformly distributed, so that the distance between cut sites is a random variable X with an exponential distribution and probability density $f(x) = \frac{1}{L}e^{-x/L}$, where L is the average distance between cut sites. Here we assume that all cut sites are indistinguishable from each other.

First we consider the case with no misaligned cuts, so that the only errors in the proposed overlap region are sizing errors. Thus our alignment data consists of two maps with fragment sizes $x_1, \ldots, x_n$ on one map that align with fragment sizes $y_1, \ldots, y_n$ on the other map, where n is the number of fragments in the overlap region. Here the quality of the alignment will be measured by a weighted squared relative sizing error, $E = \sum_1^n w_i(x_i - y_i)^2/(x_i + y_i)^2$, where w_i's are chosen as explained earlier. We need to compute $P_n = \Pr(\sum_1^n w_i(\frac{X_i-Y_i}{X_i+Y_i})^2 \le E)$, where $E \equiv \sum_1^n w_i(\frac{x_i-y_i}{x_i+y_i})^2$. By an application of the previous lemma, we have:

$$P_n \le \frac{(\frac{\pi}{4} \sum_{i=1}^n w_i(\frac{x_i-y_i}{x_i+y_i})^2)^{n/2}}{(\frac{n}{2})! \prod_{i=1}^n \sqrt{w_i}}.$$

Here, $n! \equiv \Gamma(n + 1)$. For current purposes it suffices to note that $(\frac{1}{2})! = \frac{\sqrt{\pi}}{2}$. For example, $(\frac{3}{2})! = \frac{3}{2}(\frac{1}{2})! = \frac{3\sqrt{\pi}}{4}$.

To see more clearly how this probability scales with the sizing errors, let us define the weighted RMS relative sizing error R_n, and the average weight A_n:

$$R_n \equiv \sqrt{\frac{\sum_{i=1}^n w_i(\frac{x_i-y_i}{x_i+y_i})^2}{\sum_{i=1}^n w_i}}. \tag{1}$$

$$A_n \equiv \frac{1}{n} \sum_{i=1}^n w_i. \tag{2}$$

Then we can rewrite P_n using Sterling's expression for factorials as:

$$P_n \leq \frac{(R_n/\sqrt{2/e\pi})^n}{\sqrt{n\pi}} \prod_{i=1}^{n} \sqrt{\frac{A_n}{w_i}}. \tag{3}$$

This shows that asymptotically the n-fragment false positive probability P_n will decrease with the nth power of the RMS relative error R_n provided that $R_n \leq \sqrt{2/e\pi} = 0.4839$.

To complete our computation of the False Positive Likelihood FP for a particular pair of maps D_1 and D_2, we need to consider the multiple possible choices of overlaps of n or more fragments. Let the two molecules contain N_1 and N_2 fragments with $N_1 \leq N_2$. If $n < N_1$ there will be exactly 4 possible ways of forming an overlap of n fragments. Otherwise, if $n = N_1$ there will be $2(N_2 - N_1 + 1)$ ways of forming an n fragment overlap. Each such overlap has the same independent probability P_n. Thus with 4 possible overlaps, the probability FP_n of finding at least 1 overlap of n fragments between two random maps as good as the actual alignment is bounded by the probability $FP_n \leq 1 - (1 - P_n)^4 \leq 4P_n$.

We also need to consider random overlaps of more than n fragments that are as good as the actual overlap. Under a typical Bayesian error model such as described in [2], each overlap of more than n fragments can have slightly larger sizing errors than the actual alignment with the same probability density, since the prior probability density must be biased towards larger overlaps. For an error model such as in [2] one can show that the permissible increase in relative sizing error R_{n+1} vs. R_n is given approximately by:

$$A_{n+1}R_{n+1}^2 \leq \frac{nA_nR_n^2 + K/2}{(n+1)},$$

where K is a prior bias parameter, typically in the range $1 \leq K \leq 1.4$. Hence for $n + k < N_1$ we can write FP_{n+k} as:

$$FP_{n+k} \leq 4P_n \sqrt{\frac{n}{n+k}} R_n^{k} \left(\frac{\pi A_n e^{K/2A_n R_n^2}}{2G_n} \right)^{k/2}$$

Here G_n is the geometric mean of $w_i, i = 1, \ldots, n$. If $n + k > N_1$ then $FP_{n+k} = 0$ and if $n + k = N_1$ we just need to replace the factor 4 by $2(N_2 - N_1 + 1)$.

We can now compute FP by combining overlaps of all possible number of fragments (ranging over $n, \ldots, N_1$):

$$FP \leq \sum_{k=0}^{N_1-n} FP_{n+k}$$

$$= 2P_n \left(2/(1 - Z) + \left(N_2 - N_1 - \frac{1+Z}{1-Z} \right) Z^{N_1-n} \right)$$

$$\text{where,} Z = R_n \sqrt{\frac{\pi A_n e^{K/2A_n R_n^2}}{2G_n}}$$

This result applies to the case of two maps. The generalization to a population of many maps is considered for the more general case of missing cuts in the next section.

4   False Positive Probability with Missing Cuts

When misaligned cuts are present in the actual alignment, the false positive probability becomes larger. Assuming the maps are random, we have many possible alignments for a given overlap region, greatly increasing the odds of coming up with a good alignment.

In this case our actual alignment data consists of n pairs of fragment sizes $x_1, x_2, ...,$ x_n and $y_1, y_2, ..., y_n$, as before, plus the total number of fragments m in the overlap region of the two maps, where $m = 2n + r$ and $r \geq 0$ is the number of misaligned cuts.

First consider the case where the number of misaligned cuts is fixed at $r = m - 2n$ and the number of aligned fragments is fixed at n in each map and we define the probability $P_{n,m}$ as the probability that two random maps with an overlap region of exactly m total fragments in both maps could produce an alignment of n fragments as good as the actual alignment.

The key to computing $P_{n,m}$ is a systematic way to enumerate possible alignments that can be applied to each sample of the two hypothesized random maps, then compute the probabilities that a particular enumerated alignment will have better sizing error than the actual alignment, and combine these probabilities over all enumerated alignments.

It simplifies matters if we consider random alignments between the left end of two random maps and compute the probability of finding an alignment involving the first m fragments from the left (on either of the two random maps) that is as good as the actual overlap.

We now claim that all possible alignments between the left end of two random maps involving m fragments can be enumerated, independent of the random map sample, as follows:

1. Pick any $n + 1$ numbers $s_0, s_1, ..., s_n$ and another $n + 1$ numbers $r_0, r_1, ..., r_n$ subject only to the constraints $s_i \geq 1, r_i \geq 1, \sum_{i=0}^{n}(s_i + r_i) \leq m + 2$
2. Align the two random map samples so that their left ends coincide. Then scan both maps from the left end and pick the s_0th cut site encountered on either map. Then scan further to the right on the other map until another r_0 cut sites have been encountered and align the r_0th one with the previous cut site. This defines the first aligned pair of cuts. The map is now re-aligned so that this pair of cuts coincide (rather than the left ends of the maps).
3. Repeat this step for $i = 1, ..., n$: Starting from the previous aligned cut site scan to the right on both maps until s_i sites have been seen and mark the last one, which could be on either map. Then scan to the right from that cut site on the opposite map only until r_i cut sites have been seen and align the last (r_ith) one with the previously marked cut site. This defines the ith set of aligned fragments. Realign the maps so that this pair of cut sites coincide.
4. After aligning the last (($n + 1$)th) cut site, scan right on both maps until a total of $m + 2$ cut sites have been seen (including all cut sites seen in previous steps). Mark the boundary of the aligned region anywhere between the last seen cut site and the next one.

For each enumerated alignment defined by a particular choice of $s_0, \ldots, s_n$, and $r_0, \ldots, r_n$ we can compute the probability $PA_{n,m,s,r}$ that for two random maps that particular enumerated alignment will have relative sizing errors better than the actual map alignment. We can then compute an upper bound for $P_{n,m}$ as the sum of $PA_{n,m,s,r}$ over all enumerated alignments.

First we compute E_{r_0,s_0} the probability of no overhang as a function of r_0 and s_0. An overhang occurs if the sum of r_0 random fragments add up to more than the leftmost fragment in the same molecule. Using IID random variables $\delta_1, \ldots, \delta_r$ each drawn from $\frac{1}{L}e^{-X/L}$ to represent the r intervals, and $Z_1, \ldots, Z_s$, also drawn from $\frac{1}{L}e^{-X/L}$, to represent twice the s intervals used to select the first aligned cut, we can obtain by suitable integration:

$$E_{r,s} = \frac{1}{3^{r-1}}\left(\frac{1}{3^{s-1}}\binom{r+s-2}{r-1} + \sum_{k=1}^{s-1}\frac{1}{3^k}\binom{r+k-2}{k-1}\right)$$

Using the earlier lemma 2.2, We can write $PA_{n,m,s,r}$ as follows:

$$PA_{n,m,s,r} = E_{r_0,s_0}\Pr\left(\sum_{i=1}^{n} w_i\left(\frac{X_i - Y_i}{X_i + Y_i}\right)^2 \leq E\right)$$

$$\text{where } E \equiv \sum_{i=1}^{n} w_i\left(\frac{x_i - y_i}{x_i + y_i}\right)^2 = nA_nR_n{}^2,$$

and simplify it to

$$PA_{n,m,s,r} \leq E_{r_0,s_0}P_n(2eR_n{}^2A_n)^{S/2}\left(\frac{n}{n+S}\right)^{(n+S+1)/2}\prod_{i=1}^{n}\frac{(\frac{r_i}{2})!}{(\frac{1}{2})!}\frac{\binom{s_i+r_i-1}{r_i}}{w_i^{\frac{r_i-1}{2}}}$$

$$\text{where } N \equiv \max_{1\leq i\leq n} r_i, \text{ and } S \equiv \sum_{i=1}^{n}(r_i - 1).$$

Here P_n is the result without misaligned cuts.

We will now sum up $PA_{n,m,s,r}$ over all possible choices of $s_0, \ldots, s_n$ while keeping $r_0, \ldots, r_n$ fixed subject to the constraint $\sum_{i=0}^{n}(s_i + r_i) \leq m + 2$ to produce:

$$PA_{n,m,r} \equiv \sum_{s} PA_{n,m,s,r}$$

$$\leq \frac{1}{2^{r_0}}\left(2\binom{m+1-r_0}{2n+S} + \binom{m+1-r_0}{2n+1+S}\right)$$

$$\times P_n(2eR_n{}^2A_n)^{S/2}\left(\frac{n}{n+S}\right)^{(n+S+1)/2}\prod_{i=1}^{n}\frac{(\frac{r_i}{2})!}{(\frac{1}{2})!w_i^{(r_i-1)/2}}$$

Next, we will add up $PA_{n,m,r}$ over all possible choices of r_0 then $r_1, \ldots, r_n$, where r_0 is constrained by $1 \leq r_0 \leq m - 2n - S + 1$ and $r_1, \ldots, r_n$ by $S \equiv \sum_{i=1}^{n}(r_i - 1) \leq$

$m - 2n$. Approximating w_i by its geometric mean G_n where needed we get:

$$P_{n,m} = \sum_{r_1..r_n} \sum_{r_0} PA_{n,m,r}$$

$$\leq \sum_{r_1..r_n} P_n (2eR_n^2 A_n)^{S/2} \left(\frac{n}{n+S}\right)^{(n+S+1)/2} \left(\frac{m+1}{2n+1+S}\right)^n \prod_{i=1}^{n} \frac{(\frac{r_i}{2})!}{(\frac{1}{2})! w_i^{(r_i-1)/2}}$$

$$\leq P_n \left(\frac{m+1}{2n+1}\right) \left(1 + \sum_{j=2}^{r+1} \left(\frac{(\frac{j}{2})!}{(\frac{1}{2})!}\right) \left(R_n\sqrt{\frac{2A_n}{G_n}}\right)^{j-1} \left(\frac{\binom{m+1}{2n+j}}{\binom{m+1}{2n+1}}\right)\right)^n.$$

The resulting expression diverges for large values of r but the bound is quite tight for realistic values of m, n, R_n.

As a final step in computing the False Positive probability we need to combine the $P_{n,m}$ just computed over random alignments involving fewer misaligned cuts (smaller values of r) or more aligned fragments (larger n), as well as consider the possible ways the ends of two random maps could be aligned with each other. Using the same approach as for the case without misaligned cuts to model the permissible change in sizing error we can show that the result is:

$$FP_r \leq 4P_n \binom{2n+r+2}{r} \left(1 + \sum_{j=2}^{r+1} \left(\frac{(\frac{j}{2})!}{(\frac{1}{2})!}\right) \left(R_n\sqrt{\frac{2A_n}{G_n}}\right)^{j-1} \left(\frac{\binom{2n+r+1}{r+1-j}}{\binom{2n+r+1}{r}}\right)\right)^n$$

$$\left(\frac{1}{1-Z} + \frac{1}{2}\left(N_2 - N_1 - \frac{1+Z}{1-Z}\right) Z^{N_1-n-r/2}\right)$$

where,

$$Z = R_n \sqrt{\frac{\pi A_n e^{K/2A_n R_n^2}}{2G_n}} \left(\frac{2n+r+3}{2n+3}\right)^2$$

5 False Positive Probability: Population of Maps

Finally if there are multiple maps to choose from, we need to reject the possibility that the proposed map pair is merely the best matching amongst all possible map pairs. We need not consider maps with less than $n + r/2$ fragments. Let the number of maps with at least $n + r/2$ fragments be M, with number of fragments $N_i, i = 1, \ldots, M$ arranged in ascending order. For each of the possible $M(M-1)/2$ map pairs we can compute the probability FP_r just described, but with N_1, N_2 suitably adjusted. The resulting probability FPT_r is given by:

$$FPT_r \leq \sum_{i=1}^{M-1} \sum_{j=i+1}^{M} FP_r(N_i, N_j)$$

$$\leq P_n \binom{2n+r+2}{r} \left(1 + \sum_{j=2}^{r+1} \left(\frac{(\frac{j}{2})!}{(\frac{1}{2})!}\right) \left(R_n\sqrt{\frac{2A_n}{G_n}}\right)^{j-1} \left(\frac{\binom{2n+r+1}{r+1-j}}{\binom{2n+r+1}{r}}\right)\right)^n$$

$$\left(\frac{2M(M-1)}{1-Z} + 4 \sum_{i=1}^{M-1} \sum_{j=i+1}^{M} \left(N_j + N_i - \frac{1+Z}{1-Z} \right) Z^{N_i - n - r/2} \right) \quad (4)$$

where,

$$Z = R_n \sqrt{\frac{\pi A_n e^{K/2A_n R_n^2}}{2G_n} \left(\frac{2n+r+3}{2n+3} \right)^2} \quad (5)$$

Which is to be used with the previous equations for P_n, R_n, A_n, G_n and the error model parameter K (and implicitly σ).

6   Experiment Design

In designing a shot-gun genome wide mapping experiment, one needs to ensure that the data allows correct map overlaps to be clearly distinguished from random map overlaps. If this is done using a False Positive threshold such as the FPT we have derived in this paper, the goal is to ensure that the expected FPT for correct map overlaps does not exceed some acceptable threshold (e.g. 10^{-3}). In this section we will estimate the expected value of FPT for a valid overlap based on the experimental error parameters.

In principle we just need to estimate the values of n, r, R_n, M for a correct overlap based on the experimental errors. However given the extreme sensitivity of FPT on n, the number of aligned fragments, we will compute FPT for correct map overlaps of a certain minimum size. By selecting a suitable value of θ, the minimum "overlap value", we can control the expected minimum value of n, at the cost of some reduction in effective coverage by the factor $1 - \theta$ [9].

In addition to θ assume the following experimental parameters: G = Expected Genome size. L_d = Length of each map. C = Desired coverage (before adjustment for θ). L = average distance between restriction site in Genome. $\sigma \sqrt{X}$ = Sizing error (standard deviation) for fragment of size X. P_d = The digestions rate of the restriction enzyme used.

Assuming $R \ll 1$ and $A_n \approx G_n$ we can then write FPT in terms of the experimental parameters as:

$$FPT \approx 2M^2 \left(\frac{\lceil 2nd + 2 \rceil}{\lfloor 2n(d-1) \rfloor} \right) \frac{(R\sqrt{\frac{\pi e}{8}})^n}{\sqrt{n\pi}} \left(1 + (d-1)R\sqrt{\frac{2}{\pi}} \right)^n \quad (6)$$

where we have $d = \frac{1}{P_d}, n = \frac{L_d \theta}{L P_d^2}, R = \frac{\sigma}{\sqrt{L/P_d}}$, and $M = \frac{CG}{L_d}$.

7   Simulated Experimental Results

This section provides some simulated experiments of our algorithm, to illustrate the effect of different experimental parameters. To generate simulated tests sets of the map data, we begin with a 30 Mb *in silico* map of chromosome 22 using the PAC I restriction enzyme, and sample random pieces of this map averaging 2 Mb in size, and introduce the following errors into each such map:

- : Each fragment of actual size L is re-sampled from a Gaussian distribution with mean L and variance $\sigma^2 L$ with $\sigma^2 = 0.3kB$.
- : Each restriction site is retained with an independently simulated digest rate of P_d. We ran experiments with $P_d = 0.5, 0.6, 0.7, 0.71, 0.72, 0.73, 0.74, 0.75, 0.8$ and 0.9.
- : False cuts are introduced at a uniform rate of $R_f = 1$ per Mb.
- : Missing small fragments are simulated at a rate of 0.7^L for an L Kb fragment etc.
- : The orientation of each molecule is randomly flipped with probability of 0.5.

Enough maps were generated in each experiment to produce a gross coverage of 16x (240 maps for the 30 Mb chromosome). Each simulated data set was then run on a program implementing the algorithms described in this paper, using false positive thresholds of 0.00001.

Digest-Rate	# contigs	Size of the largest contig	# singletons
0.60	9	3994 Kb	212
0.70	5	11588 Kb	110
0.71	3	13731 Kb	90
0.72	3	14385 Kb	88
0.73	1	30119 Kb	53
0.75	1	30144 Kb	47
0.80	1	30075 Kb	17
0.90	1	30000 Kb	4

The contigs generated were checked and verified to be at the correct offset (though minor local alignment errors may be present).

8   Conclusion

In this paper we derived a tight False Positive Probability bound for overlapping two maps. This can be used in the assembly of genome wide maps to reduce the search space from exponential time to sub-quadratic time with only a small increase in false negatives. The False Positive Probability bound also can be used to determine if a sequence derived map has a statistically significant match with a map.

We also showed how the False Positive Probability bound can be used to select experimental parameters for whole-genome shot-gun mapping that will allow the genome wide map to be assembled rapidly and reliably and showed that the boundary between feasible and infeasible experimental parameters is quite narrow, exhibiting a form of computational phase transition.

Our approach has certain limitations due to the assumptions underlying our model that unrelated parts of the genome will not align with each other except in a random manner. This assumption is not true for haplotypes, for example, and hence our algorithm is not sufficient to produce a haplotyped map. Similarly if some other biological process results in strong homologies over large distances our algorithm may merge homologous regions of the genome. If this turns out to be problem, explicit postprocessing of the resulting map contigs to look for merged homologous regions or haplotypes can be performed.

Acknowledgments

The research presented here was partly supported by NSF Career Grant IRI-9702071, DOE Grant 25-74100-F1799, NYU Research Challenge Grant, and NYU Curriculum Challenge Grant.

References

1. T. S. Anantharaman, B. Mishra, and D. C. Schwartz. Genomics via optical mapping ii: Ordered restriction maps. *Journal of Computational Biology*, 4(2):91–118, 1997.
2. T. S. Anantharaman, B. Mishra, and D. C. Schwartz. Genomics via optical mapping iii: Contiging genomic dna. In *Proceedings 7th Intl. Conf. on Intelligent Systems for Molecular Biology: ISMB '99*, pages 18–27. AAAI Press, 1999.
3. M. Antoniotti, B. Mishra, T. Anantharaman, and T. Paxia. Genomics via optical mapping iv: Sequence validation via optical map matching. Preprint.
4. C. Aston, B. Mishra, and D. C. Schwartz. Optical mapping and its potential for large-scale sequencing projects. *Trends in Biotechnology*, 17:297–302, 1999.
5. R. Durbin, S. Eddy, A. Krogh, and G. Mitchison. *Biological Sequence Analysis*. Cambridge University Press, 1998.
6. J. Reed et al. A quantitative study of optical mapping surfaces by atomic force microscopy and restriction endonuclease digestion assays. *Analytical Biochemistry*, 259:80–88, 1998.
7. L. Lin et al. Whole-genome shotgun optical mapping of *Deinococcus radiodurans* . *Science*, 285:1558–1562, 1999.
8. Z. Lai et al. A shotgun sequence-ready optical map of the whole *Plasmodium falciparum* genome. *Nature Genetics*, 23(3):309–313, 1999.
9. M. S. Waterman. *Introduction to Computational Biology*. Chapman and Hall, 1995.

Boosting EM for Radiation Hybrid
and Genetic Mapping

Thomas Schiex[1], Patrick Chabrier[1], Martin Bouchez[1], and Denis Milan[2]

[1] Biometry and AI Lab.
[2] Cellular Genetics Lab., INRA, Toulouse, France
{Thomas.Schiex,Denis.Milan}@toulouse.inra.fr

Abstract. Radiation hybrid (RH) mapping is a somatic cell technique that is used for ordering markers along a chromosome and estimating physical distances between them. It nicely complements the genetic mapping technique, allowing for finer resolution. Like genetic mapping, RH mapping consists in finding a marker ordering that maximizes a given criteria. Several software packages have been recently proposed to solve RH mapping problems. Each package offers specific criteria and specific ordering techniques. The most general packages look for maximum likelihood maps and may cope with errors, unknowns and polyploid hybrids at the cost of limited computational efficiency. More efficient packages look for minimum breaks or two-points approximated maximum likelihood maps but ignore errors, unknowns and polyploid hybrids.

In this paper, we present a simple improvement of the EM algorithm [5] that makes maximum likelihood estimation much more efficient (in practice and to some extent in theory too). The boosted EM algorithm can deal with unknowns in both error-free haploid data and error-free backcross data. Unknowns are usually quite limited in RH mapping but cannot be ignored when one deals with genetic data or multiple populations/panels consensus mapping (markers being not necessarily typed in all panels/populations). These improved EM algorithms have been implemented in the CAR⁴ᴴAGÈNE software. We conclude with a comparison with similar packages (RHMAP and MapMaker) using simulated data sets and present preliminary results on mixed simultaneous RH/genetic mapping on pig data.

1 Introduction

Radiation hybrid mapping [4] is a somatic cell technique that is used for ordering markers along a chromosome and estimating the physical distances between them. It nicely complements alternative mapping techniques especially by providing intermediate resolutions. This technique has been mainly applied to human cells but also used on animals, e.g. [6].

The biological experiment in RH mapping can be rapidly sketched as follows: cells from the organism under study are irradiated. The radiation breaks the chromosomes at random locations into separate fragments. A random subset of the fragments is then rescued by fusing the irradiated cells with normal rodent cells, a process that produces a collection of hybrid cells. The resulting clone may contain none, one or many chromosome fragments. This clone is then tested for the presence or absence of each of the

O. Gascuel and B.M.E. Moret (Eds.): WABI 2001, LNCS 2149, pp. 41–51, 2001.

markers. This process is performed a large number of times producing a radiated hybrid panel.

The algorithmic analysis which follows the biological experiment, based on the retention patterns of the markers observed, aims at finding the most likely linear ordering of the markers on the chromosome along with distances between markers. The underlying intuition is that the further apart two markers are, the most likely it is that the radiation will create one or more breaks between them, placing the two markers on separate chromosomal fragments. Therefore, close markers will tend to be more often co-retained than distant ones. Given an order of the markers, the retention pattern can therefore be used to estimate the pairwise physical distances between them.

Two fundamental types of approaches have been used to evaluate the quality of a possible marker's permutation. The first, crudest approach, is a non parametric approach, called "obligate chromosomal breaks" (OCB), that aims at finding a permutation that minimizes the number of breaks needed to explain the retention pattern. This approach is not considered in this paper. The second one is a statistical parametric method of maximum likelihood estimation (MLE) using a probabilistic model of the RH biological mechanism. Several probabilistic models have been proposed for RH mapping [9] dealing with polyploidy, errors and unknowns. In this paper, we are only interested in a subset of the models that are compatible with the use of the EM algorithm [5] for estimating distances between markers. According to our experience, the simplest "equal retention model" is the most frequently used in practice and also the most widely available because it is a good compromise between efficiency and realism. Such models are used in the RHMAP and RHMAPPER packages. More recently, more efficient approximated MLE versions based on two points estimation have also been used [2] but they don't deal with unknowns and won't be considered in the sequel.

The older but still widely used genetic mapping technique [10] exploits the occurrence of cross-overs during meiosis. As for RH mapping, the underlying intuition is that the further apart two markers are, the most likely it is that a cross-over will occur in between. Because cross-overs cannot be directly observed, the indirect observation of allelic patterns in parents and children is used to estimate the genetic distance between them. There is a long tradition of using EM in genetic mapping [7]. This paper will focus on RH mapping but we must mention that the improvements presented in this paper have also been applied to genetic mapping with backcross pedigree. Actually, genetic and RH data nicely complement each other for *ordering* markers. Genetic data leads to myopic ordering: set of close markers cannot be reliably ordered because usually no recombination can be observed between them. On the contrary RH data leads to hypermetropic ordering: set of closely related markers can be reliably ordered but distant groups are sometimes difficult to order because too many breaks occurred between them. Dealing with unknown is unavoidable in genetic mapping since markers may be uninformative.

In either RH or genetic mapping, the most obvious computational barrier is the shear number of possible orders. For n markers, there are $\frac{n!}{2}$ possible orders (as an order and its reverse are equivalent), which is too large to search exhaustively, even for moderate values of n. In the simplest case of error-free unknown-free data, it has been observed by several authors that the MLE ordering problem is equivalent to the famous traveling salesman problem [13, 1], an NP-hard problem. The ordering techniques used in existing packages go from branch and bound [2], to local search techniques and more or less greedy heuristics. In all cases, finding a good order requires a very large number of MLE calls. In practice, the cost of EM execution is still too heavy to make branch and bound or local search methods computationally usable on large data sets and the greedy heuristic approach remains among the most widely used in practice.

In this paper, we show how the EM algorithm for RH/genetic mapping can be sped up when it is applied to data sets where each information is either completely informative or completely uninformative. This is the case, e.g. for error-free haploid RH data with unknown or error-free backcross data with unknowns. In practice, for RH mapping, it has been observed in [9] that "analyzing polyploid radiation hybrids as if they were haploid does not compromise the ability to order markers" which makes this restriction to haploid data quite reasonable. For genetic mapping, most phase-known data can be reduced (although with some possible loss of information) to backcross data. In practice, we have applied it to diploid RH data and complex animal (pig) pedigree with very limited discrepancies (see section 5) and a speed-up factor of two orders of magnitude.

Interestingly, this boosted EM algorithm is especially adapted to the unknown/ known patterns that appear in multiple population/panels mapping when a single consensus map is built: when two or more data sets are available for the same organism (and with a similar resolution for RH), a possible way to build a consensus map is to merge the two data sets in one. markers that are not typed in one of the 2 data sets are marked as unknown in this case. In this case, we show that each iteration of the EM algorithm may be in $O(n)$ instead of $O(n.k)$, where n is the number of markers and k the number of individuals typed.

These improved EM algorithms along with several TSP-inspired ordering strategies for both framework and comprehensive mapping have been implemented in the free software package CAR$_H^T$AGÈNE [13] which allows for multiple population/panels mapping using either shared or separate distance estimation for each pair of data sets. This allows, among other, for mixed genetic/RH mapping (with separate estimations of genetic and RH distances) which nicely exploits the complementarity of genetic and RH data.

2   The EM Algorithm for RH Data

In this section, we will explain how EM can be optimized to deal with haploid RH data sets with unknowns. This optimization also applies to backcross genetic data sets with missing data. It has been implemented in the genetic/RH mapping software CAR$_H^T$AGÈNE [13] but has never been described in the literature before.

Suppose that genetic markers $M_1, \ldots M_n$ are typed on k radiation hybrids. The observations for a given hybrid, given the marker order $(M_1, \ldots, M_n)$ can be written as a vector $x = (x_1, \ldots, x_n)$ where $x_i = 1$ if the marker M_i is typed and present, $X_i = 0$ if the marker is typed and absent and $x_i = -$ if the marker could not be reliably typed. Such unknowns are relatively rare in RH mapping but are much more frequent in genetic mapping or in multiple population/panel consensus mapping.

The probabilistic HMM model for generating each sequence x in the case of error-free haploid "equal retention" data is defined by one retention probability denoted r (probability for a fragment to be retained) and $n - 1$ breakage probability denoted $b_1, \ldots, b_{n-1}$ (b_i is the probability of breakage between marker M_i and M_{i+1}). Breakage and retention are considered independent processes.

The structure of the HMM model for 4 markers ordered as M1,...,M4 is sketched as a weighted digraph $G = (V, E)$ in Figure 1. Vertices correspond to the possible state of

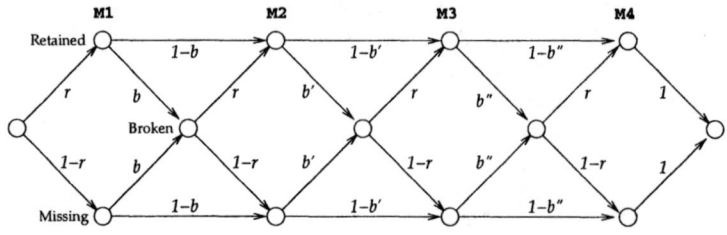

Fig. 1. A Graph Representation of the "Equal Retention Model" for RH Mapping.

being respectively retained, missing or broken. An edge $(a, b) \in E$ that connects two vertices a and b is simply weighted by the conditional probability of reaching the state b from the state a, noted $p(a, b)$. For example, if we assume that M_1 is on a retained fragment, there is a probability b_1 that a new fragment will start between M_1 and M_2 and a probability $(1 - b_1)$ that the fragment remains unbroken. In the first case, the new fragment may either be retained (r) or not ($1 - r$). In the second case, we know that M_2 is on the same retained fragment. The EM algorithm [5] is the usual choice for parameter estimation in hidden Markov models [11] where it is also known as the Forward/Backward or Baum-Welsh algorithm. This algorithm can be used to estimate the parameters $r, b_1, b_2 \ldots$ and to evaluate the likelihood of a map given the available evidence (a vector x of observation for each hybrid in the panel).

If we consider one observation $x = (0, -, 0, 1)$ on a given hybrid, the graph can be restricted to the partial graph of Figure 2 by removing vertices and edges which are incompatible with the observation (dotted in the figure). Every path in this graph corresponds to a possible reality. The path in bold corresponds to the fact that a fragment has been a breakage between each pair of markers, each fragment being successively missing, retained, missing and retained. If we define the probability of such a source-sink path as the product of all the edge probabilities, then the sum of the probabilities of all the paths that are compatible with the observation is precisely its likelihood. Although there is a worst-case exponential number of such paths, dynamic programming, embodied in the so-called "Forward" algorithm [11] may be used to compute the likelihood of a single hybrid in $\theta(n)$ time and space. For any vertex v, if we note $P_l(v)$ the sum of the probabilities of all the paths that exist in the graph from the source to the vertex v, we have the following recurrence equation:

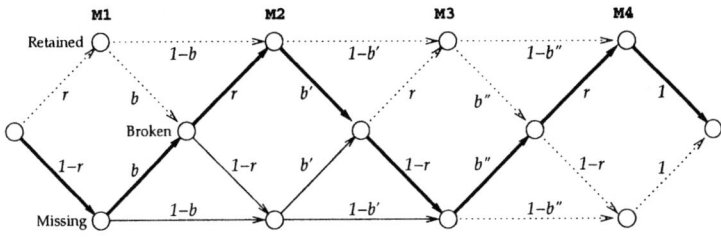

Fig. 2. The Graph Representation for $x = (0, -, 0, 1)$.

$$P_l(v) = \sum_{u \text{ s.t. } (u,v) \in E} P_l(u).p(u,v)$$

This simply says that in order to reach v from the source, we must first reach a vertex u that is directly connected to v (with probability $P_l(u)$) then go to v (with probability $p(u,v)$). We can sum up all these probabilities that correspond to an exhaustive list of distinct cases. This recurrence can simply be used by initializing the probability P_l of the source vertex to 1.0 and applying the equation "Forward" (from left to right, using a topological ordering of the vertices). Obviously, P_l for the sink vertex is nothing but the likelihood of the hybrid. One should note that the same idea can be exploited to compute for each vertex $P_r(v)$, the sum of the probabilities of all paths that connect v to the sink (simply reverse all edges and apply the "forward" version).

The EM algorithm is an iterative algorithm that starts from given initial values of the parameters $r_0, b_0, b'_0 \ldots$ and that goes repeatedly through two phases:

1. Expectation: for each hybrid $h \in H$, given the current value of the parameters, the probability $P_h(u,v)$ that a path compatible with the observation x for hybrid h uses edge (u,v) can simply be computed by:

$$P_h(u,v) = P_l(u).p(u,v).P_r(v)$$

If for a given parameter p, and given hybrid h, we note $S_h^+(p)$ the set of all edges weighted by p and $S_h^-(p)$ the set of all edges weighted by $1-p$, an expected number of occurrence of the corresponding event can be computed by:

$$E(p) = \sum_{h \in H} \frac{\sum_{(u,v) \in S_h^+(p)} P_h(u,v)}{\sum_{(u,v) \in S_h^+(p) \cup S_h^-(p)} P_h(u,v)}$$

2. Maximization: the value of each parameter p is updated by maximum likelihood under the assumption of complete data by:

$$p_{i+1} = \frac{E(p)}{k}$$

It is known that EM will produce estimates of increasing likelihood till a local maximum is reached. The usual choice is to stop iteration when the increase of log-likelihood is lower than a given tolerance. Several iterations are usually needed to reach, e.g. tolerance 10^{-4}, especially as the number of unknowns increases.

Each of the forward, backward and $E(p)$ computation of the E phase successively treat each pair of adjacent markers in constant time (there are at most 6 edges between each pair of markers). This will be called "steps" in the sequel with the aim of getting a better idea of complexity than Landau's notation can offer (and which can anyway be derived from the number of steps needed since each step is constant time). From the previous simplified presentation of EM, we can observe that each EM iteration needs $3(n + 1)k$ steps since $n + 1$ steps are required for the Forward phase, the Backward phase and the $E(p)$ computation phase. The M phase is in $\theta(n)$ only.

2.1   Speeding Up the E Phase

To make the E phase more efficient, the idea is to try to sum up the data for all hybrids in a more concise way in order to try to avoid, as far as possible the k factor in the complexity of the E phase. The crucial property that is exploited is that when a loci status is known (0 or 1), then the probabilities P_l and P_r for the corresponding "Retained" vertex are both equal to 0.0 and 1.0 respectively (1.0 and 0.0 respectively for the "Missing" vertex) and this independently of the markers around.

Any given hybrid can therefore be decomposed in segments of successive loci of 3 different types as illustrated in Figure 3:

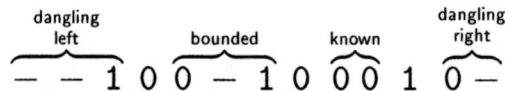

Fig. 3. Decomposing Hybrid into Segments.

- **dangling** segments are either segments that start at the first loci and are all of unknown status except the rightmost one (dangling left) or segments that end at the last loci and are all of unknown status except the leftmost one (dangling right).
- **known pairs** are segments composed from a pair of adjacent loci which are both of known status.
- **bounded** segments are segments that start and stop at loci of known status but which are separated by loci of unknown status.

Given the hybrid data set H, it is possible to precompute for all pairs of markers the number of known pairs of each type. This is done only once, when the data set is loaded. For a given loci ordering, we can precompute in a $O(n.k)$ phase the number of dangling and bounded segments of each type that occurs in the data set. Then the

EM algorithm may iterate and perform the E phase by computing expectations for each of the cases and multiplying the results by the number of occurrences. Maximizing remains unchanged.

For known pairs, the expectation computation can be done in one step and there are at most $4(n-1)$ different types of pairs which means $4(n-1)$ steps are needed. For all other segments, dangling or bounded, the expectation computation needs a number of steps equal to the length of the fragment. So, if we note u the total number of unknowns in the data set, the total length of these fragments is less than $3u$ and the expectation computation can be done in at most $9u$ steps. We get an overall number of steps of $4(n-1) + 9u$ which is usually very small compared to the $3(n+1)k$ needed before. From an asymptotic point of view, this is still in $O(nk)$ because u is in $O(nk)$ but it does improve things a lot in practice. Also, decomposing hybrids into fragments guarantees that repeated patterns occurring in two or more hybrids are only processed once.

There is a specific important case where asymptotic complexity may improve. When multiple population/panels consensus mapping is performed, a marker which is not typed in a data set will be unknown for all the hybrids/individuals of the data-set. This may induce dangling and bounded segments that are shared by all the hybrids/individuals of the data-set but that will be processed only once at each EM iteration. In this case, known pairs and all dangling/bounded segments induced by data-set merging will be handle in $O(n)$ time instead of $O(nk)$.

3 The CAR$^{\text{T}}_{\text{H}}$AGÈNE Package

This improved EM algorithms have been implemented in the CAR$^{\text{T}}_{\text{H}}$AGÈNE package [13]. Beyond RH and backcross data, CAR$^{\text{T}}_{\text{H}}$AGÈNE can also handle genetic intercross, recombinant inbred lines and phase known outbred data. Although perfectly able to handle single population mapping, CAR$^{\text{T}}_{\text{H}}$AGÈNE is oriented towards multiple population mapping: data sets can be hierarchically "merged" and the user decides which data sets share (or not) distance estimations. Markers ordering are always evaluated using a true maximum likelihood multipoint criteria. Beyond multiple population genetic mapping or multiple panels RH mapping, CAR$^{\text{T}}_{\text{H}}$AGÈNE allows to perform mixed genetic/RH mapping by merging genetic and RH data and estimating genetic/RH distances separately using a common loci order.

Considering loci ordering, CAR$^{\text{T}}_{\text{H}}$AGÈNE offers a large spectra of tools for building and validating both comprehensive and framework maps. Most of these tools have been derived from the famous Traveling Salesman Problem (TSP) solving technology. From its beginning [13], CAR$^{\text{T}}_{\text{H}}$AGÈNE has exploited the analogy between backcross genetic mapping and the TSP. As it has been observed in [1], this analogy also exists for RH mapping. CAR$^{\text{T}}_{\text{H}}$AGÈNE has been implemented in C++ and has a Tcl/Tk based graphical interface. It is available on the web at **www-bia.inra.fr/T/CarthaGene**. It runs on most Unix platforms and Windows.

4   Empirical Evaluation of the Boosted EM Algorithm

To better evaluate the gains obtained by the improved EM algorithm, we have compared it to the EM implementations available in the RHMAP RH mapping package [3] and in the MapMaker genetic mapping package [7]. Since we only wanted to compare EM efficiency and not loci ordering efficiency, all tests have been done by estimating the log-likelihood of a fixed loci ordering using the same convergence tolerance, the same starting point, on a Pentium II 400Mhz machine running Linux, using GNU compilers (respectively g77, gcc and g++ for RHMAP, MapMaker and CAR$_H^T$AGÈNE), on 1000 EM calls.

Simulated data sets have been generated using the underlying probabilistic model both for backcross and RH data:

- the first tests have been done single panel/population data. The RH data uses 50 markers evenly distributed on a 10 ray long chromosome, 100 hybrids and 5% of unknowns. The genetic data, uses 50 markers evenly distributed on a 1 Morgan long chromosome., 200 individuals and 25% of unknowns. This corresponds to typical framework mapping situations.
- the second tests have been done by merging two panels/population. The data sets have been generated using the same parameters except that each data set shares half of the markers with the other data set. When a marker is not typed in a panel/population, the corresponding hybrid/individual is marked as unknown[1]

Software package	Data type	Panels/ populations	CPU time (1000 EM)	Improvement ratio
	RH	1		
RHMAP 3.0			110"64	
CAR$_H^T$AGÈNE			5"57	19.8
	RH	2		
RHMAP 3.0			2376"48	
CAR$_H^T$AGÈNE			125"95	18.9
	BC	1		
MapMaker			60"99	
CAR$_H^T$AGÈNE			8"46	7.2
	BC	2		
MapMaker			232"27	
CAR$_H^T$AGÈNE			70"94	3.3

For radiated hybrid data, the speed-up exceeds one order of magnitude. More modest improvements are reached on genetic data. These improvements may reduce day of

---

[1] Note that for RH data, this situation corresponds to the merging of two panels that have been irradiated using a similar level of radiation. If this is not the case, one should rather perform separate distance estimations per panel. More complex models, using proportional distances, are available in RHMAP but are not compatible with the use of the EM algorithm.

computation time to hours and enable the use of more sophisticated ordering techniques, without any approximation being made. These numbers still leave room for improvements since CAR$_H^T$AGÈNE does not exploit the strategy of precomputing known pairs once and for all but recomputes them at each EM call.

5   Application to Real Genetic and Radiation Hybrid Data

The improvements obtained apply only to haploid RH data and phase known backcross genetic data. This could be considered as quite restrictive. In this section we show how RH polyploid data and complex genetic data can be reduced to these cases and evaluate the impact of these reductions. This also illustrates how genetic and RH data can be mixed in order to exploit the different resolutions for marker ordering.

Considering genetic data, the USDA porcine reference pedigree consists in a two generations backcross population of 10 parents (2 males from a White composite line and 8 F1 females) and 94 progeny [12]. The height F1 females were obtained after mating White composite females with Duroc, Fengjing, Meishan or Minzhu boars. To reduce this to backcross data, phases have been set to the most probable phase, identified using Chrompic from the CRIMAP package. Then the data is encoded as two backcross (one for the paternal allele and one for the maternal allele). The original data set can be processed by CRIMAP (but not by MapMaker).

Considering RH data, the IMpRH panel consists in 118 radiated hybrid clones produced after irradiation of porcine cells at 7000 rads [14]. Using this panel a first generation whole genome radiation map has been established using 757 markers [6] including 699 markers already mapped on the genetic map from [12].

These two data-sets have been merged and a framework consensus order built for chromosome 1 using the `buildfw` command of CAR$_H^T$AGÈNE. The resulting order contains 38 markers. This order has been validated using simulated annealing and other usual techniques (local permutations, swapping of markers...) and could not be improved, the second best order having a log-likelihood 4.48 below the best. The simulated annealing step alone took few hours on Pentium-II 400Mhz and could probably not have been done in reasonable time using standard EM algorithms.

Using CAR$_H^T$AGÈNE instead of CRIMAP/RHMAP, we made the assumptions that using an haploid model on diploid data, fixing the phase and considering outbreds as double backcross does not change *differences* in likelihood too much. In order to check these assumptions, we compared CAR$_H^T$AGÈNE to CRIMAP and RHMAP applied separately on the 4 best orders identified by CAR$_H^T$AGÈNE. We used CAR$_H^T$AGÈNE haploid model, RHMAP haploid model and RHMAP diploid model on the RH data alone. The following table indicates the log-likelihoods obtained in each case and the differences in log-likelihood with the best order. The results are consistent with our assumption.

	Order 1	Order 2	Order 3	Order 4
CAR$_H^T$AGÈNE	-927.61	-928.37	-932.65	-931.21
RHMAP haplo.	-927.61	-928.37	-932.65	-931.21
RHMAP diplo.	-926.45	-927.26	-931.36	-930.07
Dif. haplo.	0.00	-0.76	-5.04	-3.60
Dif. diplo.	0.00	-0.81	-4.91	-3.62

The same comparison was done using CRIMAP outbred model and CAR$_H^T$AGÈNE backcross model on the derived double backcross data. The results obtained are again consistent with our assumption. Note that the important change in log-likelihood is not surprising: fixing phases brings in information, while double backcross projection removes some information. The important thing is that differences in log-likelihood are not affected.

	Order 1	Order 2	Order 3	Order 4
CRIMAP	-366.54	-370.26	-366.54	-379.22
CAR$_H^T$AGÈNE	-422.29	-426.01	-422.29	-434.97
Dif. CRIMAP	0.00	-3.72	0.00	-12.68
Dif. Cartha.	0.00	-3.72	0.00	-12.68

We completed this test by a larger comparison, using more orders, and it appears that the differences in log-likelihood are well conserved: a difference of difference greater than 1.0 was observed only for orders whose log-likelihood was very far from the best one (more than 10 LOD). CAR$_H^T$AGÈNE can thus be used to build framework and comprehensive maps, integrating genetic and RH maps in reasonable time. For better final distances between markers, one can simply reestimate them with RHMAP diploid model and CRIMAP.

6   Conclusion

We introduced a simple improvement for the EM algorithm in the framework of HMM based RH and genetic maximum likelihood estimation. One of the limitation of this improvement is that it can only deal with observations that either completely determine the hidden states or leave the hidden state undetermined (unknown). It can therefore not be extended, e.g. to handle intercross genetic data, diploid RH data or to models that explicitly represent typing errors [9]. However, as we experienced on real data, these restrictions can be easily be dealt with.

This is especially attractive considering that our application to haploid radiation hybrid and backcross genetic mapping shows that the boosted EM algorithms can lead to speed-ups of more than one order of magnitude compared to the traditional EM approach, without any loss in accuracy. Ordering techniques such as simulated annealing or taboo search requires an important number of calls to the evaluation function. The efficiency of the boosted EM algorithm makes the application of such approaches practical, even in the presence of unknowns. This is crucial for multiple population/panel mapping as it has been done, e.g. in [8].

Acknowledgements

We would like to thank Gary Rohrer (USDA) for letting us use the USDA porcine reference genetic data and Martine Yerle (INRA) for the RH porcine data.

References

1. Amir Ben-Dor and Benny Chor. On constructing radiation hybrid maps. *J. Comp. Biol.*, 4:517–533, 1997.
2. Amir Ben-DOr, Benny Chor, and Dan Pelleg. RHO – radiation hybrid ordering. *Genome Research*, 10:365–378, 2000.
3. Michael Boehnke, Kathryn Lunetta, Elisabeth Hauser, Kenneth Lange, Justine Uro, and Jill VanderStoep. *RHMAP: Statistical Package for Multipoint Radiation Hybrid Mapping*, 3.0 edition, September 1996.
4. D.R. Cox, M. Burmeister, E.R. Price, S. Kim, and R.M.Myers. Radiation hybrid mapping: A somatic cell genetic method for constructing high-resolution maps of mammalian chromosomes. *Science*, 250:245–250, 1990.
5. A.P. Dempster, N.M. Laird, and D.B. Rubin. Maximum likelihood from incomplete data via the EM algorithm. *J. R. Statist. Soc. Ser.*, 39:1–38, 1977.
6. R.J. Hawken, J. Murtaugh, G.H. Flickinger, M. Yerle, A. Robic, D. Milan, J. Gellin, C.W. Beattie, L.B. Schook, and L.J. Alexander. A first-generation porcine whole-genome radiation hybrid map. *Mamm. Genome*, 10:824–830, 1999.
7. E.S. Lander, P. Green, J. Abrahamson, A. Barlow, M. J. Daly, S. E. Lincoln, and L. Newburg. MAPMAKER: An interactive computer package for constructing primary genetic linkage maps of experimental and natural populations. *Genomics*, 1:174–181, 1987.
8. E. Laurent, V.and Wajnberg, B. Mangin, T. Schiex, C. Gaspin, and F. Vanlerberghe-Masutti. A composite genetic map of the parasitoid wasp *Trichogramma brassicae* based on RAPD markers. *Genetics*, 150(1):275–282, 1998.
9. Kathryn L. Lunetta, Michael Boehnke, Kenneth Lange, and David R. Cox. Experimental design and error detection for polyploid radiation hybrid mapping. *Genome Research*, 5:151–163, 1995.
10. Jurg Ott. *Analysis of human genetic linkage.* John Hopkins University Press, Baltimore, Maryland, 2nd edition, 1991.
11. Lawrence R. Rabiner. A tutorial on hidden markov models and selected applications in speech recognition. *Proc. of the IEEE*, 77(2):257–286, 1989.
12. G.A. Rohrer, L.J. Alexander, J.W. Keele, T.P. Smith, and C.W. Beattie. A microsatellite linkage map of the porcine genome. *Genetics*, 136:231–245, 1994.
13. T. Schiex and C. Gaspin. Cartagene: Constructing and joining maximum likelihood genetic maps. In *Proceedings of the fifth international conference on Intelligent Systems for Molecular Biology*, Porto Carras, Halkidiki, Greece, 1997. Software available at http://www-bia.inra.fr/T/CartaGene.
14. M. Yerle, P. Pinton, A. Robic, A. Alfonso, Y. Palvadeau, C. Delcros, R. Hawken, L. Alexander, C. Beattie, L. Schook, D. Milan, and J. Gellin. Construction of a whole-genome radiation hybrid panel for high-resolution gene mapping in pigs. *Cytogenet. Cell Genet.*, 82:182–188, 1998.

Placing Probes along the Genome
Using Pairwise Distance Data

Will Casey[1], Bud Mishra[1], and Mike Wigler[2]

[1] Courant Institute, New York University
251 Mercer St., New York, NY 10012, USA
wcasey@cims.nyu.edu,mishra@nyu.edu
[2] Cold Spring Harbor Laboratory
P.O. Box 100, 1 Bungtown Rd.
Cold Spring Harbor, NY 11724, USA
wigler@cshl.org

Abstract. We describe the theoretical basis of an approach using microarrays of probes and libraries of BACs to construct maps of the probes, by assigning relative locations to the probes along the genome. The method depends on several hybridization experiments: in each experiment, we sample (with replacement) a large library of BACs to select a small collection of BACs for hybridization with the probe arrays. The resulting data can be used to assign a local distance metric relating the arrayed probes, and then to position the probes with respect to each other. The method is shown to be capable of achieving surprisingly high accuracy within individual contigs and with less than 100 microarray hybridization experiments even when the probes and clones number about 10^5, thus involving potentially around 10^{10} individual hybridizations.

This approach is not dependent upon existing BAC contig information, and so should be particularly useful in the application to previously uncharacterized genomes. Nevertheless, the method may be used to independently validate a BAC contig map or a minimal tiling path obtained by intensive genomic sequence determination.

We provide a detailed probabilistic analysis to characterize the outcome of a single hybridization experiment and what information can be garnered about the physical distance between any pair of probes. This analysis then leads to a formulation of a likelihood optimization problem whose solution leads to the relative probe locations. After reformulating the optimization problem in a graph-theoretic setting and by exploiting the underlying probabilistic structure, we develop an efficient approximation algorithm for our original problem. We have implemented the algorithm and conducted several experiments for varied sets of parameters. Our empirical results are highly promising and are reported here as well. We also explore how the probabilistic analysis and algorithmic efficiency issues affect the design of the underlying biochemical experiments.

1 Introduction

Genetics depends upon genomic maps. The ultimate maps are complete nucleotide sequences of the organism together with a description of the transcription units. Such maps in various degrees of completion exist for many of the microbial organisms,

O. Gascuel and B.M.E. Moret (Eds.): WABI 2001, LNCS 2149, pp. 52–68, 2001.

yeasts, worms, flies, and now humans. Short of this, genetically or physically mapped collections of objects derived from the genome under study are still of immense utility, and are often precursors to the development of complete sequence maps. These objects may be markers of any sort, DNA probes, and genomic inserts in cloning vectors.

We have been exploring the use of microarrays to assist in the development of genomic maps. We report here one such mapping algorithm, and explore its foundation using computer simulations and mathematical treatment. The algorithm uses unordered probes that are microarrayed and hybridized to an organized sampling of arrayed but unordered members of libraries of large insert genomic clones.

In the foregoing we assume some knowledge of genome organization, DNA hybridization, repetitive DNA, gene duplication, and the common forms of microarray experiments. In the proposed experimental setting, one sample at a time is hybridized to microarrayed probes, and hybridization is measured as an absolute quantity. We assume probes are of zero dimension, that is, of negligible length compared to the length of the large genomic insert clones. Most importantly, we assume that hybridization signal of a probe reflects its inclusion in one or more large genomic insert clones present in the sample, and negligible background hybridization. Our analysis is general enough to include the effects of other sources of error. *The novelty of the results reported here is in their ability to deal with ambiguities, an inevitable consequence of the use of massive parallelism in microarrays involving many probes and many clones.* Similar algorithms are reported in the literature [7], but assume only the knowledge of clone-probe inclusion information for every such combination and suggest different algorithms that do not exploit the underlying statistical structure.

One important application of our method is in measuring gene copy number in genomic DNA [10]. Such techniques will eventually have direct application to the analysis of somatic mutations in tumors and inherited spontaneous germline mutations in organisms when those mutations result in gene amplification or deletion. In contrast, low signal-to-noise ratios, due to the high complexity of genomic DNA, make other approaches such as the direct application of standard DNA microarray methods highly problematic.

2   Related Literature

The problem of physical mapping and sequencing using hybridization is relatively well studied. As shown in [6], the general problem of physical mapping is NP-complete. An approach based on traveling salesman problem (TSP) in the absence of statistics is given in [1]. The problem formalism used in this paper will be similar to the foundational work in [1–3, 5, 7, 11, 12, 14]. Our method extends the previous results by devising efficient algorithms as well as biochemical experiments capable of achieving higher resolution of probe placement within contigs. In [8] the MATRIX-TO-LINE problem is suggested as the model problem for determining Radiation Hybrid maps. Probe mapping using BACs is slightly different in that pairwise distances for probes far away cannot be resolved directly using BACs. Our design is general in that the inputs are modeled as random variables with known statistics determined a priori by our experimental designs chosen appropriately for a range of applications. Also we provide estimated probabilities of correctness for the map we produce. In this sense, this paper invokes an an experimental optimization as recommended in [2].

3   Mathematical Definitions

Given a set of P probes listed as $\{p_1, p_2, \ldots, p_P\}$ and contained in some contiguous segment of the genome we define a *probe map* to be a pair of sequences, **ordering** $= \{p_{\pi(1)}, p_{\pi(2)}, \ldots, p_{\pi(P)}\}$ and **position** $= \{x_1, x_2, \ldots, x_P\}$. The position sequence infers the positions of the probes and the ordering sequence is determined by the permutation[1] $\pi \in S_P$ that sorts the given list of probes by position.

However the underlying correct position of each probe remains unknown. We infer probe maps approximating the correct positions as best as possible from an experimental set of data which is stochastic. Experimental data sets are represented by graphs; given a set of probes $\{p_1, p_2, \ldots, p_P\}$, let V be the set of indices. Then a *pairwise distance graph* is an undirected graph $\mathcal{G} = \langle V, E \rangle$, $E \subset V \times V$ where each edge $e_{i,j}$ maps to a distance $d_{i,j}$ between probe i and probe j.

We model various experimental errors arising from the hybridization experiment used to measure probe to probe distance. With the model we can understand the distribution of pairwise distance graphs as a random variable. Under certain parameters we can implement Bayes formula to build a Maximum Likelihood Estimator (MLE) for probe map reconstruction. With the MLE established we attempt to optimize the computation involved for practical implementation.

3.1   Experimental Procedure

Consider a genome represented by the interval $[0, G]$. Take P random short sub-strings (about 200bps) which appear on the genome uniquely. Represent these strings as points $\{x_1, \ldots, x_P\}$. Assume that the probes are i.i.d. with uniform random distribution over the interval $[0, G]$. Let S be a collection of intervals of the genome, each of length L (usually ranging from few 100kbs to Mbs). Suppose the left-hand points of the intervals of S are i.i.d. uniform random variables over the interval $[0, G]$. Take a small, even in number sized subset of intervals $S' \subset S$, chosen randomly from S. Divide S' randomly into two equal-size disjoint subsets $S' = S'_R \cup S'_G$, where R indicates a red color class and G indicates a green color class. Now specify any point x in $[0, G]$ and consider the possible associations between x, and the intervals in S':

- x is not covered by any interval in S'.
- x is covered by at least one interval of S'_R but no intervals of S'_G.
- x is covered by at least one interval of S'_G but no intervals of S'_R.
- x is covered by at least one interval of S'_R and at least one interval of S'_G.

If we perform a sequence of M such experiments then for each x we get a sequence of M outcomes represented as a color vector of length M. We are interested in observing sequences of such outcomes on the set $\{x_1, \ldots, x_P\}$.

For DNA the short sub-strings can be produced with the use of restriction enzymes, or synthesized as oligoes. The collection of covering intervals may be provided by a BAC or YAC clone library. The division of a random sample taken from the clone library may be done with phosphorescent molecules added to the DNA and visible

---

[1] We denote the permutation group on P indices as S_P.

with a laser scanner. Hybridization microarrays allow us to observe such an outcome sequence for each of the 100,000 probes in a constant amount of time.

Consider an example with human. To make a set of Human Oligoe Probes we may use restriction enzymes to cut out P probe substrings of size 200bp to 1200bp from the genome and choose a low complexity representation (LCR) as discussed in [10,9]. We may arrange for a sequence of M random samples from the BAC library, suppose each sample has K BACs and coverage $c = \frac{KL}{G}$. Samples are then randomly partitioned into two color classes $\Sigma = \{R, G\}$, and then hybridized to a microarray, arrayed with P probes. If we pick one probe p_i, then the possible outcomes for one experiment are:

- p_i hybridizes to zero BACs. We say the outcome is 'B' (blank).
- p_i hybridizes to at least one red BAC and zero green BACs. We say the outcome is 'R' (red).
- p_i hybridizes to at least one green BAC and zero red BACs. We say the outcome is 'G' (green).
- p_i hybridizes to at least one green BAC and at least one red BAC. We say the outcome is 'Y' (yellow).

We call these events i_B, i_R, i_G, and i_Y respectively. We use M random samples to complete the full experiment. The parameter domain for the full experiment is $\langle P, L, K, M \rangle$, where P is the number of probes, L is the average length of the genomic material used (for BACs, $L = 160$kb), K is the sampling size, and M is the number of samples. The output is a color sequence for each probe. The sequence corresponding to probe p_j is $\mathbf{s}_j = \langle s_{j,k} \rangle_{k=1}^{M}$ with $s_{j,k} \in \{B, R, G, Y\}$.

How the Distances Are Measured. With the resulting color sequences $\mathbf{s}_j$ we can compute the pairwise Hamming distance. Let

$$H_{i,j} = \# \text{ places where } \mathbf{s}_i \text{ and } \mathbf{s}_j \text{ differ},$$
$$C_{i,j} = \# \text{ places where } \mathbf{s}_i \text{ and } \mathbf{s}_j \text{ are the same but } \mathbf{s}_i \neq B,$$
$$T_{i,j} = \# \text{ places where } \mathbf{s}_i \text{ and } \mathbf{s}_j \text{ are } B.$$

The Hamming distance defines a distance metric on the set of probes.

Lemma 31. *Consider an experiment with parameters $\langle P, L, K, M \rangle$, and $c = \frac{KL}{G}$. Let i and j be arbitrary indices from the clone set and x_{ij} is the actual distance (in number of bases) separating probe p_i from probe p_j on the genome. Let $\hat{x}_{ij} = \min\{x_{ij}, L\}$. Then:*

$$H_{i,j} \sim Bin\left(M, \frac{2ce^{(\frac{-c}{2})}\hat{x}_{ij}}{L} + O((\hat{x}_{ij})^2)\right)$$
$$C_{i,j} \sim Bin\left(M, 1 - e^{-c} + \frac{c}{2}(e^{-c} - 2e^{-\frac{c}{2}})\hat{x}_{ij} + O((\hat{x}_{ij})^2)\right)$$
$$T_{i,j} \sim Bin\left(M, (e^{-c(1+\hat{x}_{ij})})\right)$$

Proof. See appendix.

These computations for small x lead to an accurate estimator:

Corollary 32. *The estimator of x_{ij} given by $\tilde{x}_{ij} = H_{i,j}\frac{e^{\frac{c}{2}}L}{2cM}$ is good in the sense that there are values of c so that:*

$$f(\tilde{x}_{ij} = d | x_{ij}) \rightarrow \begin{cases} \dfrac{1}{\sqrt{2\pi}\sigma\sqrt{x_{ij}}}e^{-\frac{(d-x_{ij})^2}{2\sigma^2 x_{ij}}} & \text{if } x_{ij} < L; \\[2ex] \dfrac{1}{\sqrt{2\pi}\sigma\sqrt{L}}e^{-\frac{(d-x_{ij})^2}{2\sigma^2 L}} & \text{if } x_{ij} \geq L; \end{cases} \quad \text{as } M \rightarrow \infty.$$

with $\sigma^2 = \left(\dfrac{e^{\frac{c}{2}}}{2c}\right)$.

Proof. It is based on a standard approximation. We have developed Chernoff bounds to analyze tradeoff between parameters K (determining c), and M. For $x < \frac{L}{2}$ one can show that for nearly any value of c the above convergence in distribution is significantly rapid w.r.t. M. $\square$

We have developed an estimator of x_{ij} given by $\tilde{x}_{ij} = \dfrac{H_{i,j}}{H_{i,j}+2C_{i,j}}e^{\frac{H_{i,j}+2C_{i,j}}{4M}}L$ this estimator takes into account the variation of sample coverage over the genome.

Lemma 33. *The distribution for distance d is a function of x and is approximated by*

$$f(d|x) = \mathbb{I}_{0 \leq x < L}\frac{e^{-(d-x)^2/2\sigma^2 x}}{\sqrt{2\pi x}\sigma} + \mathbb{I}_{L \leq x \leq G}\frac{e^{-(d-L)^2/2\sigma^2 L}}{\sqrt{2\pi L}\sigma}.$$

Proof. Simple restatement of corollary 2.2 $\square$

Since we have assumed that any given probe is distributed uniformly randomly over the genome, the density function for the probe's position is:

$$f(x) = \frac{1}{G}$$

Our next lemma is an application of Bayes' formula to compute $f(x|d)$ from $f(x)$ and $f(d|x)$ computed above.

Lemma 34. *If $f(d|x) = \mathbb{I}_{0 \leq x < L}\dfrac{e^{-(d-x)^2/2\sigma^2 x}}{\sqrt{2\pi x}\sigma} + \mathbb{I}_{L \leq x \leq G}\dfrac{e^{-(d-L)^2/2\sigma^2 L}}{\sqrt{2\pi L}\sigma}$. Then*

$$f(x|d) \approx \mathbb{I}_{d < L}\frac{e^{-(x-d)^2/2\sigma^2 d}}{\sqrt{2\pi d}\sigma} + \mathbb{I}_{d \geq L}\,\mathbb{I}_{L \leq x \leq G}\left(\frac{1}{G-L}\right).$$

Proof. See appendix. $\square$

With conditional $f(x|d)$ we can now define the Maximum Likelihood Estimation problem:

Given an arbitrary pair-wise distance edge weighted complete graph $\mathcal{G}$ of P vertices, representing probes, and each edge (i, j) labeled with $d_{i,j}$, a sampled value of a random variable with the distribution $f(d||x_i - x_j|)$, we would like to choose an embedding of $\mathcal{G}$ (or more precisely, an embedding of the vertices of $\mathcal{G}$) into the real line:

$$\{\tilde{x}_1, \tilde{x}_2, \ldots, \tilde{x}_P\} \subset [0, G],$$

that maximizes a likelihood function $F(\tilde{x}_1, \tilde{x}_2, \ldots, \tilde{x}_P | d_{ij} : i, j \in [1, P])$. By ignoring the week dependencies, we approximate F as:

$$\prod_{1 \leq i,j \leq P} f(|\tilde{x}_i - \tilde{x}_j| | d_{ij}).$$

Hence, we can minimize a related cost function

$$\sum_{1 \leq i,j \leq P} -\ln f(|\tilde{x}_i - \tilde{x}_j| | d_{ij}).$$

Lemma 35. *The Optimization problem of finding $\tilde{x}_j$ to minimize $f(\tilde{x}_j | \{\tilde{x}_i : i < j\}, \{d_{i,j} : i < j\})$ is approximated by solving the following optimization problem:*

$$minimize \sum_{1 \leq i < j \leq P} W_{ij}(|\tilde{x}_i - \tilde{x}_j| - d_{ij})^2,$$

where W_{ij}'s are positive real valued weight functions:

$$W_{ij} = \begin{cases} \dfrac{1}{2\sigma^2 d_{ij}} & \text{if } d_{ij} < L; \\ \epsilon & \text{otherwise,} \end{cases}$$

and $\epsilon = O\left(\dfrac{1}{(G-L)^2}\right)$.

Proof.

$$-\ln f(x|d) \approx \begin{cases} \dfrac{(x-d)^2}{2\sigma^2 d} + \ln\left(\sqrt{2\pi d}\sigma\right) & \text{if } d < L; \\ \ln(G - L) - \ln \mathbb{I}_{L \leq x \leq G} & \text{otherwise.} \end{cases}$$

Hence

$$\sum_{1 \leq i,j \leq P} -\ln f(|\tilde{x}_i - \tilde{x}_j| | d_{ij}) = \sum_{1 \leq i < j \leq P} W_{ij}(|\tilde{x}_i - \tilde{x}_j| - d_{ij})^2.$$

Note that $\epsilon = \dfrac{1}{2\sigma_M^2 d_{ij}} \leq \dfrac{1}{2\sigma_M^2 L} \leq \dfrac{1}{2(G-L)^2 L}$ as σ_M being the maximum variance is bounded by $(G - L)$ $\square$.

3.2   Simple Algorithm

The simplest algorithm to place probes proceeds as follows: Initially, every probe occurs in just one singleton contig, and the relative position of a probe $\tilde{x}_i$ in contig C_i is at the position 0. At any moment, two contigs $C_p = [\tilde{x}_{p_1}, \tilde{x}_{p_2}, \ldots, \tilde{x}_{p_l}]$ and $C_q = [\tilde{x}_{q_1}, \tilde{x}_{q_2}, \ldots, \tilde{x}_{q_m}]$ may be considered for a "join" operation: the result is either a failure to join the contigs C_p and C_q or a new contig C_r containing the probes from the constituent contigs. Without loss of generality, assume that $|C_p| \geq |C_q|$, and that the probe corresponding to the right end of the first contig (x_{p_l}) is closet to the left end of the other contig (x_{q_1}). That is the estimated distance d_{p_l, q_1} is smaller than all other estimated distances: d_{p_1, q_1}, d_{p_1, q_m} and d_{p_l, q_m}.

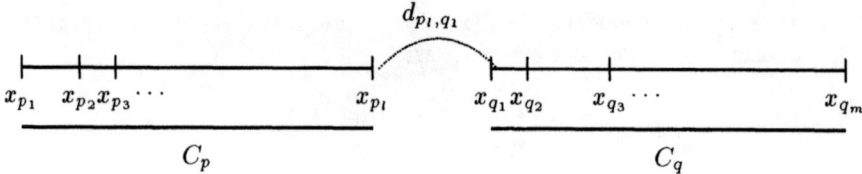

Let $0 < \theta \le 1$ be a parameter to be explored further later, and $L' = L\theta \le L$. If $d_{p_l,q_1} \ge L'$ then the join operation fails. Otherwise, the join operation succeeds with the probes of C_p placed to the left of the probes of C_q, with all the relative positions of the probes of each contig left undisturbed. We will estimate the distance between the probes in C_p and the probe x_{q_1} by minimizing the function:

$$\text{minimize} \sum_{i \in \{p_1, \ldots, p_l\}: d_{i,q_1} < L'} \frac{(\tilde{x}_{q_1} - \tilde{x}_i - d_{i,q_1})^2}{2\sigma^2 d_{i,q_1}},$$

where $\tilde{x}_i$'s ($i \in \{p_1, \ldots, p_l\}$) are fixed by the locations assigned in the contig C_p. Thus taking a derivative of the expression above with respect to $\tilde{x}_{q_1}$ and equating it to zero, we see that the optimal location for x_{q_1} in C_r is

$$d^* = \max \left[\tilde{x}_{p_l}, \frac{\sum_{i \in \{p_1, \ldots, p_l\}: d_{i,q_1} < L'} (\tilde{x}_i + d_{i,q_1}) / \sigma^2 d_{i,q_1}}{\sum_{i \in \{p_1, \ldots, p_l\}: d_{i,q_1} < L'} 1 / \sigma^2 d_{i,q_1}} \right].$$

Once the location of x_{q_1} is determined in C_r at d^*, the locations of all other probes of C_q in the new contig C_r are computed by shifting them by the value d^*. Thus

$$C_r = [\tilde{x}_{r_1}, \ldots, \tilde{x}_{r_l}, \tilde{x}_{r_{l+1}}, \ldots, \tilde{x}_{r_{l+m}}],$$

where $r_i = p_i$ and $\tilde{x}_{r_i} = \tilde{x}_{p_i}$, for $1 \le i \le l$; $r_{l+i} = q_i$ and $\tilde{x}_{r_{l+i}} = d^* + \tilde{x}_{q_i}$, for $1 \le i \le m$. Note that when the join succeeds, the distance between the pair of consecutive probes $\tilde{x}_{r_l}$ and $\tilde{x}_{r_{l+1}}$ is

$$0 \le \tilde{x}_{r_{l+1}} - \tilde{x}_{r_l} \le L',$$

and the distances between all other consecutive pairs are exactly the same as what they were in the original constituent contigs. Thus, in any contig, the distance between every pair of consecutive probes takes a value between 0 and L'. Note that one may further simplify the distance computation by simply considering the k nearest neighbors of $\tilde{x}_{q_1}$ from the contig C_p: namely, $\tilde{x}_{p_l-k+1}, \ldots, \tilde{x}_{p_l}$.

$$d_k^* = \max \left[\tilde{x}_{p_l}, \frac{\sum_{i \in \{p_l-k+1, \ldots, p_l\}: d_{i,q_1} < L'} (\tilde{x}_i + d_{i,q_1}) / \sigma^2 d_{i,q_1}}{\sum_{i \in \{p_l-k+1, \ldots, p_l\}: d_{i,q_1} < L'} 1 / \sigma^2 d_{i,q_1}} \right].$$

In the greediest version of the algorithm $k = 1$ and

$$d_1^* = \tilde{x}_{p_l} + d_{p_l,q_1},$$

as one ignores all other distance measurements.

At any point we can also improve the distances in a contig, by running an "adjust" operation on a contig C_p with respect to a probe $\tilde{x}_{p_j}$, where

$$C_p = [\tilde{x}_{p_1}, \ldots, \tilde{x}_{p_{j-1}}, \tilde{x}_{p_j}, \tilde{x}_{p_{j+1}}, \ldots, \tilde{x}_{p_l}.]$$

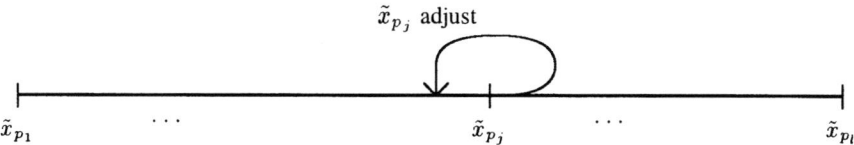

We achieve this by minimizing the following cost function:

$$\text{minimize} \quad \sum_{i \in \{p_1, \ldots, p_l\} \setminus \{p_j\} : d_{i,p_j} < L'} \frac{(|\tilde{x}_{p_j} - \tilde{x}_i| - d_{i,p_j})^2}{2\sigma^2 d_{i,p_j}},$$

where $\tilde{x}_i$'s $(i \in \{p_1, \ldots, p_l\} \setminus \{p_j\})$ are fixed by the locations assigned in the contig C_p. Let:

$$I_1 = \{i_1 \in \{p_1, \ldots, p_{j-1}\} : d_{i_1,p_j} < L'\}$$
$$I_2 = \{i_2 \in \{p_{j+1}, \ldots, p_l\} : d_{i_2,p_j} < L'\}$$
$$x^* = \frac{\sum_{i_1 \in I_1} (\tilde{x}_{i_1} + d_{i_1,p_j})/\sigma^2 d_{i_1,p_j} + \sum_{i_2 \in I_2} (\tilde{x}_{i_2} - d_{i_2,p_j})/\sigma^2 d_{i_2,p_j}}{\sum_{i_1 \in I_1} 1/\sigma^2 d_{i_1,q_1} + \sum_{i_2 \in I_2} 1/\sigma^2 d_{i_2,p_j}}.$$

At this point, if $x^* \neq \tilde{x}_{p_j}$, then the new position of the probe $\tilde{x}_{p_j}$ in the contig C_p is x^*. As before, one can use various approximate version of the update rule, where only k probes from the left and k probes from the right are considered and in the greediest version only the two nearest neighbors are considered. Note that the "adjust" operation always improves the quadratic cost function of the contig locally and since it is positive valued and bounded away from zero, the iterative improvement operations terminate.

4   Implementation of the k-Neighbor Algorithm

INPUT

The input domain is a probe set V, and a symmetric positive real-valued distance weight matrix $D \in \mathbb{R}_+^{P \times P}$, where $P = |V|$.

PRE–PROCESS

Construct a graph $\mathcal{G}' = \langle V, E' \rangle$, where $E' = \{e_k = (x_i, x_y) | d_{i,j} < L'\}$. The edge set of the graph $\mathcal{G}'$ is sorted into an increasing order as follows: $e_1, e_2, \ldots, e_Q$, with $Q = |E'|$ such that for any two edges $e_{k_1} = [x_{i_1}, x_{j_1}]$ and $e_{k_2} = [x_{i_2}, x_{j_2}]$, if $k_1 < k_2$ then $d_{i_1,j_1} \leq d_{i_2,j_2}$. $\mathcal{G}'$ can be constructed in $O(|V|^2)$ time, and its edges can be sorted in $O(|E'|\log(|V|))$ time. In a simpler version of the algorithm it will suffice to sort the edges into an "approximate" increasing order by a parameter $H_{i,j}$ (related to $d_{i,j}$) that takes values between 0 and M. Such a simplification would result in an algorithm with $O(|E'|\log M)$ runtime.

MAIN ALGORITHM

Data-structure: Contigs are maintained in a modified union-find structure designed to encode a collection of disjoint unordered sets of probes which may be merged at any time. Union-find supports two operations, *union* and *find* [13], union merges two sets into one larger set, find identifies the set an element is in. At any instant, a is represented by the following:

- Doubly linked list of probes giving left and right neighbor with estimated consecutive neighbor distances.
- Boundary probes: each contig has a reference to left and right most probes.

In the kth step of the algorithm consider edge $e_k = [x_i, x_j]$: if find(x_i) and find(x_j) are in distinct contigs C_p and C_q, then join C_p and C_q, and update a single distance to neighbor entry in one of the contigs.

At the termination of this phase of the algorithm, one may repeatedly choose a random probe in a randomly chosen contig and apply an "adjust" operation.

OUTPUT

A collection of probe contigs with probe positions relative to the anchoring probe for that contig.

4.1   Time Complexity

First we estimate the time complexity of the main algorithm implementing the k−neighbor version: For each $e \in E'$ there are two find operations. The number of union operations cannot exceed the number of probes $P = |V|$, as every successful join operation leading to a union operation involves a boundary vertex of a contig. Any vertex during its life time can appear at most twice as a boundary vertex of a contig, taking part in a successful join operation. The time cost of a single find operation is at most $\gamma(P)$, where γ is the inverse of Ackermann's function. Hence the time cost of all union-find operations is at most $O(|E'|\gamma(P))$. The join operation on the other hand requires running the k−neighbor optimization routine which is done at a cost $O(k)$. Thus the main algorithm has a worst case time complexity of:

$$O\Big(|E'|\gamma(|V|) + k|V|\Big)$$

The Full Algorithm including preprocessing is:

$$O\Big(|E'|\log(|V|) + |V|^2\Big)$$

In a slightly more robust version the contigs may be represented by a dynamic balanced binary search tree which admit find and implant operations. Each operation has worst case time complexity of $O(\log(|V|))$. Thus after summing over all $|E'|$ operations the worst case runtime for the main algorithm is:

$$O\Big(|E'|\log(|V|) + k|V|\Big)$$

and for the full algorithm is:

$$O\Big(|E'|\log(|V|) + |V|^2\Big)$$

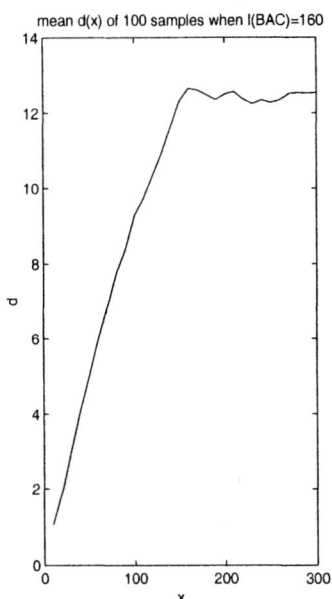
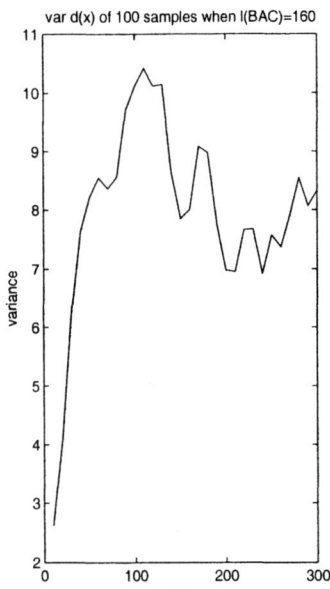

5   What Do the Simulations Tell?

5.1   Simulation: Observed Distance

The sample mean and variation of the distance function are computed with a simple simulation done in-silico. BACs are 160Kb in length, we generate $1,200$ BACs and place them randomly on a genome of size $G = 32,000$Kb, This gives a $6\times$ BAC set. In this experiment 100 random points are chosen on the genome and for each point we compute the Hamming distance compared to points $10, 20, 30, \ldots 300$ Kb to the right on the Genome. Color sequences are computed by using 20 samples of 130 randomly chosen BACs of which half are likely to be red and the other half green.

5.2   Simulation: Full Experiment

Below we describe an in-silico experiment for a problem with 150 probes. On a Genome of size 5,000 Kb we randomly place 150 probes, there positions are graphed as a monotone function in the probe index. Next we construct a population of 500 randomly placed

BACs. From the population we repeat a sampling experiment using a sample size of 32 BACS 16 are colored red, and 16 are colored green. Each sample is hybridized in-silico to the probe set. Here we assume a perfect hybridization so there are no cross hybridizations or failures in hybridizations associated with the experiment. We repeat the sample experiment 130 times. This produces the observed distance matrix, whose distribution we modeled earlier. This is the input for the algorithm presented in this paper. In the distance vs. observed data plot we see that using a large $M = 130$ (suggested by the Chernoff bounds) has its benefits in cutting down the rate of the false positives. The observed distance matrix is input into the $(10-\text{neighbor}, \theta = \frac{11}{16})$ algorithm without the use of the adjust operation, the result is 7 contigs. The order within contigs had five mistakes. We look at the the 4th contig and plot the relative error in probe placement.

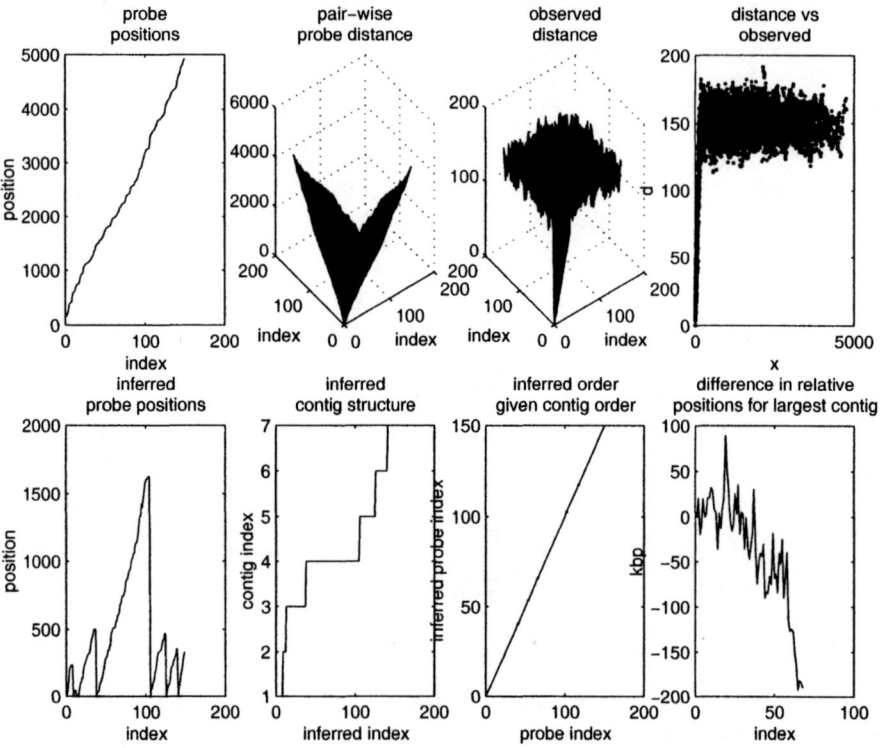

6   Future Work

The more robust variation of the algorithm based on a dynamically balanced binary search tree will be studied in more detail. A comparison with Traveling Salesman heuristics and an investigation of an underlying relation to the heat equation will show why this algorithm works well. We will work on a probabilistic analysis for the statistics of contigs. A model incorporating failure in hybridization and cross hybridization will be developed. We are able to prove that, if errors are not systematic, then a slight modification of the Chernoff bounds presented here can be applied to ensure the same results. We shall also consider the choice of probes to limit the cross-hybridization error and a choice of melting points to further add to the goal of decreasing experimental noise. A set of experimental designs will be presented for the working biologists. More extensive simulations, and results on real experiments shall report the progress of what appears to be a promising algorithm.

Acknowledgments

The research presented here was supported in part by a DOE Grant, an NYU Research Challenge Grant, and an NIH Grant.

Appendix

A Proof of Lemma 2.1

Lemma A1. $H_{i,j} \sim Bin\left(M, \frac{2c\exp(\frac{-c}{2})x}{L} + O(x^2)\right), C_{i,j} \sim Bin\left(M, 1 - e^{-c} + \frac{c}{2}(e^{-c} - 2e^{-\frac{c}{2}})x + O(x^2)\right), T_{i,j} \sim Bin\left(M, (e^{-c(1+\frac{x}{L})})\right)$ *with parameters* $\langle P, L, K, M \rangle$ *as above and* $c = \frac{KL}{G}$, i, j *are arbitrary indices from the Clone Set and* x *is the actual distance as number of bases separating the probe positions on the Genome.*

Proof. Since the M samples are done independently the proof reduces to showing that when $M = 1$ the probabilities are Bernoulli with respective parameters. Let us define events $T = (i_B \wedge j_B)$, $C = ((i_R \wedge j_R) \vee (i_G \wedge j_G) \vee (i_Y \wedge j_Y))$, and $H = (\neg T \wedge \neg C)$.

Given a set of K BACs on a genome $[0, G]$ the probability that none start in an interval of length l is $(1 - \alpha)^l \approx e^{-\alpha l}$ where $\alpha = \frac{K}{G}$.

Shown below is a diagram that is helpful in computing the probabilities for events C, H, T when $x < L$. The heavy dark bar labeled a represents a set of BACs which covers probe p_i but not p_j; the bar labeled b represents a set of BACs that covers probe p_i and p_j; finally, the bar labeled c represents a set of BACs that covers p_j but not p_i.

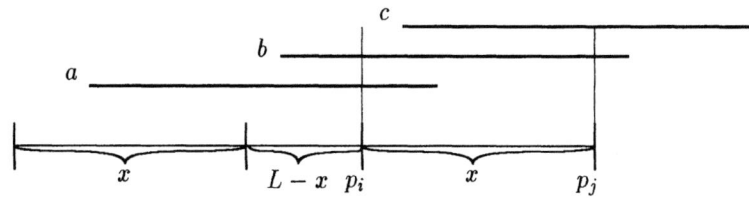

Hence we derive:

$$P(T|x \leq L) = \exp(-(\alpha_R + \alpha_G)(L + x))$$
$$P(i_R \wedge j_R|x < L) = e^{-\alpha_G(L+x)}\{(1 - e^{-\alpha_R(L-x)})$$

$$+(1 - e^{-\alpha_R x})(e^{-\alpha_R(L-x)})(1 - e^{-\alpha_R x})\}$$
$$= e^{-\alpha_G(L+x)}\{1 - 2e^{-\alpha_R L} + e^{-\alpha_R(L+x)}\}$$
$$P(i_G \wedge j_G | x \le L) = e^{-\alpha_R(L+x)}\{1 - 2e^{-\alpha_G L} + e^{-\alpha_G(L+x)}\}$$
$$P(i_Y \wedge j_Y | x \le L) = (1 - 2e^{-\alpha_R L} + e^{-\alpha_R(L+x)})(1 - 2e^{-\alpha_G L} + e^{-\alpha_G(L+x)})$$
$$P(C | x \le L) = P(i_R \wedge j_R | x \le L) + P(i_G \wedge j_G | x \le L) + P(i_Y \wedge j_Y | x \le L)$$
$$P(H | x \le L) = 1 - [P(T | x \le L) + P(C | x \le L)]$$

When $x \ge L$ the probabilities are:

$$P(T | x \ge L) = \exp(-(\alpha_R + \alpha_G)(2L))$$
$$P(i_R \wedge j_R | x \ge L) = e^{-\alpha_G(2L)}\{(1 - e^{-\alpha_R L})^2\}$$
$$P(i_G \wedge j_G | x \ge L) = e^{-\alpha_R(2L)}\{(1 - e^{-\alpha_G L})^2\}$$
$$P(i_Y \wedge j_Y | x \ge L) = (1 - e^{-\alpha_R L})^2(1 - e^{-\alpha_G L})^2$$
$$P(C | x \ge L) = P(i_R \wedge j_R | x \ge L) + P(i_G \wedge j_G | x \ge L) + P(i_Y \wedge j_Y | x \ge L)$$
$$P(H | x \ge L) = 1 - [P(T | x \ge L) + P(C | x \ge L)]$$

Because $\alpha_R = \alpha_G$, $\alpha_R L = \alpha_G L = \frac{c}{2} = \frac{KL}{2G}$. Let $q = q(x) = P(H)$ and $p = p(x) = P(C)$. In general $q(x)$ and $p(x)$ are complicated function of x, below we derive a first order approximation of $x(q)$ to be used as a biased estimator.

$$P(H) = (1 - (1 - 2e^{\frac{-c}{2}} + 2e^{\frac{-c}{2}(1+\frac{x}{L})})^2) = \frac{2c\exp(\frac{-c}{2})x}{L} + O(x^2)$$
$$P(T) = \left(e^{-c(1+\frac{x}{L})}\right)$$
$$P(C) = 1 - e^{-c} + \frac{c}{2}(e^{-c} - 2e^{-\frac{c}{2}})x + O(x^2)$$

With independent sampling:

$$P(H_{i,j}) \sim \text{Bin}\left(M, \frac{2c\exp(\frac{-c}{2})x}{L} + O(x^2)\right)$$
$$P(C_{i,j}) \sim \text{Bin}\left(M, 1 - e^{-c} + \frac{c}{2}(e^{-c} - 2e^{-\frac{c}{2}})x + O(x^2)\right)$$
$$P(T_{i,j}) \sim \text{Bin}\left(M, (e^{-c(1+\frac{x}{L})})\right) \quad \square$$

B Proof of Lemma 2.3 Using Bayes' Formula

Lemma B1. *If* $f(d|x) = \mathbb{I}_{0 \le x < L}\frac{e^{-(d-x)^2/2\sigma^2 x}}{\sqrt{2\pi x}\sigma} + \mathbb{I}_{L \le x \le G}\frac{e^{-(d-L)^2/2\sigma^2 L}}{\sqrt{2\pi L}\sigma}$. *Then*

$$f(x|d) \approx \mathbb{I}_{d < L}\frac{e^{-(x-d)^2/2\sigma^2 d}}{\sqrt{2\pi d}\sigma} + \mathbb{I}_{d \ge L}\,\mathbb{I}_{L \le x \le G}\left(\frac{1}{G-L}\right),$$

Proof.

$$f(x|d) = \frac{f(d|x)f(x)}{\int_0^G f(d|x)f(x)\,dx}$$

$$= \frac{\frac{1}{G}\left(\mathbb{I}_{0 \le x < L}\frac{e^{-(d-x)^2/2\sigma^2 x}}{\sqrt{2\pi x}\sigma} + \mathbb{I}_{L \le x \le G}\frac{e^{-(d-L)^2/2\sigma^2 L}}{\sqrt{2\pi L}\sigma}\right)}{\frac{1}{G}\int_0^G\left(\mathbb{I}_{0 \le x < L}\frac{e^{-(d-x)^2/2\sigma^2 x}}{\sqrt{2\pi x}\sigma} + \mathbb{I}_{L \le x \le G}\frac{e^{-(d-L)^2/2\sigma^2 L}}{\sqrt{2\pi L}\sigma}\right)\,dx}$$

For small values of σ^2 the denominator in the above expression can be approximated as follows[2]:

$$f(d) = \left(\frac{1}{G}\right)\int_0^L \frac{e^{-(d-x)^2/2\sigma^2 x}}{\sqrt{2\pi x}\sigma}\,dx + \left(\frac{G-L}{G}\right)\frac{e^{-(d-L)^2/2\sigma^2 L}}{\sqrt{2\pi L}\sigma}$$

$$\approx \frac{1}{G}\mathbb{I}_{d<L} + \left(1 - \frac{L}{G}\right)\delta_{d=L}.$$

Thus, we make further simplifying assumptions and choose the following likelihood function:

$$f(x|d) \approx \mathbb{I}_{d<L}\frac{e^{-(x-d)^2/2\sigma^2 d}}{\sqrt{2\pi d}\sigma} + \mathbb{I}_{d \ge L}\,\mathbb{I}_{L \le x \le G}\left(\frac{1}{G-L}\right), \qquad \square$$

C   How Good Are the Results?

C.1   False Positives, False Negatives

We treat the problem of false positives, and false negatives with Chernoff's tail bounds. We find upper bounds on the probability of getting a false positive or false negative in terms of the parameters $\theta, M, c = \frac{KL}{G}, 0 \le \theta \le 1, L' = L\theta \le L$.

A *false positive* is a pair of probes that appear to be close by the Hamming distance but are actually far apart on the genome. We denote the event as:

$$\text{F.P.} = (d < L') \wedge (x > L)$$

A *false negative* is a pair of probes that appear to be far by the Hamming distance but are actually close on the genome. We denote the event as:

$$\text{F.N.} = (x < L') \wedge (d > L)$$

In the following picture the volume of data which are false positives and false negatives are indicated by the squares noted F.P. and F.N. respectively.

---

[2] The   Dirac   Delta   Function   is   distribution   defined   by   the   equations $\left\{\begin{array}{l}\delta_{x=0} = 0 \quad\quad \text{if } x \ne 0 \\ \int_x \delta_{x=0}\,dx = 1\end{array}\right\}.$

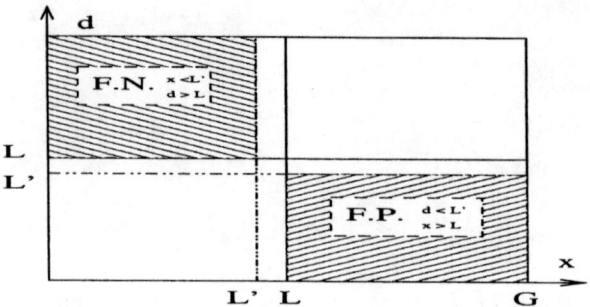

We develop a Chernoff bound to bound the probability that the volume of false positive data is greater than a specified size.

The Chernoff bounds for a binomial distribution with parameters (M, q) are given by:

$$P(H > (1 + v)Mq) < \left(\frac{e^v}{(1 + v)^{(1+v)}}\right)^{Mq} \quad \text{with } v > 0$$

$$P(H < \theta Mq) < e^{\frac{-Mq(1-\theta)^2}{2}} \quad \text{with } 0 \leq \theta < 1$$

Let $H(M)$ be the Hamming distance when M phases are complete. Let $q(L) = P(H|x \geq L) \approx \frac{2\alpha L}{e^{\frac{c}{2}}} = \frac{2c}{e^{\frac{c}{2}}}$ We start by noting equivalent events:

$$(d < \theta L|x > L) = (\sigma^2 H(M) < \theta L|x > L)$$
$$= (H(M) < \theta \frac{L}{\sigma^2}|x > L)$$
$$\subset (H(M) < \theta \frac{2cM}{e^{\frac{c}{2}}})$$
$$= (H(M) < \theta M q(L))$$

Using the Chernoff bound we have:

$$P(d < \theta L|x > L) \leq P(H(M) < \theta Mq_L) < e^{\frac{-Mc(1-\theta)^2}{e^{\frac{c}{2}}}}$$

For the False Negatives we begin by noting that:

$$(d > L|x \leq \theta L) = (\sigma^2 H(x) > L|x < L')$$
$$= (\sigma^2 H(x) > (1 + v)L'|x < L') \text{ where } v = (\frac{1}{\theta} - 1)$$
$$= (H(x) > \frac{(1 + v)L'}{\sigma^2}|x < L')$$
$$\subset (H(x) > (1 + v)Mq(x))$$

The last event inclusion is because:

$$(x \leq L') \Rightarrow \left(\frac{2cMx}{e^{\frac{c}{2}}L} \leq \frac{2cML'}{e^{\frac{c}{2}}L}\right) \Rightarrow \left(Mq(x) \leq \frac{1}{\sigma^2}L'\right)$$

Applying the Chernoff bound we get:

$$P(\text{ F.N. }) \leq P(H > (1+v)Mq(x)) < \left(\frac{e^v}{(1+v)^{(1+v)}}\right)^{Mq(x)} < \left(e^{(\frac{1}{\theta}-1)}\theta^{\frac{1}{\theta}}\right)^{Mq_L}$$

$$= \left(e^{(\frac{1}{\theta}-1)}\theta^{\frac{1}{\theta}}\right)^{M\frac{2c}{e^{\frac{c}{2}}}}$$

Chernoff bounds are:

$$P(\text{ F.P. }) < e^{\frac{-Mc(1-\theta)^2}{e^{\frac{c}{2}}}}$$

$$P(\text{ F.N. }) < \left(e^{(\frac{1}{\theta}-1)}\theta^{\frac{1}{\theta}}\right)^{M\frac{2c}{e^{\frac{c}{2}}}}$$

The Chernoff bounds for typical parameters are shown below.

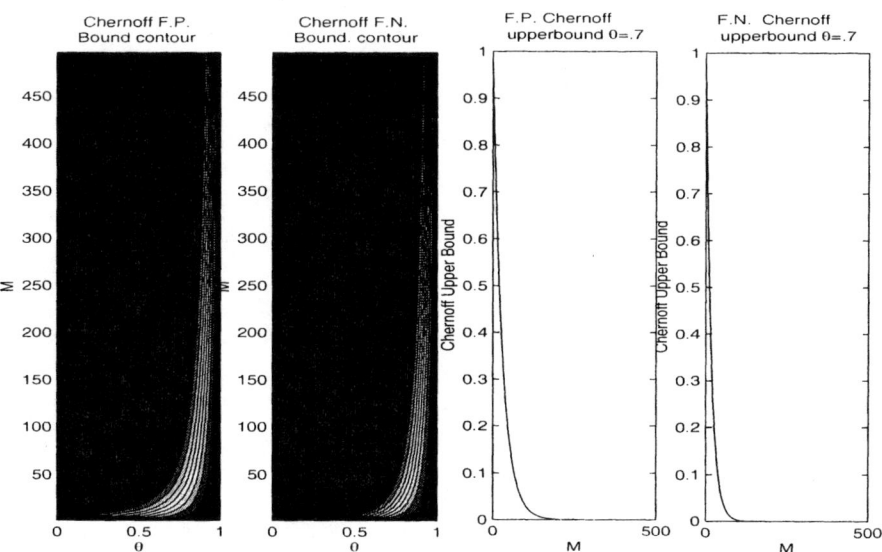

References

1. F. Alizadeh,R.M. Karp,D.K. Weisser, and G. Zweig. "Physical Mapping of Chromosomes Using Unique Probes," **Journal of Computational Biology, 2(2)**:159–185, 1995.
2. E. Barillot, J. Dausset, and D. Cohen. "Theoretical Analysis of a Physical Mapping Strategy Using Random Single-Copy Landmarks," **Journal of Computational Biology, 2(2)**:159–185, 1995.

3. A. Ben-Dor and B. Chor. "On constructing radiation hybrid maps," **Proceedings of the First International Conference on Computational Molecular Biology,** 17–26, 1997

4. H. Chernoff. "A measure of asymptotic efficiency for tests of a hypothesis based on the sum of observations," **Annals of Mathematical Statistics, 23**:483–509,1952.

5. R. Drmanac *et al.,* "DNA sequence determination by hybridization: a strategy for efficient large-scale sequencing," **Science, 163(5147)**:596, Feb 4, 1994.

6. P.W. Goldberg, M.C. Golumbic, H. Kaplan, and R. Shamir. "Four Strikes Against Physical Mapping of DNA," **Journal of Computational Biology, 2(1)**:139–152, 1995.

7. D. Greenberg and S. Istrail. "Physical mapping by STS hybridization: Algorithmic strategies and the challenge of software evaluation," **Journal of Computational Biology, 2(2)**:219–273, 1995.

8. J. Håstad, L. Ivansson, J. Lagergren, "Fitting Points on the Real Line and its Application to RH Mapping," **Lecture Notes in Computer Science, 1461**:465–467, 1998.

9. N. Lisitsyn and M. Wigler, "Cloning the differences between two complex genomes," **Science, 258**:946–951,1993.

10. R. Lucito, J. West, A. Reiner, J. Alexander, D. Esposito, B. Mishra, S. Powers, L. Norton, and M. Wigler, "Detecting Gene Copy Number Fluctuations in Tumor Cells by Microarray Analysis of Genomic Representations," **Genome Research, 10(11)**: 1726–1736, 2000.

11. M.J. Palazzolo, S.A. Sawyer, C.H. Martin, D.A. Smoller, and D.L. Hartl "Optimized Strategies for Sequence-Tagged-Site Selection in Genome Mapping," **Proc. Natl. Acad. Sci. USA, 88(18)**:8034–8038, 1991.

12. D. Slonim, L. Kruglyak, L. Stein, and E. Lander, "Building human genome maps with radiation hybrids," **Journal of Computational Biology, 4(4)**:487–504, 1997.

13. R. E. Tarjan. **Data Structures and Network Algorithms**, CBMS 44 SIAM, Philadelphia, 1983.

14. D.C. Torney, "Mapping Using Unique Sequences," **J Mol Biol, 217(2)**:259–264, 1991.

Comparing a Hidden Markov Model and a Stochastic Context-Free Grammar

Arun Jagota[1], Rune B. Lyngsø[1,*], and Christian N.S. Pedersen[2,**]

[1] Baskin Center for Computer Science and Engineering
University of California at Santa Cruz
Santa Cruz, CA 95064, U.S.A.
{jagota,rlyngsoe}@cse.ucsc.edu
[2] Basic Research in Computer Science (BRICS)
Department of Computer Science
University of Aarhus, Ny Munkegade, DK-8000 Århus C, DK
cstorm@brics.dk

Abstract. Stochastic models are commonly used in bioinformatics, e.g., hidden Markov models for modeling sequence families or stochastic context-free grammars for modeling RNA secondary structure formation. Comparing data is a common task in bioinformatics, and it is thus natural to consider how to compare stochastic models. In this paper we present the first study of the problem of comparing a hidden Markov model and a stochastic context-free grammar. We describe how to compute their co-emission—or collision—probability, i.e., the probability that they independently generate the same sequence. We also consider the related problem of finding a run through a hidden Markov model and derivation in a grammar that generate the same sequence and have maximal joint probability by a generalization of the CYK algorithm for parsing a sequence by a stochastic context-free grammar. We illustrate the methods by an experiment on RNA secondary structures.

1 Introduction

The basic chain-like structure of the key biomolecules, DNA, RNA, and proteins, allows an abstract view of these as strings, or sequences, over finite alphabets, obviously of finite length. Furthermore, these sequences are not completely random, but exhibit various kinds of structures in different contexts. E.g. a family of homologous proteins is likely to have similar amino acid residues in "equivalent" positions; an RNA sequence will have pairs of complementary subsequences to form base pairing helices. Hence, it is natural to consider applying models from formal language theory to model different classes of biological sequences.

Though not completely random, biological sequences can still possess inherent stochastic traits, e.g., due to mutations in a family of homologous sequences or a lack

* Supported by grants from Carlsbergfondet and the Program in Mathematics and Molecular Biology
** Partially supported by the IST Programme of the EU under contract number IST-1999-14186 (ALCOM-FT)

O. Gascuel and B.M.E. Moret (Eds.): WABI 2001, LNCS 2149, pp. 69–84, 2001.

of knowledge (and computing power) to correctly model all aspects of RNA secondary structure formation. Thus, it is often better to use stochastic models giving a probability distribution over all sequences, where a high probability reflects a sequence likely to belong to the class of sequences being modeled, instead of formalisms only distinguishing sequences as either belonging to the class being modeled or not. The two most widely used grammatical models in bioinformatics are hidden Markov models [1, 3, 8, 9, 16] and stochastic context-free grammars [7, 14, 15], though other models have also been proposed [13, 18]. These two types of stochastic models were originally developed as tools for speech recognition (see [2, 12]). One can identify hidden Markov models as a stochastic version of regular languages and stochastic context-free grammars as a stochastic version of context-free languages (see [17] for an introduction to formal languages). A more in-depth treatment of biological uses of hidden Markov models and stochastic context-free grammars can be found in [5, Chap. 3–6 and 9–10].

As stochastic models are commonly used to model families of biological sequences, and as a common task in bioinformatics is that of comparing data, it is natural to ask how to compare two stochastic models. In [10] we described how to compare two hidden Markov models, by computing the co-emission—or collision—probability of the probability distributions of the two models, i.e., the probability that the two models independently generate the same sequence. Having the co-emission probability for a pair of probability distributions as well as for each of the distributions with itself, it is easy to compute the L_2- and the Hellinger-distance between the two distributions. In this paper we study the problem of comparing a hidden Markov model and a stochastic context-free grammar. We develop recursions for the co-emission probability of the distributions of the model and the grammar, recursions that lead to a set of quadratic equations. Though quadratic equations are generally hard to solve, we show how to find an approximate solution by a simple iteration scheme. Furthermore, we show how to solve the equivalent maximization problem, the problem of finding a run through the hidden Markov model and derivation in the grammar that generate the same sequence and have maximal joint probability. This is in essence parsing the hidden Markov model by the grammar, so the algorithm can be viewed as a generalization of the CYK algorithm for parsing a sequence by a stochastic context-free grammar. Indeed, in most cases the complexity of our algorithm will be identical to the complexity of the CYK algorithm. Finally we discuss the undecidability of some natural extensions of our results.

The structure of this paper is as follows. In Sect. 2 and 3 we briefly introduce hidden Markov models and stochastic context-free grammars and the terminology we use. In Sect. 4 we consider the problem of computing the co-emission probability of the probability distributions of a hidden Markov model and a stochastic context-free grammar, and in Sect. 5 we develop the algorithm for parsing a hidden Markov model by a stochastic context-free grammar. In Sect. 6 we present an illustrative experiment and in Sect. 7 we discuss the problems occurring when trying to extend the methods presented in Sect. 4 and 5.

2   Hidden Markov Models

A hidden Markov model M is a generative model consisting of n states and m transitions between states, where each state is either silent or non-silent. The model generates a string s over a finite alphabet Σ with probability $P_M(s)$ such that $\sum_s P_M(s) = 1$, i.e., M describes a probability distribution over the set of finite strings over Σ. A hidden Markov model is a *left-right model* if the states can be numbered $1, 2, \ldots, n$ such that all transitions $i \to j$ from state i to state j satisfy $i \leq j$.

A run in M begins in a special *start-state* and continues from state to state according to the *state transition probabilities*, where $a_{q,q'}^M$ is the probability of a transition from state q to q', until a special *end-state* is reached. A state is either a *silent* or a *non-silent* state. Each time a non-silent state is entered, a symbol is emitted according to the *symbol emission probabilities*, where $e_{q,\sigma}^M$ is the probability of emitting symbol $\sigma \in \Sigma$ in state q. When entering a silent state, nothing is emitted. A run thus follows a Markovian path $\pi = (\pi_0, \pi_1, \ldots, \pi_k)$ of states and generates a string $s \in \Sigma^*$ which is the concatenation of the emitted symbols. The probability $P_M(s)$ of M generating s is the probability of following any path and generating s. The probability $P_M(\pi, s)$ of following path $\pi = (\pi_0, \pi_1, \ldots, \pi_k)$ and generating s depends on the subsequence $(\pi_{i_1}, \pi_{i_2}, \ldots, \pi_{i_\ell})$ of non-silent states on the path π. If the length of s is different from the number of non-silent states along π, the probability $P_M(\pi, s)$ is zero. Otherwise:

$$P_M(\pi, s) = P_M(\pi) \cdot P_M(s \mid \pi) = \prod_{i=1}^{k} a_{\pi_{i-1}, \pi_i}^M \cdot \prod_{j=1}^{\ell} e_{\pi_{i_j}, s_j}^M \; ; \quad P_M(s) = \sum_{\pi} P_M(\pi, s).$$

A *partial run* in M is a run that starts in a state p and ends in a state q, which are not necessarily the special start- and end-states. To ease the presentation of the methods in Sect. 4, we introduce the concept of a partial run from p to q *semi-including* q meaning that if q is a non-silent state then no symbol has yet been emitted from q. The probability of a partial run from p to q semi-including q and generating s is thus the probability of generating s on the path from p to the immediate predecessor of q and taking the transition to q. Given a string s and a model M, efficient algorithms for determining $P_M(s)$ and the run in M of maximal probability generating s are known, see [5, 12].

3   Stochastic Context Free Grammars

A context-free grammar G describes a set of finite strings over a finite alphabet Σ, also called a language over Σ. It consists of a set V of non-terminals, a set $T = \Sigma \cup \{\epsilon\}$ of terminals, where ϵ is the empty string, and a set P of production rules $\alpha \to \beta$, where $\alpha \in V$ and $\beta \in (V \cup T)^+$. A production rule $\alpha \to \beta$ means that α can be rewritten to β by applying the rule. A string $s \in \Sigma^*$ can be *derived* from a non-terminal U, $U \stackrel{*}{\Rightarrow} s$, if U can by rewritten to s by a sequence of production rules; s is in the language described by G if it can be derived from a special start non-terminal S. A derivation D of s in G is a sequence of production rules which rewrites S to s. A derivation of s in G is also called a parsing of s in G.

A *stochastic* context-free grammar G is a context-free grammar in which all production rules $\alpha \rightarrow \beta$ are assigned probabilities $P_G(\alpha \rightarrow \beta)$ such that the sum of the probabilities of all possible production rules from any given non-terminal is one. The resulting stochastic grammar describes a probability distribution over the set of finite string over the finite alphabet Σ, where the probability $P_G(s)$ of deriving s in G (sometimes also referred to as the probability of G generating s) is the sum of the probabilities of all possible derivations of s from S. The probability of a derivation D of s in G, i.e., $P_G(D) = P_G(S \overset{D}{\Rightarrow} s)$, is the product of the probabilities of the production rules in the sequence D applied to derive s from S.

Any (stochastic) context-free grammar can be transformed to an equivalent (stochastic) context-free grammar in Chomsky normal form which describes the same language but where the production rules can be split into two sets; a set P_n of non-terminal production rules on the form $U \rightarrow XY$, and a set P_t of terminal production rules on the form $U \rightarrow \sigma$ and $U \rightarrow \epsilon$, where U, X and Y are non-terminals, σ is a symbol in Σ, and ϵ is the empty string. For a stochastic context-free grammar G in Chomsky normal form, we can compute $P_G(s)$ by the inside algorithm, and the most likely parse of s in G by the CYK algorithm, in time $O(|P| \cdot |s|^3)$ where $|P|$ is the number of production rules in G, i.e., $|P_n| + |P_t|$, and $|s|$ is the length of s, see [5].

4   Comparing a SCFG and a HMM

In this section we will consider the problem of comparing a stochastic context-free grammar G (in Chomsky normal form) and a hidden Markov model M which generate strings over the same alphabet Σ. More precisely, we will consider the problem of computing the co-emission probability, $C(G, M)$, of the probability distributions of G and M over all strings, i.e., the quantity

$$C(G, M) = \sum_{s \in \Sigma^*} P_G(s) \cdot P_M(s),$$

which is similar to the definition of the co-emission probability of two hidden Markov models in [10]. This quantity is also often referred to as the collision probability of the probability distributions of G and M, as it is the probability that strings picked at random according to the two probability distributions collide, i.e., are identical. We initially assume that M is an acyclic hidden Markov model, i.e., a left-right hidden Markov model where no states have self-loop transitions. By itself, this is not a very interesting class of hidden Markov models, e.g., a model of this type cannot generate strings of arbitrary length, but the ideas of our approach for computing the co-emission probability for this class of models are also applicable to left-right and general hidden Markov models.

For acyclic hidden Markov models we can use an approach closely mimicking the inside algorithm for computing the probability that a stochastic context-free grammar generates a given string. In the inside algorithm, when computing the probability that a string s is derived in a stochastic context-free grammar, track is being kept of the probability that a substring of s is derived from a non-terminal of the grammar. In our algorithm for computing the co-emission probability of G and M we keep track of the

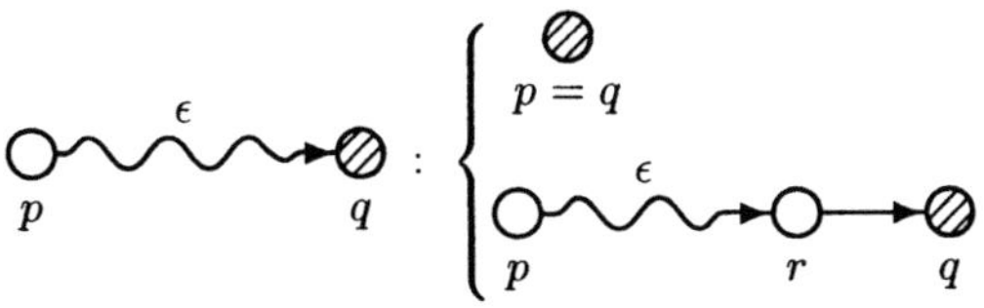

(a) The A array holds the probabilities of getting from one state p to another state q in M without emitting any symbols.

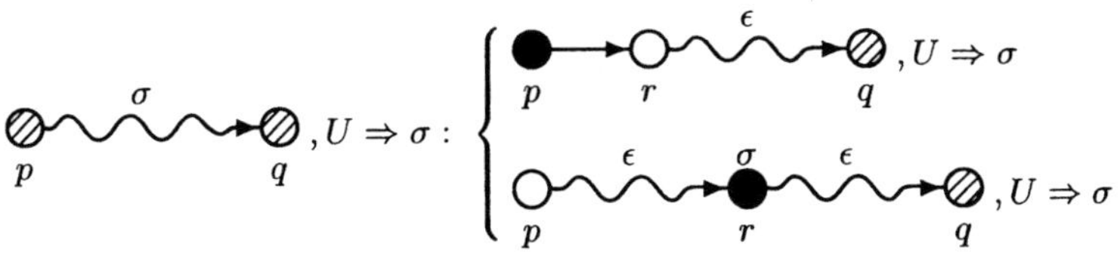

(b) The B array holds the probabilities of getting from one state p to another state q in M while emitting only one symbol and at the same time generating the same symbol by a terminal production rule for the non-terminal U in G.

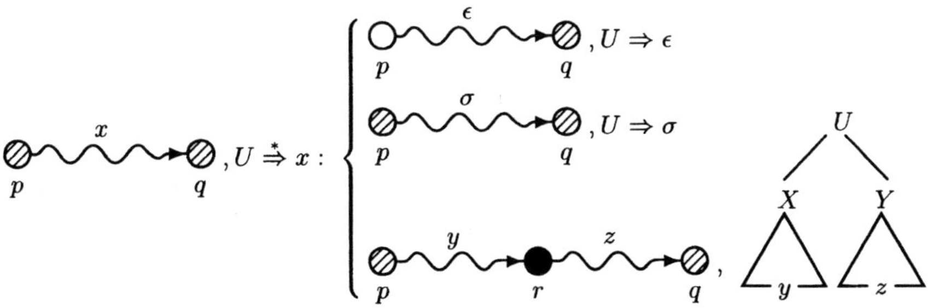

(c) The C array holds the probabilities of getting from one state p to another state q in M while emitting any string and at the same time generating the same string from a non-terminal U in G.

Fig. 1. Illustration of the individual purposes and recursions of the three arrays used. Hollow circles denote silent states, solid circles denote non-silent states, and hatched circles denote states of any type. Squiggle arrows indicate partial runs of arbitrary length and straight arrows indicate single transitions between states.

probability of deriving the same string from a non-terminal that is generated on a partial run from a state p to a state q semi-including q in M. In our dynamic programming based algorithm we maintain three arrays, A, B, and C, for the following purposes.

– $A(p, q)$ will be the sum of the probabilities of all partial runs from p to q semi-including q that does not emit any symbols, illustrated in Fig. 1(a). I.e. all states on the partial runs, except possibly for q, are silent states.

- $B(U, p, q)$ is the probability of independently deriving a single symbol string from the non-terminal U and generating the same single symbol string on a partial run from p to q semi-including q, illustrated in Fig. 1(b).
- $C(U, p, q)$ is the probability of independently deriving a string from the non-terminal U and generating the same string on a partial run from p to q semi-including q, illustrated in Fig. 1(c).

The purpose of the A and B arrays is to deal efficiently with partial runs consisting of silent states. As all symbols of a string are in a sense "non-silent", this is the main new problem encountered when modifying the inside algorithm to "parse" acyclic hidden Markov models.

The C array is similar to the array maintained in the inside algorithm. The array of the inside algorithm tells us the probability of deriving any substring of the string being parsed from any of the non-terminals of G. Similarly, the C array will tell us the probability of independently generating a sequence on a partial run between any pairs of states and at the same time deriving it from any of the non-terminals of G. It is evident that $C(G, M) = C(S, start, end)$, where S is the start symbol of G and $start$ and end are the start- and end-states of M, as the co-emission probability of G and M is the probability of deriving the same string from S that is generated on a (genuine) run from $start$ to end. This is assuming that the end-state of M is silent, so that any partial run from $start$ to end semi-including end in M is also a genuine run in M.

Having described the required arrays, we must specify recursions for computing them—and argue that these recursions split the computations into ever smaller parts, i.e., that the dependencies of the recursions are acyclic—to obtain a dynamic programming algorithm computing $C(G, M)$. The A array are the probabilities for getting from the state p to the state q in M along a path that only consists of silent states. In general such a path can be broken down into the last transition of the path and a preceding path only going through silent states. Hence, we obtain the following recursion where the first case takes care of the initialization.

$$A(p, q) = \begin{cases} 1 & \text{if } p = q \\ \displaystyle\sum_{\substack{p \leq r < q, \\ r \text{ silent}}} A(p, r) \cdot a_{r,q}^M & \text{otherwise} \end{cases} \qquad (1)$$

The ordering of states referred to in the summation is any ordering consistent with the (partial) ordering of states by the acyclic transition structure. One immediately observes that each entry of the A array requires time at most time $O(n)$ to compute. Thus the entire A array can be computed in time $O(n^3)$. However, one can observe that we actually only need to sum over states r with a transition to q in the summation of (1). This observation reduce the time requirements to $O(nm)$ as each transition is part of $O(n)$ of the above recursions.

The B array holds the probabilities for getting from the state p to the state q in M by a partial run generating exactly one symbol, and at the same time generating the same symbol from U by a terminal production rule. In general we can partition the partial run into an initial path consisting only of silent states and the remaining partial run starting with the non-silent state emitting the symbol. Hence, we obtain the following recursion where "initialization" is handled by paying special attention to the special case where the initial path of silent states is empty.

$$B(U, p, q) = \sum_{(U \to \sigma) \in G} P_G(U \to \sigma) \cdot$$

$$\left(e_{p,\sigma}^M \cdot \sum_{p < r \leq q} a_{p,r}^M \cdot A(r, q) + \sum_{\substack{p < r < q, \\ r \text{ non-silent}}} A(p, r) \cdot B(U, r, q) \right) \quad (2)$$

To find the time requirements for computing the B array one observes that each non-terminal of G occurs in $O(n^2)$ entries. Hence, each terminal production rule of G is part of $O(n^2)$ of the above recursions. Each of these recursions requires time $O(n)$ to compute. Thus, computing the B array requires time $O(|P_t| \cdot n^3)$.

The C array should hold the probabilities for getting from the state p to the state q in M by a partial run generating any string, and at the same time generating the same string from a non-terminal U in G. The purpose of the A and B arrays is to handle the special cases where at most a single symbol is emitted on a partial run from p to q. In these cases we do not really recurse on a non-terminal U from G but either ignore G completely or only consider terminal production rules. In the general case where a string with more than one symbol is generated, we need to apply a non-terminal production rule $U \to XY$ from G. Part of the string is then derived from X and part of the string is derived from Y. This leads to the following recursion.

$$C(U, p, q) = P_G(U \to \epsilon) \cdot A(p, q) + B(U, p, q) +$$

$$\sum_{(U \to XY) \in G} P_G(U \to XY) \cdot \sum_{\substack{p < r < q \\ r \text{ non-silent}}} C(X, p, r) \cdot C(Y, r, q) \quad (3)$$

The reason for requiring r to be a non-silent state in the last sum is to ensure a unique decomposition of the partial runs from p to q. If we allowed r to be silent we would erroneously include some partial runs from p to q several times in the summation. As we did with the computation of the B array, we can observe that each non-terminal production rule is part of $O(n^2)$ of the above recursions, and that each recursion requires time $O(n)$ to compute. Hence, we obtain total time requirements of $O(|P_n| \cdot n^3)$ for computing the C array. Adding the time requirements for computing the three arrays leads to overall time requirements of $O(|P| \cdot n^3)$ for computing the co-emission probability of G and M. This is in correspondence with the time requirements of $O(|P| \cdot |s|^3)$ for parsing a string s by the grammar G cf. Sect. 3.

As previously stated, in order for these recursions to be used for a dynamic programming algorithm, we need to argue that the recursions only refer to elements that in some sense are smaller. I.e. that no recursion for any of the entries of the three arrays depends cyclically on itself. But this is an easy observation as all pairs of states indexing an array on the right-hand side of the recursions are closer in the ordering of states than the pair of states indexing the array on the left-hand side of the recursions.

The point of always recursing by smaller elements is exactly where we run into problems when trying to extend the above method to allow cycles in M. Even when only allowing the self-loops of left-right models this problem crops up. We can easily modify (1), (2), and (3) to remain valid as equations for the entries of the A, B and C arrays. All that is needed is to change some of the strict inequalities in the summation boundaries to include equality. More specifically, if we assume that only non-silent states can have self-loop transitions (self-loop transitions for silent states are rather meaningless and can easily be eliminated), we need to change (2) and (3) to

$$B(U, p, q) = \sum_{(U \to \sigma) \in G} P_G(U \to \sigma) \cdot$$

$$\left(e_{p,\sigma}^M \cdot \sum_{p \leq r \leq q} a_{p,r}^M \cdot A(r, q) + \sum_{\substack{p < r \leq q, \\ r \text{ non-silent}}} A(p, r) \cdot B(U, r, q) \right), \quad (4)$$

$$C(U, p, q) = P_G(U \to \epsilon) \cdot A(p, q) + B(U, p, q) +$$

$$\sum_{(U \to XY) \in G} P_G(U \to XY) \cdot \sum_{p \leq r \leq q} C(X, p, r) \cdot C(Y, r, q). \quad (5)$$

For the B array, (4) still refers only to smaller elements. Either the A array or an entry of the B array where the two states are closer in the ordering as r has to be strictly larger than p in the ordering in the last sum. But for the C array we might choose r equal to either p or q in the last sum. Hence, to compute $C(U, p, q)$ we need to know $C(X, p, q)$ and $C(Y, p, q)$ which might in turn depend on (or even be) $C(U, p, q)$.

But, as stated above, (5) still holds as equations for the entries of the C array. For each pair of states, p and q, in M we thus have a system of equations with one variable and one equation (and the restriction that all variables have to attain non-negative values) for each non-terminal of G. Assume that we solve these systems in an order where the distance between the pair of states in the ordering is increasing, i.e., we first consider systems with $p = q$, then systems with q being the successor to p, etc. Then most of the systems will be systems of linear equations. In the equation for a given entry $C(U, p, q)$ the only unknown quantities are the occurrences of $C(X, p, q)$ and $C(Y, p, q)$ corresponding to production rules $U \to XY$ of G. These occurrences have coefficients $P_G(U \to XY) \cdot C(Y, q, q)$ and $P_G(U \to XY) \cdot C(X, p, p)$, respectively, coefficients that have known values if $p < q$. But if $p = q$ the last sum of (5) will lead to a number of terms of the form $C(X, p, p) \cdot C(Y, p, p)$. I.e. the system of equations is quadratic. Hence, for each state with a self-loop transition we need to solve a system of quadratic equations with one variable and one equation for each non-terminal of G. General systems of quadratic equations are hard to solve, see [6], but the construction proving this requires equations with all terms having coefficients with the same sign. One can immediately observe that in a system of equations based on (5) the left-hand side terms will have coefficients that have opposite sign of the right-hand side terms. Hence, the hardness proof does not relate to the system of quadratic equations we obtain from (5). We have not been able to find any literature on algorithms solving systems of the type derivable from (5). But as all the dependencies are positive we can approximate a solution simply by initializing all entries to the terms depending only on the A and B arrays and then iteratively update each entry in turn. This process will converge to the true solution. We conjecture that this convergence will be very rapid for all realistic grammars and models.

For general hidden Markov models we do not have an ordering of the states of M. Hence, to modify (1), (2), and (3) to hold for general hidden Markov models, the ordering constraints on states should be removed from all the summations. When we do no longer have any ordering of the states, all entries might depend on each other. Thus, we cannot separate the systems of equations for the entries of the C array into independent blocks based on the pair of states indexing the entry of the array. This means that we obtain just one aggregate system of quadratic equations with $|V| \cdot n^2$ variables and equations for the entries of the C array, where $|V|$ is the number of non-terminals in G and n is the number of states in M. However, for the entries of the A and B arrays we still only get systems of linear equations. Actually, the B array can even be computed by simple dynamic programming.

5   Parsing an HMM by a SCFG

Given a stochastic context-free grammar G modeling RNA secondary structures and a hidden Markov model M representing a family of RNA sequences, we can use the method of the previous section to determine the likelihood that the family has a common secondary structure. This being established, one might want to find the most likely structure. In other words, an equivalent to the CYK algorithm for finding the most likely parse of a string, but finding the most likely "parse" of a hidden Markov model, is desirable. Hence, we want to compute

$$\max\{P_G(D) \cdot P_M(r) \mid D \text{ a derivation in } G, r \text{ a run in } M, \text{ and } S \stackrel{D}{\Rightarrow} s_r\}, \quad (6)$$

where s_r is the string generated by the run r, preferably in a way that allows us to easily extract a derivation D and a run r witnessing the maximum.

The basic principles we will use to compute the value of (6) are similar to those used for computing the co-emission probability in the previous section. Thus, the general technique is to find the probabilities of optimal combinations of a derivation from a non-terminal U in G and a partial run between a pair of states p and q in M yielding the same string. Furthermore, this optimal combination is split into a pair of optimal combinations of a derivation from a non-terminal and a partial run between states as illustrated in Fig. 1(c). However, computing the maxima is easier than computing the sums. The main reason for this is that we do not have to consider combinations of derivations and partial runs of arbitrary lengths. The probability of an optimal combination for a particular choice of a non-terminal and pair of states cannot depend cyclicly on itself—the product of two or more probabilities can never be larger than any of these probabilities. A similar principle is used in algorithms for finding shortest paths in graphs with only positive edge weights, cf. [4, chapters 25–26]. Not surprisingly, the methods described in this section will bear strong resemblances to such algorithms, though the particular nature of the problem does prevent formulating it simply as a shortest path problem in a well chosen graph.

The description of our method will employ three arrays $A_{\max}$, $B_{\max}$, and $C_{\max}$, similar to the A, B, and C arrays used in previous section. The $A_{\max}(p, q)$ entries hold the maximum probability of any partial run from state p to state q semi-including q not emitting any symbols. But this is just the path of maximum weight in the graph defined by the transitions not leaving a non-silent state in M. We can thus compute the $A_{\max}$ array by standard all-pairs shortest paths graph algorithms in time $O(n^2 \log n + nm)$, cf. [4, p. 550].

The $B_{\max}(U, p, q)$ entries hold the maximum probability of a combination of a derivation consisting of only a terminal production rule $U \to \sigma$ and a partial run from p to q generating a string consisting only of the symbol σ. This imposes a restriction on the path from p to q preventing the use of standard graph algorithms. Still, we only need to combine transitions from non-silent states with preceding and succeeding partial runs not emitting any symbols, i.e., with entries of $A_{\max}$. However, if we compute the $B_{\max}$ entries directly from the equation

$$B_{\max}(U, p, q) = \max_{\substack{(U \to \sigma) \in G \\ p \to q \in M}} \left\{ P_G(U \to \sigma) \cdot e^M_{r,\sigma} \cdot a^M_{r,s} \cdot A_{\max}(p, r) \cdot A_{\max}(s, q) \right\} \quad (7)$$

we use time $O(|P_t| \cdot m \cdot n^2)$. This could possibly be the dominating term in the overall time requirements for computing the value of (6), as we will see later.

Hence, we will specify a more efficient way to compute the $B_{\max}(U, p, q)$ entries. If p is a non-silent state the optimal choice of a preceding partial run not emitting any symbols must be the empty run. Thus

$$B_{\max}(U, p, q) = \max_{\substack{(U \to \sigma) \in G \\ p \to r \in M}} \left\{ P_G(U \to \sigma) \cdot e^M_{p,\sigma} \cdot a^M_{p,r} \cdot A_{\max}(r, q) \right\} \quad (8)$$

for all entries of $B_{\max}$ with p non-silent. Having computed these entries, we can now proceed to compute the entries for p silent by

$$B_{\max}(U, p, q) = \max_{\substack{(U \to \sigma) \in G \\ r \in M, \, r \text{ non-silent}}} \left\{ A_{\max}(p, r) \cdot B_{\max}(U, r, q) \right\} . \quad (9)$$

Computing the $B_{\max}$ array this way reduces the time requirements to $O(|P_t| \cdot n^3)$.

We are now ready to specify how to determine the value of (6), i.e., how to compute the entries of the $C_{\max}$ array. An entry $C_{\max}(U, p, q)$ holds the maximum probability of a combination of a derivation from U in G and a partial run from p to q in M yielding the same string. The following equation for computing an entry of $C_{\max}$ closely follows Fig. 1(c).

$$C_{\max}(U, p, q) = \max \begin{cases} P_G(U \to \epsilon) \cdot A_{\max}(p, q) \\ B_{\max}(U, p, q) \\ \max \left\{ P_G(U \to XY) \cdot C_{\max}(X, p, r) \cdot C_{\max}(Y, r, q) \right. \\ \qquad \left. \mid (U \to XY) \in G, r \in M \right\} \end{cases} \quad (10)$$

The only exception is that there is no harm in considering the same combination of a derivation and a partial run several times when working with maxima instead of sums. Hence, there is no restriction on the type of the state (denoted by r) that we maximize over in the general case. In an actual implementation one might want to retain the type restriction to speed up the program, though.

```
/* Initialization */
PQ = ∅, S = ∅
for all U ∈ G, p, q ∈ M do
    PQ.insert((U, p, q); min{P_G(U → ε) · A_max(p, q), B_max(U, p, q)})
/* Main loop where one entry is fixed at a time */
while PQ is not empty do
    /* Fix the entry with highest probability not yet fixed */
    (U, p, q); x = PQ.deletemax
    S.insert((U, p, q); x)
    /* Combine this entry with all feasible, fixed entries */
    for all X, Y ∈ G with (X → UY) ∈ G and all r ∈ M with (Y, q, r); y ∈ S do
        PQ.increasekey((X, p, r); x · y)
    for all X, Y ∈ G with (X → YU) ∈ G and all r ∈ M with (Y, r, p); y ∈ S do
        PQ.increasekey((X, r, q); y · x)
```

Algorithm 1: The algorithm for computing an optimal parse of a hidden Markov model M by a stochastic context-free grammar G.

Again, this is slightly beyond a shortest path problem. However, (10) gives us a number of inequalities for each of the entries of C_{max}, similar to [4, Lemma 25.3]; $C_{max}(U, p, q)$ must be at least as large as any of the terms on the right-hand side of (10). Thus, we can use a technique very similar to the relaxation technique discussed in [4, pp. 520–521]; if at any time $C_{max}(U, p, q) < P_G(U → XY) · C_{max}(X, p, r) · C_{max}(Y, r, q)$ for any $(U → XY) ∈ G$ and $r ∈ M$ we can increase $C_{max}(U, p, q)$ to this value. This means that we could start by initializing each $C_{max}(U, p, q)$ to $max\{P_G(U → ε) · A_{max}(p, q), B_{max}(U, p, q)\}$ and then keep updating C_{max} by iterating over all possible relaxations until no more changes occur. This process will eventually terminate as no entry can depend cyclicly on itself, as discussed above.

The worst-case time requirements of this scheme are quite excessive, though. Thus, the question of how to order the relaxations so as to compute C_{max} most efficiently still remains. We propose an approach very similar to Dijkstra's algorithm for computing single-source shortest paths in a graph, cf. [4, Sect. 25.2]. Assume that S is the set of entries for which we have already determined the correct value, and that we have performed all relaxations combining the correct values of these entries (and initialized all entries using the A_{max} and B_{max} arrays as mentioned in the preceding paragraph). Let $C_{max}(U, p, q)$ be an entry of maximum value not in S. We claim that the current value of $C_{max}(U, p, q)$ must be correct, i.e., that it cannot be increased. In fact, no entry not in S can have a correct value larger than $C_{max}(U, p, q)$. The reason for this is that any relaxation not combining two entries both in S—which we assumed have already been performed—will involve an entry not in S, and thus with current value at most $C_{max}(U, p, q)$. As the other entry used in the relaxation can have a value of at most 1, no future relaxations can lead to values above $C_{max}(U, p, q)$. Hence, we can insert $C_{max}(U, p, q)$ in S and perform all relaxations combining $C_{max}(U, p, q)$ with an entry from S. This idea is formalized in algorithm 1.

So what are the time requirements of algorithm 1? To some extent this of course depends on the choice of data structures implementing the priority queue PQ and the set S, but a key observation is that each possible relaxation, i.e., combination of two particular entries, is only performed once, namely when the entry with the smaller value is inserted in S. Hence, algorithm 1 performs $O(|P_n| \cdot n^3)$ relaxations. One can observe that for each relaxation we need to perform the operation increasekey on PQ, while all other operations on PQ are performed at most $O(|V| \cdot n^2)$ times. Thus, increasekey is the most critical operation, why we will assume that PQ is implemented by a Fibonacci heap. This limits the time requirements for all operations on PQ to $O(|P_n| \cdot n^3 + |V| \cdot n^2 \cdot \log(|V| \cdot n))$.

For the set S we need to be able to insert an element efficiently, and to efficiently iterate through all elements with a particular non-terminal and state. But having already set aside time $\Omega(|P|_n \cdot n^3)$ for the priority queue operations, the operations on S do not need to be that efficient. As it turns out, it is actually sufficient to maintain S simply as a three-dimensional boolean array indexed by (U, p, q). This makes insertion a constant time operation. However, it does not allow for an efficient way to iterate over all elements in S with a particular non-terminal and state, short of iterating over all elements with that non-terminal and state and test the membership of each individual element. However, this turns out to be sufficiently efficient. In some situations, especially in the beginning when there are only a few elements in S, we might test the membership of numerous elements not in S. But each membership test can be associated with a relaxation involving a particular pair of elements, namely the relaxation that is performed if the test succeeds. Furthermore, for each relaxation we will only test membership twice, once for each element that the relaxation combines. Hence, the total time we spend iterating over elements in S is $O(|P_n| \cdot n^3)$. Thus, the time requirements for algorithm 1 is $O(|P_n| \cdot n^3 + |V| \cdot n^2 \cdot \log(|V| \cdot n))$. Combined with the time complexity of computing the $A_{\max}$ and $B_{\max}$ arrays, this leads to an overall time complexity of $O(|P| \cdot n^3 + |V| \cdot n^2 \cdot \log(|V| \cdot n))$ for determining the optimal parse of a general hidden Markov model by a stochastic context-free grammar (having computed $A_{\max}$, $B_{\max}$, and $C_{\max}$ it is easy to find the optimal parse by standard backtracking techniques). This should be compared with the time requirements of $O(|P| \cdot |s|^3)$ for finding the optimal parse of a string s by the CYK algorithm.

In the above description we did not use any assumptions about the structure of the hidden Markov model M. A natural question to ask is thus how much can be gained with respect to time requirements, by restricting our attention to left-right models. That it is not a lot should not be surprising, considering that we are already close to the complexity of the CYK algorithm. However, we can observe that $C_{\max}(U, p, q)$ can only depend on entries $C_{\max}(U', p', q')$ with $p \leq p' \leq q' \leq q$ (where the ordering is with respect to the implicit partial ordering of states in a left-right model). Thus, we can separate (10) into $O(n^2)$ systems, one for each choice of p and q, that can be solved one at a time in a predefined order. Hence, the priority queue only needs to hold at most $|V|$ elements at any time, reducing the time complexity for finding the optimal parse to $O(|P| \cdot n^3 + |V| \cdot n^2 \cdot \log |V|)$. More importantly, though, as we only need a priority queue with at most $|V|$ elements and $|V|$ will usually be very small, it might be feasible to replace the Fibonacci heap implementation with implementations that have worse asymptotic complexities but smaller overhead. Thus, if we just implement PQ with an array, scanning the entire array each time a deletemax operation is performed,

```
      (((((((..((((...  .. ..)))) (((((((...))))))).........(((((.......)))))))))))    )).
seq1 GGGGAUGUAGCUCAGU--GG-UAGAGCGCAUGCUUCGCAUGUAUGAGGCCCCGGGUUCGAUCCCCGGCAUCU--
      -CCA
      (((((((...((((((((((( (.....)))......)))))))))......(((((.......)))))))))))))....
seq2 CGGCACGUAGCGCAGCCUGG-UAGCGCACCGUCCUGGGGUUGCGGGGGUCGGAGGUUCAAAUCCUCUCGUGCCGACCA

      .((((((( ((((...  ..       ))))))))))) (((((((((((((((((((...)))))))))))))))))))....
max  GGGCAUGU-GCGCAGU--GG---GCGCACAUGCCUCGGGAUGAGGGGGUCCGAGGUUCGGACCCCCUCAUCCCGACCA
                                                    ↑                          ↑
```

Fig. 2. Alignment and predicted secondary structure of the two sequences, *seq1* and *seq2*, used to construct the `trna2` model, and the sequence and secondary structure of the maximal parse of `trna2` aligned according to the states of `trna2` emitting the symbols. In the two positions indicated with an ↑ the sequence of the maximal parse does not match any of the two other sequences.

the complexity of parsing a left-right model only increases to $O(|P| \cdot n^3 + |V|^2 \cdot n^2)$ with the involved constants being very small.

6   Results

Algorithm 1 has been implemented as the program **CStoRM** (Comparison of Stochastic (Random) Models) and we are currently working on adding computation of the co-emission probability to the implementation. The implementation is available at http://www.cse.ucsc.edu/~rlyngsoe/cstorm.tar.gz. As an illustrating experiment we have used the program to parse the `trna2` profile hidden Markov model that is part of the test suite for the SAM software distribution available at http://www.cse.ucsc.edu/research/compbio/sam.html. This model is built from the alignment of *seq1* and *seq2* shown in Fig. 2, with the symbol emission probabilities in each position being a little less than 0.1 for symbols not present in that position in the alignment and the rest of the probability distributed evenly among the symbols present in that position in the alignment. Each match-state has a probability of at least 0.85—and for positions without gaps in the alignment more than 0.97—for choosing the transition to the next match-state. The grammar used is the stochastic context-free grammar for general RNA secondary structure presented in [7]. This grammar was also used for predicting the secondary structure of *seq1* and *seq2* shown in Fig. 2.

As is evident from Fig. 2, the maximal parse does a poor job at finding the common structural elements of the two sequences. This might in part be explained by the fact that only about half the base pairs of the structures of each of the sequences is shared with the structure of the other sequence. But not even any of the shared base pairs are present in the structure found by the maximal parse. Instead, one can observe that the structure of the maximal parse is much more dense than the structures of *seq1* and *seq2*, with only a few bases not being part of one of the two uninterrupted helices constituting the structure. This is probably an indication of the main problem of using the maximal parse to predict the secondary structure. In each position we can choose a symbol so as to construct a sequence with a structure of very high probability, i.e., the maximal parse seems to be a question of highly probable structures finding matching sequences instead of what we are really looking for, highly probable sequences finding a matching structure. This is further supported by the two positions where the maximal parse disagrees with *seq1* and *seq2*, especially as *seq1* and *seq2* agree in these positions—the sequence obtained by changing the symbols in these two positions to the symbols shared by *seq1* and *seq2* would be roughly 80 times as probable in the hidden Markov model.

Another problem is that the maximum is not very good for discriminating states that exhibit complementarity from states that do not exhibit complementarity. E.g. a state that has probabilities 0.5 for emitting either a C or a G gets a lower probability if paired with a state identical to itself, than if paired with a state that has probability 0.51 for emitting a C and 0.49 for emitting a A. However, having the framework of algorithm 1 it is easy to modify the details to accommodate a scoring of combinations of pairs of states and derivations introducing base pairs that captures complementarity better. Furthermore, the co-emission probability will indeed capture that the two C/G-emitting states exhibit better complementarity than the C/G-C/A pair. Thus, a better idea of a common structure might be obtained by looking at the probability that two states emit symbols that are base paired for all pairs of non-silent states, similar to [11, 19]. Indeed, as the dependencies of the energy rules commonly used for RNA secondary structure prediction can be captured by a context-free grammar, one can also combine the computation of the co-emission probability as discussed in this paper with the computation of the equilibrium partition function presented in [11] to obtain probabilities for base pairing of positions including both the randomness of base pairing captured by the partition functions as well as the variability of a family of sequences captured by a (profile) hidden Markov model.

7   Discussion

In this paper we have considered the problem of comparing a hidden Markov model with a stochastic context-free grammar. The methods presented can be viewed as natural generalizations of methods for analyzing strings by means of stochastic context-free grammars, or of the idea of comparing two hidden Markov models in [10]. A natural question is thus whether we can further extend the results to comparing two stochastic context-free grammars. If we could determine the co-emission probability of—or just the maximal joint probability of a pair of parses in—two stochastic context-free grammars, we could also determine whether the languages of two context-free grammars are disjoint, simply by assigning a uniform probability distribution to the derivations of each of the variables and asking whether the computed probability is zero. However, a well-known result in formal language theory states that it is undecidable whether the languages of two context-free grammars are disjoint [17, Theorem 11.8.1]. Hence, we cannot generalize the methods presented in this paper to methods comparing two stochastic context-free grammars with a precision that allows us to determine whether the true probability is zero or not.

In [10] we use the co-emission probability to compute the L_2-distance between the probability distributions of two hidden Markov models. Having demonstrated how to compute the co-emission probability between a hidden Markov model and a stochastic context-free grammar, the only thing further required to compute the L_2-distance is the co-emission probability of the grammar with itself. As stated above, the problem of computing the co-emission probability between two stochastic context-free grammars is undecidable, but one could hope that computing the co-emission probability of a stochastic context-free grammar with itself would be easier. However, given two stochastic context-free grammars G_1 and G_2 we can construct an aggregate grammar G' where the start symbols of G_1 and G_2 are chosen with equal probability one half. It is easy to see that the co-emission probability of G' with itself is the sum of the co-emission probabilities of G_1 and G_2 with themselves plus twice the co-emission probability between G_1 and G_2. Hence, computing the co-emission probability of a stochastic context-free grammar with itself, or the L_2- or Hellinger-distances between the probability distributions of a context-free grammar and a hidden Markov model, is as hard as computing the co-emission probability between two stochastic context-free grammars.

In this paper we have presented the co-emission probability as a measure for comparing two stochastic models. However, the co-emission probability has at least two other interesting uses. First, it allows us to use one model as a prior for training the other model, e.g., using the distribution over sequences of a hidden Markov model as our prior belief about the distribution over sequences for a stochastic context-free grammar we want to construct. Secondly, it allows us to compute the probability that two stochastic models have independently generated a sequence s given the two models generate the same sequence. I.e. we can combine two models under the assumption of independence.

References

1. K. Asai, S. Hayamizu, and K. Handa. Prediction of protein secondary strucuture by the hidden markov model. *Computer Applications in the Biosciences (CABIOS)*, 9:141–146, 1993.
2. J. K. Baker. Trainable grammars for speech recognition. In *Speech Communications Papers for the 97th Meeting of the Acoustical Society of America*, pages 547–550, 1979.
3. G. A. Churchill. Stochastic models for heterogeneous DNA sequences. *Bulletin of Mathematical Biology*, 51:79–94, 1989.
4. T. H. Cormen, C. E. Leiserson, and R. L. Rivest. *Introduction to Algorithms*. The MIT Press, 1990.
5. R. Durbin, S. R. Eddy, A. Krogh, and G. Mitchison. *Biological Sequence Analysis: Probalistic Models of Proteins and Nucleic Acids*. Cambridge University Press, 1998.
6. J. Håstad, S. Phillips, and S. Safra. A well characterized approximation problem. *Information Processing Letters*, 47(6):301–305, 1993.
7. B. Knudsen and J. Hein. RNA secondary structure prediction using stochastic context-free grammars and evolutionary history. *Bioinformatics*, 15:446–454, 1999.
8. A. Krogh. Two methods for improving performance of an HMM and their application for gene finding. In *Proceedings of the 5th International Conference on Intelligent Systems for Molecular Biology (ISMB)*, pages 179–186, 1997.
9. A. Krogh, M. Brown, I. S. Mian, K. Sjölander, and D. Haussler. Hidden markov models in computational biology: Applications to protein modeling. *Journal of Molecular Biology*, 235:1501–1531, 1994.

10. R. B. Lyngsø, C. N. S. Pedersen, and H. Nielsen. Metrics and similarity measures for hidden Markov models. In *Proceedings of the 7th International Conference on Intelligent Systems for Molecular Biology (ISMB)*, pages 178–186, 1999.

11. J. S. McCaskill. The equilibrium partition function and base pair binding probabilities for RNA secondary structure. *Biopolymers*, 29:1105–1119, 1990.

12. L. R. Rabiner. A tutorial on hidden markov models and selected applications in speech recognition. In *Proceedings of the IEEE*, volume 77, pages 257–286, 1989.

13. E. Rivas and S. R. Eddy. The language of RNA: A formal grammar that includes pseudo-knots. *Bioinformatics*, 16(4):334–340, 2000.

14. Y. Sakakibara, M. Brown, R. Hughey, I. S. Mian, K. Sjölander, R. C. Underwood, and D. Haussler. Stochastic context-free grammars for tRNA modeling. *Nucleic Acids Research*, 22:5112–5120, 1994.

15. D. B. Searls. The linguistics of DNA. *American Scientist*, 80(579–591), 1992.

16. E. L. L. Sonnhammer, G. von Heijne, and A. Krogh. A hidden Markov model for predicting transmembrane helices in protein sequences. In *Proceedings of the 6th International Conference on Intelligent Systems for Molecular Biology (ISMB)*, 1998.

17. T. A. Sudkamp. *Languages and Machines*. Computer Science. Addison-Wesley Publishing Company, Inc., 1998.

18. Y. Uemura, A. Hasegawa, S. Kobayashi, and T. Yokomori. Tree adjoining grammars for RNA structure prediction. *Theoretical Computer Science*, 210:277–303, 1999.

19. M. Zuker. On finding all suboptimal foldings of an RNA molecule. *Science*, 244:48–52, 1989.

Assessing the Statistical Significance of Overrepresented Oligonucleotides

Alain Denise[1], Mireille Régnier[2], and Mathias Vandenbogaert[2,3]

[1] LRI, UMR CNRS 8623 and IGM,UMR CNRS 8621
Université Paris-Sud, 91405 Orsay cedex, France
Alain.Denise@lri.fr
[2] INRIA-BP 105. 78 153 Le Chesnay, France
Mireille.Regnier@inria.fr
[3] LaBRI, Université Bordeaux I
351, Cours de la Libération, 33405 Talence, France
Mathias.Vandenbogaert@inria.fr

Abstract. Assessing statistical significance of over-representation of exceptional words is becoming an important task in computational biology. We show on two problems how large deviation methodology applies. First, when some oligomer H occurs more often than expected, e.g. may be overrepresented, large deviations allow for a very efficient computation of the so-called p-value. The second problem we address is the possible changes in the oligomers distribution induced by the over-representation of some pattern. Discarding this noise allows for the detection of weaker signals. Related algorithmic and complexity issues are discussed and compared to previous results. The approach is illustrated with three typical examples of applications on biological data.

1 Introduction

Putative DNA recognition sites can be defined in terms of an idealized sequence that represents the bases most often present at each position. Conservation of only very short consensus sequences is a typical feature of regulatory sites (such as promoters) in both prokaryotic and eukaryotic genomes. Structural genes are often organized into clusters that include genes coding for proteins whose functions are related. Data from the *Arabidopsis* genome project suggest that more than 5% of the genes of this plant encode transcription factors. The necessity for the use of genomic analytical approaches becomes clear when it is considered that less than 10% of these factors have been genetically characterized. Transcription-factor genes comprise a substantial fraction of all eukaryotic genomes, and the majority can be grouped into a handful of different, often large, gene families according to the type of DNA-binding domain that they encode. Functional redundancy is not unusual within these families; therefore the proper characterization of particular transcription-factor genes often requires their study in the context of a whole family. The scope of genomic studies in this area is to find cis-acting regulatory elements from a set of co-regulated DNA sequences (e.g. promoters). The basic assumption is that a cluster of co-regulated genes is regulated by the same transcription factors and the genes of a given cluster share common regulatory motifs.

O. Gascuel and B.M.E. Moret (Eds.): WABI 2001, LNCS 2149, pp. 85–97, 2001.

On the other hand, the importance of whole-genome studies is highlighted by the fact that approximately 50% of the *Arabidopsis* genes have no proposed function, and that the *Arabidopsis* genome, like those of *Saccharomyces cerevisiæ, Caenorhabditis elegans* and *Drosophila*, contains extensive duplications. These include many tandem gene duplications as well as large-scale duplications on different chromosomes. The prevalence of gene duplication in Arabidopsis implies that redundancy is a problem that will have to be dealt with in the functional analysis of genes. In order to find long approximate repeats, some approaches consist in searching for short exact motifs that appear in the sequences.

In both problems, there is a need for identification of over-represented motifs in the considered sequences. Research is very active in this area [4, 23, 15, 21, 24, 25, 22, 14, 7, 1, 9]. In these works, one searches for exceptional patterns in nucleotidic sequences, using various tools to assess the significance of such rare events.

Large deviation is a mathematical area that deals with rare events; to our knowledge, it has not been used in computational biology, although the extremal statistics on alignments [6] can be viewed as large deviation results. Nevertheless, our recent results in [3], that extend preliminary results in [17] show it may be a very powerful method to assess statistical significance of very rare events.

The first problem we address is the following. One considers a *candidate*, e.g. a word that occurs more often than expected. One needs to quantify this difference between the observation and the expectation. Among the classical statistical tools, the so-called p-values are much more precise than the Z-scores (or the χ-scores). The drawback is that their computation is considered as much harder. Large deviations provide a very efficient way to compute them in some cases.

As a second problem, we consider some consequences of the over-representation of a word on a sequence distribution. In particular, it has been observed that, whenever a word is overrepresented, its subwords or the words that contain it, look overrepresented. Such words are called below *artifacts* [1]. It is a desirable goal to choose the best element in the set composed of a word and its artifacts. It is also important to discard automatically the "noise" created by the artifacts, in order to detect other words that are potentially overrepresented. An important example is the noise introduced by the *Alu* sequences. Another one is the χ-sequence GNTGGTGG in *H. influenzae* [11]. We provide some mathematical results and the algorithmic consequences.

The efficiency of this approach comes from the existence of explicit formulae for the (conditioned) distribution. Large deviations allow for a very fast computation. Moreover, due to the "simplicity" of the result -if not of the proof-, their implementation is easy and provides numerically stable and guaranteed computations. The reason is that the problem reduces to the numerical computation of the root of a polynomial. Other approaches need the numerical computation(s) of exp, log functions. The implementation is delicate and machine dependent. Hence, the large deviation approach occasionally corrects commonly used approximations. Interestingly, they apply with the same cost to self-overlapping patterns. This is a great improvement on the approaches based on binomial or related formulae [26, 23], where autocorrelation effects are neglected. Still, our computation that takes into account the correlations is much faster and precise than computing the approximation. Approach is valid for various counting models. For a sake of clarity, we present it for the most commonly used, the overlapping model [27].

In Section 2, we present and discuss the statistical criteria commonly used for evaluating over-representation of patterns in sequences. Section 3 is devoted to our mathematical results. In Sections 4, 6 and 5, we validate our approach by a comparison with published results derived by other methods that are computationally more expensive. Finally, in Section 7, we discuss possible improvements and present further work.

2   Statistical Tools in Computational Biology

In the present section, we present basic useful definitions for statistical criteria and we briefly discuss their limits, e.g the validity domains and the computational efficiency. Below, we denote by $O(H)$ the number of observations of a given pattern H in a given sequence. Depending of the application, it may be either the number of occurrences [14,7] or the number of sequences where it appears [1,23].

Z-Scores. Many definitions of this parameter can be found in the literature. Other names can be used: see for instance the so-called *contrast* used in [14]. A common feature is the comparison of the observation with the expectation, using the variance as a normalization factor. A rather general definition is

$$Z(H) = \frac{E(H) - O(H)}{\sqrt{V(H)}} \tag{1}$$

where H is a given pattern or word, $O(H)$ is the *observed* number of occurrences, $E(H)$ the expectation and $V(H)$ the variance. Many recent works allow for a fast computation of E and V, hence Z. Relevant approximations are discussed in [16], notably the Poisson approximation $V = E$. Nevertheless, if Z-scores are a very efficient filter to detect potential candidates, they are not precise enough. Notably, this parameter is not stable enough for very exceptional words, e.g. when the expectation is much smaller than 1. This will be detailed in Section 4. Moreover, it is relevant only for large sequences, and does not adapt easily to the search in several small sequences.

p-Values. For each word that occurs r times in a sequence or in a set of N (related) sequences, one computes the probability that this event occurs just "by chance":

$$pval(H) = P(O(H) \geq r) \ . \tag{2}$$

When the expectation of a given word is much smaller than 1, a single occurrence is a rare event. In this case, the p-value is defined as: $P(O(H) \geq r$ knowing that $O(H) \geq 1)$, e.g.:

$$pval(H) = \frac{P(O(H) \geq r)}{P(O(H) \geq 1)} \ . \tag{3}$$

The computation is performed in two steps. First, the probability that H occurs in a given sequence, e.g. $P(O(H) \geq 1)$, is known. An exact formula is provided in [17] and used in [7]. An approximated formula is often used, for instance in software RSA-tools (http://www.ucmb.ulb.ac.be/bioinformatics/rsa-tools/) or in [1]. Then two different cases occur.

Set of small sequences. The p-value in (2) is the probability that r sequences out of N contain H; when they are independent, it is given by a binomial formula:

$$pval(\mathrm{H}) = \binom{N}{r}(P(O(\mathrm{H}) \geq 1))^r(1 - P(O(\mathrm{H}) \geq 1))^{N-r} \qquad (4)$$

Large sequences. One needs to compute $P(O(\mathrm{H}) \geq r)$, through the exact formulae in [17] or an approximation.

This p-value is evaluated in [23] through a *significance coefficient*. Given a motif H, the significance coefficient is defined as

$$\mathrm{Sig} = -log_{10}[P(O(\mathrm{H}) \geq r) * D] \,,$$

which takes into account the number of different oligonucleotides D. The number of distinct oligonucleotides depends whether one counts on a single strand or on both strands.

3   Main Results

3.1   Basic Notations

The model of random text that we handle with is the *Bernoulli model*: one assumes the text to be randomly generated by a memoryless source. Each letter s of the alphabet has a given probability p_s to be generated at any step. Generally, the p_s are not equal.

Definition 1. *Given a pattern* H *of length* m *on the alphabet* S *and a Bernoulli distribution on the letters of* S, *the probability of* H *is defined as*

$$P(\mathrm{H}) = \prod_{i=1}^{m} p_{h_i}$$

where h_i *denotes the* i-*th character of* H. *By convention, empty string* ϵ *has probability* 1.

Finding a pattern in a random text is, in some sense, correlated to the previous occurrences of the same or other patterns [13]. Hence for example, the probability of finding $H_1 = \mathtt{ATT}$ knowing that one has just found $H_2 = \mathtt{TAT}$ is - intuitively - rather good since a T right after H_2 is enough to give H_1. *Correlation polynomials* and *correlation functions* give a way to formalize this intuition.

Definition 2. *The* correlation set *of two patterns* H_i *and* H_j *is the set of words* w *which satisfy: there exists a non-empty suffix* v *of* H_i *such that* $vw = H_j$. *It is denoted* $A_{i,j}$. *If* $H_i = H_j$, *then the correlation set is called the* autocorrelation set *of* H_i.

Thus for example, the correlation set of $H_1 = \mathtt{ATT}$ and $H_2 = \mathtt{TAT}$ is $A_{1,2} = \{AT\}$; the autocorrelation set of H_1 is $\{\epsilon\}$, while the autocorrelation set of H_2 is $\{\epsilon, \mathtt{AT}\}$. Empty string always belong to the autocorrelation set of any pattern.

Definition 3. *The* correlation polynomial *of two patterns* H_i *and* H_j *of length* m_i *and* m_j *is defined as:*

$$A_{i,j}(z) = \sum_{w \in A_{i,j}} P(w)z^{|w|} \ ,$$

where $|w|$ *denotes the length of word* w. *If* $H_i = H_j$, *then this polynomial is called the* autocorrelation polynomial *of* H_i. *The* correlation function *is:*

$$D_{i,j}(z) = (1 - z)A_{i,j}(z) + P(H_j)z^{m_j} \ .$$

. *When* $H_i = H_j$, *the correlation function can be written* D_i.

The most common counting model is the *overlapping model*: overlapping occurrences of patterns are taken into account. It is as follows. For example, consider two oligonucleotides $H_1 = $ ATT, $H_2 = $ TAT and a sequence TTATTATATATT. This sequence contains 2 occurrences of H_1 and 4 occurrences of H_2, as shown below:

$$
\begin{array}{cccc}
 & H_1 & H_2 & H_1 \\
 & \overbrace{} & \overbrace{} & \overbrace{} \\
T\,T & A\,T\,T\,A & T\,A\,T & A\,T\,T\,T\,T \\
 & \underbrace{} & \underbrace{} & \underbrace{} \\
 & H_2 \quad H_2 & & H_2
\end{array}
$$

It turns out [3] that our main results rely on the computation of the (real) roots of a polynomial equation:

Definition 4. *Let* a *be a real number such that* $a > P(H_1)$. *Let* (E_a) *be the* fundamental equation:

$$D_1(z)^2 - (1 + (a - 1)z)D_1(z) - az(1 - z)D_1'(z) = 0 \ . \tag{5}$$

Let z_a *be the largest real positive solution of Equation* (E_a) *that satisfies* $0 < z_a < 1$. *The number* z_a *is called the* fundamental root *of* (E_a).

3.2 p-Value for a Single Pattern

The main result of this section is the theorem below, proven in [3], that provides the probability for the observed number of occurrences to be much greater than the expectation.

Theorem 1. *Let* H_1 *be a given pattern, and* k *be its observed number of occurrences in a random sequence of length* n. *Denote* $a = \frac{k}{n}$ *and assume that* $a > P(H_1)$. *Then:*

$$pval(H_1) = Prob(O(H_1) \geq k) \approx \frac{1}{2\sigma_a \sqrt{n}} e^{-nI(a) + \delta_a} \tag{6}$$

where

$$I(a) = a \ln\left(\frac{D_1(z_a)}{D_1(z_a) + z_a - 1}\right) + \ln z_a \ , \tag{7}$$

$$\sigma_a^2 = a(a - 1) - a^2 z_a \left(\frac{2D_1'(z_a)}{D_1(z_a)} - \frac{(1 - z_a)D_1''(z_a)}{D_1(z_a) + (1 - z_a)D_1'(z_a)}\right) \ , \tag{8}$$

$$\delta_a = \log\left[\frac{P(H)z^m}{D_1(z_a) + (1 - z_a)D_1'(z_a)}\right] \ , \tag{9}$$

and z_a is the fundamental root of E_a. $I(a)$ is called the rate function. Additionally:

$$Prob(O(\mathrm{H}_1) = k) \approx \frac{1}{\sigma_a \sqrt{2\pi n}} e^{-nI(a)+\delta_a} \quad . \tag{10}$$

Remark 1. When $a = P(\mathrm{H}_1)$, the number of H_1-occurrences is equal to its expected value. Conditional variance σ_a in (8) becomes: $\sigma = P(\mathrm{H}_1)(2A_1(1) - 1 + (1 - 2m) P(\mathrm{H}_1))$ (where m denotes the length of H_1),, e.g. the unconditional variance computed by various authors [27, 19].

Remark 2. The two probabilities $Prob(O(\mathrm{H}_1) \geq k)$ and $Prob(O(\mathrm{H}_1) = k)$ appear to be very similar in magnitude.

These results turn out to be very precise. It is shown in [3] that the *relative error* is $O(1/n)$; that is to say, the *neglected term* is upper bounded by $e^{-nI(a)} \frac{1}{\sigma_a n^{3/2}}$, which is exponentially small. A numerical comparison with the exact computation implemented, for the Bernoulli model, in [7] is given in Section 4. It appears that, when the random sequence is large, the formulae above provide an attractive alternative to the exact computation. Moreover, they also hold for the Markov model [3].

Another approach [5, 18] is the approximation of the word counting distribution by a compound Poisson distribution with the same mean, e.g. $nP(\mathrm{H})$ in the overlapping counting model. For the compound Poisson distribution defined in [18], the variance is not asymptotically equal to the variance of the process. As a consequence, the validity domain is restricted to the domain where the difference is small. More important, the normalization factor is improper. This implies an error on the *rate function* $I(a)$, and the neglected term in the validity domain, is not exponentially small [18]. Numerical evaluation of the compound Poisson distribution can be found in [12].

3.3 Conditioning by an Overrepresented Word

In this subsection, we assume that a pattern H_1 has been detected as an overrepresented word and we provide mathematical results to investigate the changes induced on the sequence distribution. Intuitively, the artifacts of an overrepresented word should look overrepresented. For example, if $\mathrm{H}_1 = AATAAA$, any word $\mathrm{H2} = ATAAAN$ is an artifact. A rough approximation of its expected value is $O(\mathrm{H}_1) \times \frac{P(N)}{P(A)}$. As $O(\mathrm{H}_1) \gg E(\mathrm{H}_1)$, this is much greater than unconditioned expectation $E(\mathrm{H}_1) \times \frac{P(N)}{P(A)}$. The theorem below, proven in [3], establishes the precise formulae:

Theorem 2. *Given two patterns H_1 and H_2, assume the number of H_1-occurrences, $O(\mathrm{H}_1)$, is known and equal to k, with $a = \frac{k}{n} \geq P(\mathrm{H}_1)$. Then, the conditional expectation of $O(\mathrm{H}_2)$ is:*

$$E(O(\mathrm{H}_2)/O(\mathrm{H}_1) = k) \sim n\alpha \tag{11}$$

where α is a function of the autocorrelation functions, the probabilities and a:

$$\alpha = a \frac{D_{1,2}(z_a) \times D_{2,1}(z_a)}{D_1(z_a)(D_1(z_a) + z_a - 1)} \tag{12}$$

and z_a is the fundamental root of Equation (5). Moreover, the variance is a linear function of n.

Remark 3. In the central region, e.g. $k = nP(H_1)$, substitutions $a = P(H_1)$ and $z_a = 1$ in (11) yield $\alpha = P(H_2)$, if H_1 and H_2 do not overlap.

Once a dominating signal has been detected, one looks for a weaker signal by a comparison of the number of observed occurrences of patterns with their conditional expectations. This procedure automatically eliminates artifacts. An example is provided in Section 6. It also allows for a choice of the best candidate between a word and its artifacts.

Computational Complexity. Another approach is used in *Regexpcount* [11]. Although our formal proof of Theorem 2 relies on similar mathematical tools, our *explicit formulae* allow for skipping the expensive intermediate computations (bivariate generating functions, recurrences,...), hence provide a much faster algorithm.

4 Tandem Repeats in B. Subtilis and A. Thaliana

In [7], authors search for localized repeats with a statistical filter. Software *Except* relies on a simple basic idea: long approximate repeats are likely to contain multiple exact occurrences of shorter words. DNA sequences are divided into overlapping fragments of size n. This size n is a parameter of the algorithm chosen for each run. Typically, n ranges from 250 to 5000. In each window, the p-value is computed for any pattern that occurs more than once. As the total number of occurrences, r, remains relatively small (typically 3 to 5), exact computation through generating functions is (theoretically) possible. Nevertheless, this approach, chosen by the authors, is computationally expensive. Typically, r repeated multiplications of polynomials of degree n. This gives a time complexity $O(n \log n \log r)$, if a Fast Fourier Transform is used, and numerical stability is rather delicate. Large deviation computation for rare events reduces to the numerical computation of real roots of a polynomial equation, namely Equation (5). Hence, it is easier to program, faster and much more stable numerically. This was efficiently implemented in *Maple* and compared with the results published in [7]; Table 1 gives the results. The results in [7] are given for one 2008 nucleotides long fragment

Table 1. Measures on the 7 Oligonucleotides Considered in [7].

Oligomer	Obs.	p-val. (large dev.)	p-val. [7]	Z-sc.
AATTGGCGG	2	8.059×10^{-4}	8.343×10^{-4}	48.71
TTTGTACCA	3	4.350×10^{-5}	4.611×10^{-5}	22.96
ACGGTTCAC	3	2.265×10^{-6}	1.458×10^{-6}	55.49
AAGACGGTT	3	2.186×10^{-6}	2.780×10^{-6}	48.95
ACGACGCTT	4	1.604×10^{-9}	0.982×10^{-9}	74.01
ACGCTTGG	4	5.374×10^{-10}	4.391×10^{-10}	84.93
GAGAAGACG	5	0.687×10^{-14}	1.180×10^{-14}	151.10

in *A. thaliana* where 5 approximate tandem repeats of a 40-uple were found. For all patterns, the occurrence probability $nP(H)$ ranges between 10^{-6} and 10^{-7}. For each oligonucleotide, the first value is the number of occurrences in the window and the second one is the p-value computed by our large deviation formulae, where the correcting term δ_a has been neglected. The third one is the p-value computed in [7] with a generating function method and the last one is the Z-score. We notice that, for any pattern, the p-values computed with two different methods are of the same magnitude order (this is illustrated in Figure 1, where the logs of p-values are plotted.) However, they can differ up to a factor 1.72. This is due to the approximation done in our calculations. When a increases, the difference $z_a - 1$ also increases as well as the contribution of $e^{-\delta_a}$. Nevertheless, it is worth noticing that the p-value order is almost the same. One inversion occurs between patterns ACGGTTCAC and AAGACGGTT, that have similar p-values.

On the other hand, the last column of the table confirms that Z-score is not adequate for very rare events. Patterns AAGACGGTT and AATTGGCGG have the same Z-score 48, while p-values have a ratio 100. For patterns ACGACGCTT and ACGCTTGG, the two parameters define a different order. The same inversion appears between AATTGGCGG and TTTGTACCA.

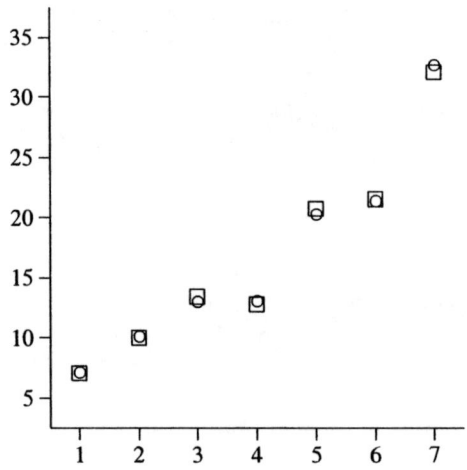

Fig. 1. Graphical comparison of $-$logs of p-values of Table 1. Abscissae refer to the ordering of motifs in the table. Circles denote large deviations formula and squares the results of [7].

5   Oligonucleotide Frequencies from Yeast Upstream Regulatory Regions

In [23], van Helden *et al.* study the frequencies of oligonucleotides extracted from regulatory sites from the upstream regions of yeast genes. Statistical significance of the oligonucleotides occurring in the 800 bp upstream sequences of regulatory regions is assessed by evaluating the probability of observing r or more occurrences of the oligonucleotide in the regulatory sequence, using the binomial formula. In [23], the probabilities are not computed in the Bernoulli model. For a given oligonucleotide, the authors count the number f of its occurrences in the non-coding sequences of yeast. Then, an approximate formula for $P(O(H) \geq 1)$ is given and p-value follows through a computation of binomial formula (4). It is observed in [23] that these binomial statistics prove to be appropriate, except for self-overlapping patterns such as AAAAAA, ATATAT, ATGATG. As a matter of fact, auto-correlation does not affect the expected occurrence number, but increases the variance [8].In other words, the probability to observe either very high or very low occurrence values is increased for auto-correlated patterns.

Table 2 compares the results of several methods to compute the significance coefficient defined above. Figure 2 presents a graphical view of this comparison. The se-

Table 2. Computations of significance coefficient (Sig) of some hexanucleotides according to several methods, in the 800 bp upstream region of ORF YGR022c of *Saccharomyces cerevisiæ* chromosome VII. *O(H)*: number of observed occurrences. *BF1*: Binomial formula with expected occurrences computed as in [23]. *BF2*: Binomial formula, Bernoulli probabilities. *LD*: Large deviations, without considering overlaps. *LDo*: Large deviations considering overlaps. *GF*: Generating function approach [7]. Column *GF* was computed using EXCEP software [7], based on formulas of [20, 17].

Motif	O(H)	BF1	BF2	LD	LDo	GF
TGATGA	22	20	30.13	31.31	23.79	23.92
GATGAT	20	20	26.31	27.25	20.89	21.02
ATGATG	19	10.03	24.43	25.26	19.46	19.59
GATGAG	12	10.21	15.18	15.41	15.01	15.14
GGATGA	11	20	13.26	13.43	13.43	13.56
ATGAGG	11	20	13.26	13.43	13.43	13.56
TGAGGA	10	10.27	11.37	11.50	11.50	11.62
AGGATG	9	9.18	9.53	9.61	9.61	9.73
GAGGAT	9	9.05	9.53	9.61	9.61	9.73
TGAAGA	8	4.54	5.64	5.68	5.68	5.79
AAGATG	6	2.55	2.75	2.72	2.72	2.83
GAAGAT	6	2.39	2.75	2.72	2.72	2.83
GATGAA	6	2.35	2.75	2.72	2.72	2.83

quence considered is the 800 bp upstream region of ORF YGR022C of *Saccharomyces cerevisiæ* chromosome VII. By comparing *BF2*, *LDo* and *GF* (this last one representing the "more exact" result), we see that the large deviations values are very close to the GF ones (while their computation is much faster). The table also confirms that the overlaps must be taken into account when counting the significance coefficients : the three first patterns, for which the difference between *BF2* and *GF* is huge, are the periodic ones.

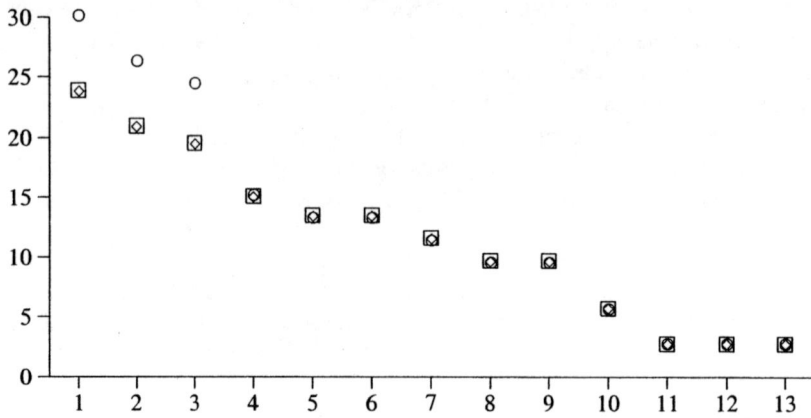

Fig. 2. Graphical comparison of significance coefficients (Sig) of Table 2. Abscissae refer to the ordering of motifs in the table. Circles denote the binomial formulas for Bernoulli probabilities (BF2), diamonds the large deviations considering overlaps, and boxes the generating function approach (GF).

6 Polyadenylation Signals in Human Genes

In [1], Beaudoing et al. study polyadenylation signals in mRNAs of human genes. One of their aims is to find several variants of the well known AAUAAA signal. For this purpose, they select 5646 putative mRNA 3' ends of length 50 nucleotides and seek for overrepresented hexamers. Pattern AAUAAA is clearly the most represented: it occurs in 3286 sequences, for a total number of 3456 occurrences. Seeking for other (weaker) signals involves searching for other overrepresented hexanucleotides. Nevertheless, it is necessary to avoid *artifacts*, e.g. patterns that appear overrepresented because they are similar to the first pattern. The algorithm designed by Beaudoing et al. consists in canceling all sequences where the overrepresented hexamer has been found. Hence, they search for the most represented hexamer in the 2780 sequences which do not contain the strong signal AAUAAA.

Here we show how Theorem 2 gives a procedure for dropping the artifacts of a given pattern without canceling the sequences where it appears. Table 3 presents the 15 most represented hexamers in the sequences considered in [1]. Columns 2 and 3 respectively give the observed number of occurrences and the rank according to this criteria. Columns 4, 5 and 6 present the (non-conditioned) expected number of occurrences, the corresponding Z-score and the rank of the hexamer according to this Z-score. Here, the variance has been approximated by the expectation; this is possible as stated in [16]. Remark that rankings of columns 3 and 6 are quite similar: only patterns UAAAAA and UAAAUA do not belong to both rankings. A number of motifs look like the canonical one: they may be artifacts. This is confirmed by the three last columns which present, respectively, the expected number of occurrences conditioned by the observed number of occurrences of AAUAAA, the corresponding conditioned Z-score and the rank ac-

Table 3. Table of the most frequent hexanucleotides. *Obs*: number of observed occurrences. *Rk*: Rank. *Exp.*: (non-conditional) expectation. *Cd.Exp.*: Expectation conditioned by number of occurrences of AAUAAA.

Hexamer	Obs.	Rk	Exp.	Z-sc.	Rk	Cd.Exp.	Cd.Z-sc.	Rk
AAUAAA	3456	1	363.16	167.03	1			1
AAAUAA	1721	2	363.16	71.25	2	1678.53	1.04	1300
AUAAAA	1530	3	363.16	61.23	3	1311.03	6.05	404
UUUUUU	1105	4	416.36	33.75	8	373.30	37.87	2
AUAAAU	1043	5	373.23	34.67	6	1529.15	-12.43	4078
AAAAUA	1019	6	363.16	34.41	7	848.76	5.84	420
UAAAAU	1017	7	373.23	33.32	9	780.18	8.48	211
AUUAAA	1013	8	373.23	33.12	10	385.85	31.93	3
AUAAAG	972	9	184.27	58.03	4	593.90	15.51	34
UAAUAA	922	10	373.23	28.41	13	1233.24	-8.86	4034
UAAAAA	922	11	363.16	29.32	12	922.67	9.79	155
UUAAAA	863	12	373.23	25.35	15	374.81	25.21	4
CAAUAA	847	13	185.59	48.55	5	613.24	9.44	167
AAAAAA	841	14	353.37	25.94	14	496.38	15.47	36
UAAAUA	805	15	373.23	22.35	21	1143.73	-10.02	4068

cording to this criteria. It is clear that artifacts are dropped out, generally very far away in the ranking. It is worth noticing that some patterns which seemed overrepresented are actually avoided: this is the case for AUAAAU which goes down from 5th to last place (among the 4096 possible hexamers, only 4078 are present in the sequences). As AUAAAU is an artifact of the strong signal, this means that U is rather avoided right after this signal.

The case of UUUUUU in rank 2 is particular: this pattern is effectively overrepresented, but was not considered by Beaudoing *et al.* as a putative polyadenylation signal because its position does not match with observed requirements (around -15/-16 nucleotides upstream of the putative polyadenylation site.) It should also be stated that the approximation of the variance by the expectation that we do here for all patterns is not as good for periodic patterns like UUUUUU as for others [16]. By this way, variance of UUUUUU is under-evaluated; so its actual Z-score is significantly lower than the one given in the table.

Now over-representation of AUUAAA (rank 3) is obvious; this is the known first variant of the canonical pattern. We remark that the following hexamer, UUAAAA, is an artifact of AUUAAA. It suggests to define a conditional expectation, or, even better, a p-value that takes into account the over-representation of two or more signals instead of one: in this example, AAUAAA *and* AUUAAA. This extension of Theorem 2 is the subject of a future work.

As it is mentioned above, the Z-score is not precise enough, and this remark also holds for conditioned Z-scores. In a second step, the authors of [1] computed a p-value defined by formula (4). This formula is approximated by the incomplete β-function. Nevertheless, any computation is rather delicate, and machine dependent due to numerous call to exp and log functions. The numerical stability necessitates a very careful use of real precision. It is worth noticing that large deviation principle applies for a Bernoulli process, with explicit values for the rate function and σ_a [3].

7   Conclusion and Perspectives

In this paper, we illustrated a possible use of large deviation methods in computational biology. These results allow, in some cases, a very fast computation of p-values that is numerically stable. These preliminary results are quite appealing and should be extended in several directions. First, it may be necessary to eliminate several strong independent signals [1]. A second task is the simplification of our formulae for artifacts: this would allow to achieve automatically the choice between a word and its subwords. A third task is the extension to the computation of the p-value for the conditioned case. Finally, regulatory sites may also be associated with structured motifs [10] or spurious motifs [2] and extension to this case should be realized.

Acknowledgments

We thank Eivind Coward for making the EXCEP software available for us and Jacques Van Helden for fruitful electronic discussions. This research was partially supported by the IST Program of the EU under contract number 99-14186 (ALCOM-FT), the REMAG Action from INRIA and IMPG French program.

References

1. E. Beaudoing, S. Freier, J. Wyatt, J.M. Claverie, and D. Gautheret. Patterns of Variant Polyadenylation Signal Usage in Human Genes. *Genome Research.*, 10:1001–1010, 2000.
2. J. Buhler and M. Tompa. Finding Motifs Using Random Projections. In *RECOMB'01*, pages 69–76. ACM-, 2001. Proc.RECOMB'01, Montréal.
3. A. Denise and M. Régnier. Word statistics conditioned by overrepresented words, 2001. in preparation; http://algo.inria.fr/regnier/index.html.
4. M.S. Gelfand and E.V. Koonin. Avoidance of palindromic words in bacterial and archaeal genomes: a close connection with restriction enzymess. *Nucleic Acids Research*, 25(12):2430–2439, 1997.
5. M. Geske, A. Godbole, A. Schafner, A. Skolnick, and G. Wallstrom. Compound Poisson Approximations for Word Patterns Under Markovian Hypotheses. *J. Appl. Prob.*, 32:877–892, 1995.
6. R. Karlin and S.F. Altschul. Applications and statistics for multiple high-scoring segments in molecular sequences. *Proc. Natl. Acad. Sci. U.S.A.*, 90:5873–5877, 1993.
7. Maude Klaerr-Blanchard, Hélène Chiapello, and Eivind Coward. Detecting localized repeats in genomic sequences: A new strategy and its application to *B. subtilis* and *A. thaliana* sequences. *Comput. Chem.*, 24(1):57–70, 2000.
8. J. Kleffe and M. Borodovsky. First and second moment counts of words in random texts generated by Markov chains. *Comput. Appl. Biosci.*, 8, 433-441, 1992.
9. X. Liu, D.L. Brutlag, and J. Liu. Bioprospector: Discovering conserved dna motifs in upstream regulatory regions of co-expressed gene. In *6-th Pacific Symposium on Biocomputing*, pages 127–138, 2001.

10. L. Marsan and M.F. Sagot. Extracting structured motifs using a suffix tree-algorithms and application to promoter consensus identification. In *RECOMB'00*, pages 210–219. ACM-, 2000. Proceedings RECOMB'00, Tokyo.

11. P. Nicodème. The symbolic package Regexpcount. In *GCB'00*, 2000. presented at GCB'00, Heidelberg, October 2000; available at
    http://algo.inria.fr/libraries/software.html.

12. G. Nuel. *Grandes déviations et chaines de Markov pour l'étude des mots exceptionnels dans les séquences biologiques*. Phd thesis, Université René Descartes, Paris V, 2001. to be defended in July,2001.

13. P.A. Pevzner, M. Borodovski, and A. Mironov. Linguistic of Nucleotide sequences:The Significance of Deviations from the Mean: Statistical Characteristics and Prediction of the Frequency of Occurrences of Words. *J. Biomol. Struct. Dynam.*, 6:1013–1026, 1991.

14. E.M. Panina, A.A. Mironov, and M.S. Gelfand. Statistical analysis of Complete Bacterial Genomes:Avoidance of Palindromes and Restriction-Modification Systems. *Genomics. Proteomics. Bioinformatics*, 34(2):215–221, 2000.

15. F.R. Roth, J.D. Hughes, P.E. Estep, and G.M. Church. Finding DNA regulatory motifs within unaligned noncoding sequences clustered by whole-genome mRNA quantitation. *Nature Biotechnol.*, 16:939–945, 1998.

16. M. Régnier, A. Lifanov, and V. Makeev. Three variations on word counting. In *GCB'00*, pages 75–82. Logos-Verlag, 2000. Proc. German Conference on Bioinformatics, Heidelberg; submitted to BioInformatics.

17. M. Régnier and W. Szpankowski. On Pattern Frequency Occurrences in a Markovian Sequence. *Algorithmica*, 22(4):631–649, 1998. preliminary draft at ISIT'97.

18. G. Reinert and S. Schbath. Compound Poisson Approximation for Occurrences of Multiple Words in Markov Chains. *Journal of Computational Biology*, 5(2):223–253,

19. G. Reinert, S. Schbath, and M. Waterman. Probabilistic and Statistical Properties of Words: An Overview. *Journal of Computational Biology*, 7(1):1–46, 2000.

20. S. Robin and J. J. Daudin. Exact distribution of word occurrences in a random sequence of letters. *J. Appl. Prob.*, 36(1):179–193, 1999.

21. E. Rocha, A. Viari, and A. Danchin. Oligonucleotides bias in *bacillus subtilis*: general trands and taxonomic comparisons. *Nucl. Acids Research*, 26:2971–2980, 1998.

22. A.T. Vasconcelos, M.A. Grivet-Mattoso-Maia, and D.F. de Almeida. Short interrupted palindromes on the extragenic DNA of Escherichia coli K-12, Haemophilus influenzae and Neisseria meningitidis. *BioInformatics*, 16(11):968–977, 2000.

23. J. van Helden, B. Andre, and J. Collado-Vides. Extracting regulatory sites from the upstream region of yeast genes by computational analysis of oligonucleotide frequencies. *J. Mol. Biol.*, 281:827–842, 1998.

24. Martin Tompa. An exact method for finding short motifs in sequences, with application to the ribosome binding site problem. In *ISMB'99*, pages 262–271. AAAI Press, 1999. Seventh International Conference on Intelligent Systems for Molecular Biology, Heidelberg,Germany.

25. A. Vanet, L. Marsan, and M.-F. Sagot. Promoter sequences and algorithmical methods for identifying them. *Res. Microbiol.*, 150:779–799, 1999.

26. M.S. Waterman, R. Arratia, and D.J. Galas. Pattern recognition in several sequences: consensus and alignment. *Bull. Math. Biol.*, 45, 515-527, 1984.

27. M. Waterman. *Introduction to Computational Biology*. Chapman and Hall, London, 1995.

Pattern Matching and Pattern Discovery Algorithms for Protein Topologies

Juris Vīksna[1] and David Gilbert[2]

[1] Institute of Mathematics and Computer Science
The University of Latvia
Rainis Boulevard 29, Rīga LV 1459, Latvia
juris@cclu.lv

[2] Department of Computing, City University
Northampton Square, London EC1V 0HB, UK
drg@soi.city.ac.uk

Abstract. We describe algorithms for pattern-matching and pattern-learning in TOPS diagrams (formal descriptions of protein topologies). These problems can be reduced to checking for subgraph isomorphism and finding maximal common subgraphs in a restricted class of ordered graphs. We have developed a subgraph isomorphism algorithm for ordered graphs, which performs well on the given set of data. The maximal common subgraph problem then is solved by repeated subgraph extension and checking for isomorphisms. Despite its apparent inefficiency, this approach yields an algorithm with time complexity proportional to the number of graphs in the input set and is still practical on the given set of data. As a result we obtain fast methods that can be used for building a database of protein topological motifs and for the comparison of a given protein of known secondary structure against a motif database.

1   Biological Motivation

Once the structure of a protein has been determined, the next task for the biologist is to find hypotheses about its function. One possible approach is a pairwise comparison of its structure with the structures of proteins whose functions are already known. There are several tools that allow such comparisons, for example DALI [7] or CATH [11]. However there are two weaknesses with these approaches. Firstly, as the number of proteins with a given structure is growing, the time needed to do such comparisons is also growing. Currently there are about 15,000 protein structure descriptions deposited in the Protein Data Bank [1], but in the future this number may grow significantly. Secondly, even if a similarity with one or more proteins has been found, it may not be apparent whether this may also imply functional similarity, especially if the similarity is not very strong.

Another possibility is to try a similar approach at a structural level similar to that used for sequences in the PROSITE database [6]. That is, precompute a database of motifs for proteins with known structures—i.e., structural patterns which are associated with some particular protein function. This effectively requires computing the maximal

O. Gascuel and B.M.E. Moret (Eds.): WABI 2001, LNCS 2149, pp. 98–111, 2001.
© Springer-Verlag Berlin Heidelberg 2001

common substructure for a set of structures. One such approach is that of CORA [10], based on multiple structural alignments of protein sequences for given CATH families.

Both approaches have been successfully used for protein comparison at the sequence level. The main difficulty in adapting them to the structural level is the complexity of the necessary algorithms—while exact sequence comparison algorithms work in linear time, exact structure comparison algorithms may require exponential time and the situation only gets worse with algorithms for finding maximal common substructures. Another aspect of the problem is that it is far from clear which is the best way to define structure similarity. There are many possible approaches, which require different algorithmic methods and are likely to produce different results.

Our work is aimed at the development of efficient comparison and maximal common substructure algorithms using TOPS diagrams for structural topology descriptions, at the definition of structure similarity in a natural way that arises from such formalisation, and at the evaluation of usefulness of such an approach. The drawback of our approach is that TOPS diagrams are not very rich in information; however it has the advantage that it is still possible to design practical algorithms for this level of abstraction.

2   TOPS Diagrams

At a comparatively simple level, protein structures can be described using TOPS cartoons (see [4,13,14]). A sample cartoon for 2bopA0 is shown in Figure 1(a); for comparison a Rasmol-style picture is given in Figure 1(b). The cartoon shows the secondary

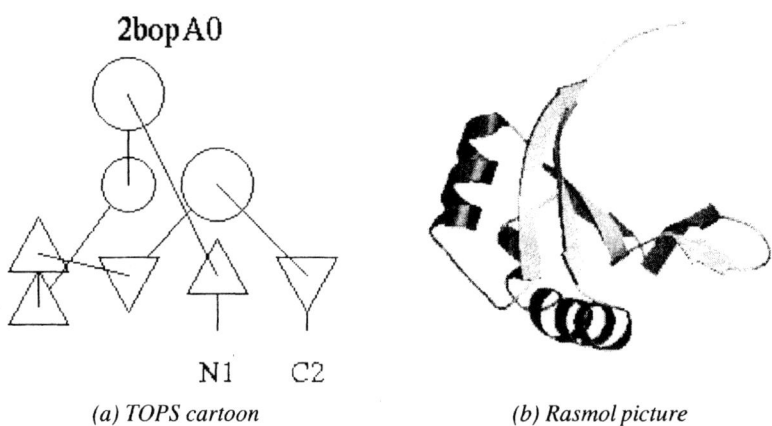

(a) TOPS cartoon   (b) Rasmol picture

Fig. 1. TOPS Cartoon and Rasmol Picture of 2bopA0.

structure elements (SSEs)—β-strands (depicted by triangles) and α-helices (depicted by circles)—, how they are connected in a sequence from amino to carboxyl terminus, and their relative spatial positions and orientations. Such representations have been used by biologists for some time. However the graphical images do not explicitly represent all topological information implied by such descriptions and there are no strict rules governing the appearance of a TOPS cartoon for a given protein.

TOPS *diagrams*, developed by Gilbert *et al.* [5], are a more formal description of protein structural topology and are based on TOPS cartoons. Instead of representing spatial positions by element positions in a plane, a TOPS diagram contains information about the grouping of β-strands in β-sheets (two adjacent elements in a β-sheet are connected by an H-bond, which can be either parallel or anti-parallel) and some information about relative orientation of elements (any two SSEs can be connected by either left or right chirality). Note that, in the topological sense, we reduce the set of atomic hydrogen bonds between a pair of strands to a single H-bond relationship between the strands. In principle chiralities can be defined between any two SSEs; however only a subset of the most important chiralities is included in TOPS diagrams—this subset roughly corresponds to the implicit position information in TOPS cartoons. A TOPS diagram can be regarded as a graph with four different types of vertices (corresponding to up- or down-oriented strands and up- or down-oriented helices) and four different types of edges (corresponding to parallel or antiparallel H-bonds and left or right oriented chiralities). Moreover, the corresponding graph is ordered—each vertex is assigned a unique number from 1 to n, where n is the total number of vertices. In Figure 2 the ordering is also indicated by placing the vertices in the order of increasing numbers (looking from left to right).

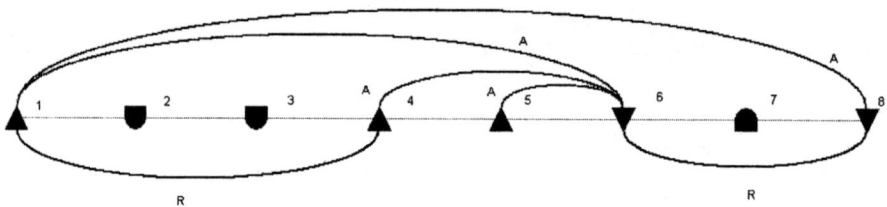

Fig. 2. TOPS Diagram of 2bopA0.

3   Pattern Matching and Pattern Discovery in TOPS

If we describe protein secondary structure by TOPS diagrams, a natural way to characterise the similarity of two proteins is by using patterns. In general, we can define patterns using the same type of graphs as for TOPS diagrams. We say that a given pattern matches a given TOPS diagram if and only if the corresponding pattern graph is a subgraph of the corresponding TOPS diagram graph. Here we assume that subgraph relation also preserves the order of vertices—i.e., there is a mapping F of pattern graph vertices to target graph vertices such that, for any pair of vertices v and w in a pattern graph:

- if the number of v is larger than the number of w, then also the number of $F(v)$ is larger than the number of $F(w)$,; and

– if there is an edge between v and w, then there is an edge (of the same type) between $F(v)$ and $F(w)$.

Figure 3 shows one of the possible patterns that matches the diagram for 2bopA0 by mapping vertices with numbers 1, 2, 3, 4, 5, and 6, corresponding to vertices with numbers 1, 2, 4, 6, 7, and 8. In practice, however, it might be useful to make the pattern

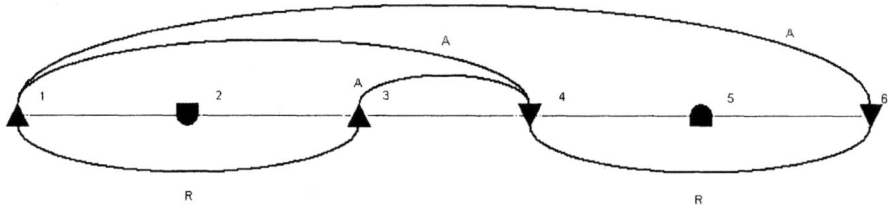

Fig. 3. TOPS Pattern.

definition more complicated. There might be reasons to require that close vertices in pattern (i.e., vertices with close numbers) are to be mapped to close vertices in the target diagram (for some natural notion of close). Alternatively it might be useful to require that the target graph does not contain extra edges between vertices to which pattern graph vertices are mapped (in this case the pattern graph must be an induced subgraph of the target graph).

If we want to compare a target TOPS diagram to a set of diagrams, we can do this by pairwise comparisons between the target and each of the comparison sets; each such comparison can be made by finding a largest common pattern for two diagrams and assigning a similarity measure based on the size of the pattern and the sizes of the two diagrams. Alternatively, if we want to use a motif-based approach, we can find the largest common patterns for a given set of proteins, consider these patterns as motifs, and check whether a pattern for some motif matches the diagram of a target protein. In practice the definition of a motif may be more complicated—for example, it may include several patterns or some additional information.

Several algorithms for protein comparison based on the notion of patterns have already been developed and implemented by David Gilbert. The system is available at http://www3.ebi.ac.uk/tops/; it permits searching for proteins that match a given pattern or to perform pattern-based comparisons of TOPS descriptions of proteins. Our current task is to implement the more efficient algorithms that we describe here. These algorithms will permit the fast generation of motif databases, which we plan to make available on the web.

4   Experimental Results

4.1   Methodology and Databases

In experiments that we have performed to date we have tried to estimate the usefulness of the pattern-based protein motifs, i.e., what is the probability that the fact that a protein matches a given motif implies that protein has also some real similarity with other proteins characterised by the same motif. To do this, we have tried to compare our approach against the existing CATH protein classification database. CATH [11] is a hierarchical classification of protein domain structures, which clusters proteins at four major levels—Class (C), Architecture (A), Topology (T) and Homologous superfamily (H). There are four different C classes—mainly alpha (class 1), mainly beta (class 2), alpha-beta (class 3) and low secondary structure content (class 4). In most cases C classes are assigned automatically. The architecture level describes the overall shape of the domain structure according to orientations of the secondary structures; classes in this level are assigned manually. Classes in the topology level depend on both the overall shape and connectivity of the secondary structures and are assigned automatically by the SSAP algorithm. Classes in the homologous superfamily level group together protein domains which are thought to share a common ancestor and can therefore be described as homologous. They are assigned automatically from the results of sequence comparisons and structure comparisons (using SSAP).

Our comparisons are based on the assumption that identical CATH numbers will also imply some similarity of the TOPS diagrams for the corresponding proteins. The TOPS Atlas database [13], containing 2853 domains and based on clustering structures from the protein data bank [1] using the standard single linkage clustering algorithm at 95% sequence similarity, was selected as the data set for this investigation. Structures with identical CATH numbers (to a given level) have been placed in one group and a maximal common pattern for this group has been computed. Then the pattern was matched against all structures in the selected subset and the quality q of the pattern, corresponding to positive predictive value, computed as follows:

q = *number of proteins in a given group / number of successful matches*

Thus, $q = 1$ corresponds to a good pattern (no false positives) and the value of q is lower for less good patterns.

4.2   Results

The experiments were performed using the CATH number identity at levels A, T, and H. The CATH number identity at the A level was clearly insufficient to guarantee any similarity at the TOPS diagram level; somewhat more surprising was the fact that identity at the T (topological) level still produced noticeably weaker results than identity at the H level. Results for the latter are shown in Figure 4. Here the values of q for all domains from the data set (in lexicographical order by CATH numbers) are shown. The first 527 structures correspond to CATH class 1 (mainly α), the next 1048 to class 2 (mainly β), the following 1151 to class 3 ($\alpha - \beta$) and the last 124 to class 4 (weak secondary structure contents). As can be expected q values are small for class 4, since there is very little secondary structure information and also for class 1, since in mainly alpha domains there are few H-bonds and the corresponding TOPS diagrams contain little information about topology. Better q values can be observed for classes 2 and 3. Figure 5 shows q values (in light-grey) for class 3. Here the proteins have been reordered according to increasing q values. As can be seen, in about 36% of cases the q value is 1, i.e., the CATH number is uniquely defined by a TOPS pattern. Also, there are not many proteins with q values close to, but less than 1. Therefore, if a pattern has been shown

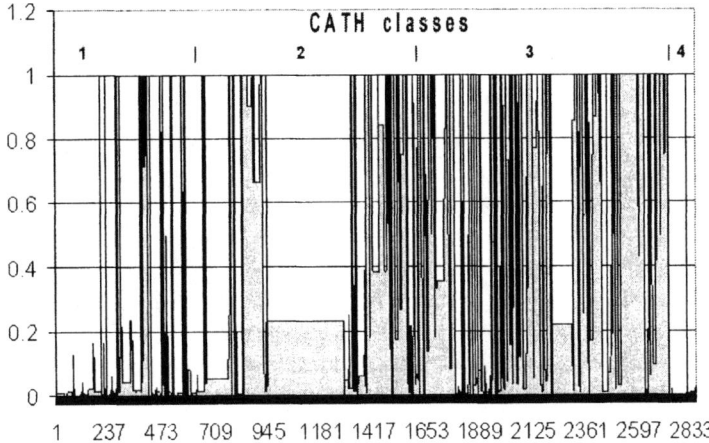

Fig. 4. Quality of TOPS Patterns at CATH H Level.

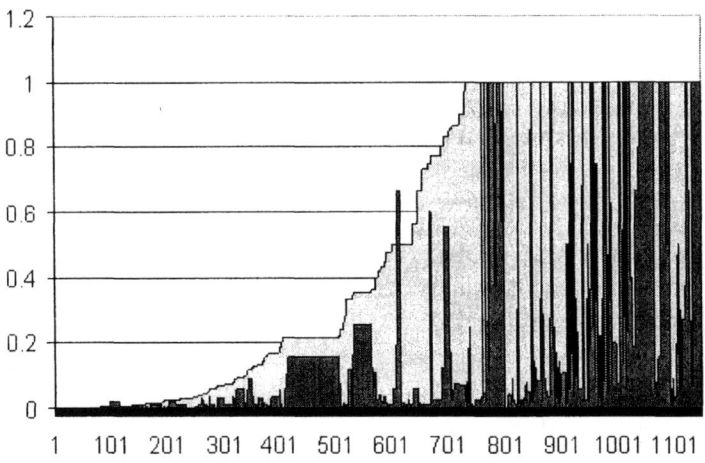

Fig. 5. Quality of TOPS Patterns for CATH Class 3.

to be good for known proteins, it is likely that it will remain good for new, as yet un-classified, proteins. For comparison the figure also contains values (in dark-grey) where q values have been computed using only secondary-structure sequence patterns instead of complete TOPS diagrams. This demonstrates that good sequence patterns only exist for approximately 8% of structures. The superiority of sequence patterns for one group is caused by different definitions of the largest pattern.

Figure 6 contains the same data as Figure 5, but initially ordered by pattern size as computed by the number of SSEs in the pattern, and then by q values. It can be seen that we start to get good q values when the number of SSEs reaches 7 or 8 (proteins with numbers from 459 or 531 on horizontal axis), and that q values are good in most cases

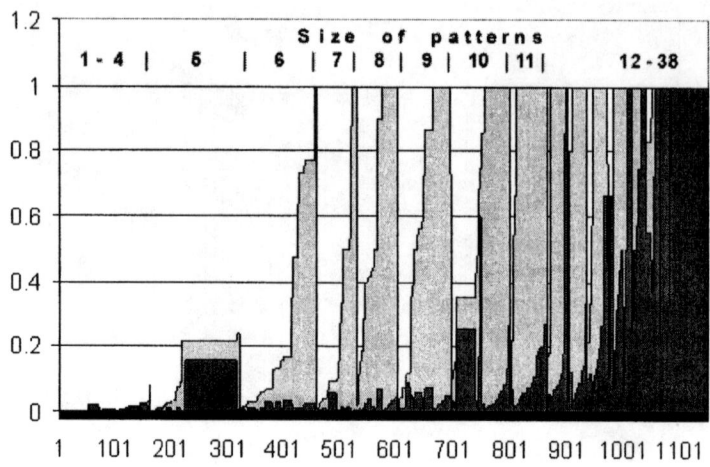

Fig. 6. Quality of TOPS Patterns for CATH Class 3 Ordered by the Size of Patterns.

when the number of SSEs reaches 11 (proteins with numbers from 800 on horizontal axis). Therefore, if a protein contains 7 or more SSEs, there is a good chance that it will have a good pattern and, if it contains 11 or more SSEs, then in most cases it will have a good pattern.

Thus, the results obtained so far suggest that a database of pattern motifs could be quite useful for comparison of those proteins that have sufficiently rich secondary-structure content and especially for proteins with a large number of strands. This is not the largest subgroup of all proteins; however for this subgroup there are good chances that comparison with TOPS motifs will give biologically useful information. Of course, TOPS diagrams contain limited information about secondary structure; thus we can expect that motifs based on richer secondary structure models may give better results. At the same time the TOPS formalism has the advantage that all computations can be performed comparatively quickly. The exact computation times are very dependent on the given data, but in general we have observed that the comparison of a given protein against a database of about 1000 motifs requires less than 0.1 second on an ordinary 600 MHz PC workstation. The discovery of motifs and associated evaluation via pattern matching over the TOPS Atlas has been done in about 2 hours on the same equipment.

5   TOPS Patterns and Related Graph Problems

The basic problems that arise in the TOPS formalism, namely, pattern matching and pattern discovery, can be easily reduced to that of subgraph isomorphism and maximal common subgraph problems in ordered graphs. We consider the following types of vertex-ordered labelled graphs.

5.1   Definitions

A given graph $G = (V, E)$ is *vertex-ordered* if there is a one-to-one mapping between the set of numbers $\{1, 2, \ldots, |V|\}$ and the set of vertices V. Let us call the number that corresponds to $v \in V$ the *vertex position* and denote it by $p(v)$. We consider undirected graphs, thus we can assume that edges are defined by ordered pairs (v, w) with $p(v) < p(w)$. Given a vertex-ordered graph, we define the *edge order* in the following way:

$$p((v_1, w_1)) < p((v_2, w_2)) \iff p(v_1) < p(v_2) \text{ or } p(v_1) = p(v_2) \wedge p(w_1) < p(w_2)$$

and assign to edges numbers $\{1, 2, \ldots, |E|\}$ according to this order. We call these numbers *edge positions* and denote them by $p(e)$.

A graph $G = (V, E)$ is *vertex- (edge-) labelled* with set of labels S, if there is given a function $l_v : V \to S$ (and, for edge labels, $l_e : E \to S$). We denote the label of a vertex $v \in V$ by $l(v)$ and the label of an edge $e \in E$ by $l(e)$. For given vertex-ordered and vertex- and edge-labelled graphs $G_1 = (V_1, E_1)$ and $G_2 = (V_2, E_2)$, we say that G_1 is isomorphic to a subgraph of G_2 if there is an injective mapping I from V_1 to V_2 such that:

- $\forall v, w \in V_1, p(v) < p(w) \implies p(I(v)) < p(I(v))$
- $\forall v, w \in V_1, (v, w) \in E_1 \implies (I(v), I(w)) \in E_2$
- $\forall v \in V_1, l(v) = l(I(v))$
- $\forall (v, w) \in E_1, l((v, w)) = l((I(v), I(w)))$

Since each edge is uniquely determined by two vertices we can extend the isomorphism I to edges by defining $I((v, w)) = (I(v), I(w))$. Then I preserves edge order just like it preserves vertex order, i.e., $\forall e_1, e_2 \in E_1, p(e_1) < p(e_2) \implies p(I(e_1)) < p(I(e_2))$.

5.2   Graph Representation of TOPS Diagrams

We can consider a TOPS diagram as a vertex ordered and vertex and edge labelled graph with the set of vertex labels $S_V = \{e+, e-, h+, h-\}$ (up- or down-oriented strand or up- or down-oriented helix) and the set of edge labels $S_E = \{P, A, L, R, PL, PR, AL, AR\}$ (parallel or antiparallel H-bonds or left- or right-oriented chiralities or a combination of H-bonds and chiralities). In practice, P edges are only permitted between $e+$ and $e+$ or $e-$ and $e-$ vertices, and A edges are allowed only between $e+$ and $e-$ or $e-$ and $e+$ vertices, but here for us this is not essential.

For practical purposes it is also worth noting the complexity of graphs that have to be dealt with in TOPS formalism—the maximal number of vertices is around 50 and the number of edges is comparatively small and similar to the number of vertices.

Let P be a TOPS pattern, D_1 and D_2 be TOPS diagrams, and $G(P)$, $G(D_1)$ and $G(D_2)$ be the graphs corresponding to these patterns or diagrams. Then the problem of checking whether TOPS pattern P will match diagram D_1 is equivalent to checking whether $G(P)$ is isomorphic to a subgraph of $G(D_1)$. Similarly, the problem of finding a largest common pattern P of D_1 and D_2 is equivalent with finding a largest common subgraph $G(P)$ of $G(D_1)$ and $G(D_2)$.

5.3   Complexity and Relation to Other Work

First, it is easy to see that the subgraph isomorphism problem for vertex-ordered graphs remains **NP**-complete, since the maximal clique problem is **NP** complete, and this is not altered by vertex ordering. Also, the relatively small number of edges cannot be exploited to obtain polynomial algorithms, since in [3] and [15] similar graph structures are considered that are even simpler (the vertex degree is 0 or 1) and for such graphs the subgraph isomorphism problem is proved to be **NP** -complete. In [3] an algorithm is given that is polynomial with respect to the number of overlapping edges—however in TOPS this number tends to be quite large.

There are several good non-polynomial algorithms for subgraph isomorphism, the two most popular being by Ullmann [12] and McGregor [9]. Although these are not easily adaptable to vertex-ordered graphs, the vertex ordering seems to be the property that could considerably improve the algorithm efficiency. Our algorithm can be regarded as a variant of a method based on constraint satisfaction [9]; however there is an additional mechanism for periodically recomputing constraints. A very similar class of graphs has also been considered by I. Koch, T. Lengauer and E. Wanke in [8] where the authors describe a maximal common subgraph algorithm based on searching for maximal cliques in a vertex product graph. This method seems to be applicable also for TOPS; however it is only practical for finding maximal common subgraphs for two graphs and is not directly useful for finding motifs for larger sets of proteins.

6   Subgraph Isomorphism Algorithm for Ordered Graphs

We have developed a subgraph isomorphism algorithm that exploits the fact that the graphs are vertex oriented. Initially, let us assume that we are dealing with graphs that are connected and do not contain isolated vertices (this set is also the most important in practice). Then an isomorphism mapping I is uniquely determined by defining the mapping of edges.

The algorithm tries to match edges in the increasing order of edge positions and backtracks if for some edge match can not be found. Since the graphs are ordered, the positions in the target graph to which a given edge may be mapped and which have to be checked can only increase. Two additional ideas are used to make this process more efficient. Firstly, we assign a number of additional labels to vertices and edges. Secondly, if an edge e can not be mapped according to the existing mapping for previous edges, then the next place where this edge can be mapped according to the labels is found, and the minimal match positions of previous edges are advanced in order to be compatible with the minimal position of e.

6.1   Labelling

By definition vertices and edges are already assigned labels l_v and l_e correspondingly that must be preserved by isomorphism mapping. Additionally we use an another kind of label for both vertices and edges, which we call *Index*. For a vertex v, Index(v) is a 16-tuple of integers (containing twice as many elements as there are edge labels). The

ith element of Index(v) is the number of edges (x, v) with $l_e((x, v))$ equal to the ith possible value of l_e (according to some initially fixed order of labels). Similarly, the $(k + i)$th element of Index(v) is the number of edges (v, x) with $l_e((v, x))$ equal to the ith possible value of l_e. Thus, the value Index(v) encodes the numbers of incoming and outgoing edges of all possible types for a given vertex v. For an edge $e = (v, w)$, Index(e) is a 4-tuple of integers $\langle S_1, S_2, E_1, E_2 \rangle$, where S_1 is the number of edges (v, x) with $p(x) < p(w)$, S_2 is the number of edges (v, x) with $p(x) > p(w)$, E_1 is the number of edges (y, w) with $p(y) < p(w)$, and E_2 is the number of edges (y, w) with $p(y) > p(w)$. The edge index describes how many shorter or longer other edges are connected to the endpoints of a given edge. For both vertices and edges we define Index$(x) \leq$ Index(y) if the inequality holds between the all corresponding pairs of 16-tuples (or 4-tuples). It is easy to see that for any vertex or edge x we must have Index$(x) \leq$ Index$(I(x))$.

6.2 Algorithm

We assume that graphs are given as arrays *PV*, *PE*, *TV* and *TE*, where *PV* is an array of vertices in the pattern graph with *PV[i]* being the vertex v with $p(v) = i$, *PE* is an array of edges in the pattern graph with *PE[i]* being the edge e with $p(e) = i$, and *TV* and *TE* are similar arrays for the target graph. For an edge e of the pattern graph, list Matches(e) contains all possible positions (in increasing order) in target graph to which e can be matched according to vertex and edge labels and Index values. By Matches$(e)[i]$ we denote the ith element from this list. The number Next(e) is the first position in Matches(e) list to which it still may be worth to try to match the edge. Initially for all edges we have Next$(e) = 1$. For vertex v the number Pos(v) is the position in target graph to which vertex v is matched, otherwise we have Pos$(v) = 0$.

Algorithm 1 shows the main loop. Starting from the first edge the algorithm tries to find matches for all edges in increasing order and returns an array Pos of vertex mappings, if it succeeds. If for some edge a match consistent with matches for previous edges can not be found a procedure *AdvanceEdgeMatchPositions* is invoked, which tries to increase the values Next(e) for some of already matched edges and the matching process is continued starting from the first edge for which the value Next(e) has been changed.

Procedure *AdvanceEdgeMatchPositions* uses a variant of depth-first search to find edges for which Next(e) can be increased. Alternative strategies are of course possible,

6.3 Correctness

The informal motivation why the algorithm correctly finds an isomorphic subgraph (or gives the answer that no isomorphic subgraph exists) is the following. First, as already noted above, for connected oriented graphs the isomorphism mapping is completely defined by defining the mapping for edges. For an isomorphism mapping it is sufficient to satisfy the labelling constraints on edge endpoints, preserve edge order and connectivity. If the *AdvanceEdgeMatchPositions* procedure is not used, the algorithm simply performs an exhaustive search of all mappings satisfying these constraints and either

procedure *SubgraphIsomorphismInOrderedGraphs*(PV, PE, TV, TE);
begin
    foreach *vertex e in PE* **do**
        Compute the list Matches(e);
        Next(e) $\leftarrow$ 1;
        if *Matches(e)* $= \emptyset$ **then return** *Not Isomorphic*;
    end
    foreach *vertex v in PV* **do**
        Pos(v) $\leftarrow$ 0;
    end
    $k \leftarrow 1$;
    while $k \leq |PE|$ **do**
        edge $= (v, w) \leftarrow PE[k]$;
        if *Next(edge)* $> |Matches(edge)|$ **then return** *Not Isomorphic*;
        else
            Find the smallest $i \geq$ Next(edge) such that, for the target graph edge
            $(vt, wt) =$ Matches(edge)[i] and for both vertices v and w, either Pos(v) $=$
            0 or Pos(v) $= vt$ and either Pos(w) $= 0$ or Pos(w) $= wt$ (and
            $p((vt, wt)) >$ Matches($PE[k-1]$)—if $k > 1$, use Next($PE[k-1]$));
            if *such an i is found* **then**
                Next(edge) $\leftarrow i$;
                Pos(v) $\leftarrow vt$; Pos(w) $\leftarrow wt$;
                $k \leftarrow k + 1$;
            end
            else
                Find the smallest $j \geq$ Next(edge) such that (if $k > 1$) for the tar-
                get graph edge $(vt, wt) =$ Matches(edge)[j] we have $p((vt, wt)) >$
                Matches($PE[k-1]$)[Next($PE[k-1]$)]
                (take $j \leftarrow$ Next(edge), if $k = 1$);
                Next(edge) $\leftarrow j$;
                for *all edges e in PE with $p(e) < k$* **do**
                    Moved(e) $\leftarrow$ **false** ;
                end
                $(vt, wt) \leftarrow$ Matches(edge)[j];
                AdvanceEdgeMatchPositions(v,vt);
                Set k to be the smallest value for which there is an edge $e = (v2, w2)$ with
                $p(e) = k$ and either Pos($v2$) $= 0$ or Pos($w2$) $= 0$;
            end
        end
    end
    return Pos (array of vertex mappings);
end

Algorithm 1: Main Loop.

```
procedure AdvanceEdgeMatchPositions(v,vt);
begin
    PatternVertexStack ← ∅; push (PatternVertexStack,v);
    TargetVertexStack ← ∅; push (TargetVertexStack,vt);
    while PatternVertexStack ≠ ∅ do
        pvert ← pop (PatternVertexStack); tvert ← pop (TargetVertexStack);
        foreach edge e with p(e) < k, Moved(e) = false, and with endpoint pvert do
            Moved(e) ← true ;
            Find the smallest i ≥ Next(e) such that, for (vt2, wt2) = Matches(e)[i], we
            have wt2 ≥ tvert (or vt2 ≥ tvert, if pvert is the rightmost endpoint of e);
            if such an i is found then
                Next(e) ← i;
                Let newpvert be the other endpoint of e;
                newtvert ← vt2 (or newtvert ← wt2, if pvert is the rightmost endpoint of
                e);
                if Pos(newpvert) ≠ 0 then
                    Pos(newpvert) ← 0;
                    push (PatternVertexStack,newpvert);
                    push (TargetVertexStack,newtvert);
                end
            end
            else
                return Not Isomorphic
            end
        end
    end
end
```

Algorithm 2: The Depth-First Search for Edges.

finds one, or returns an answer that no such mapping exists. If the *AdvanceEdgeMatch-Positions* procedure is included, then when invoked it receives a vertex v in pattern graph and the first vertex vt in target to which v may be mapped according to search performed so far. The constraints on edge mappings are then narrowed down to be consistent with the mapping requirement for vertex v.

6.4   General Case of Disconnected Graphs

To deal with graphs that may be disconnected (but do not have isolated vertices) we additionally have to check that the vertex positions are preserved by isomorphism mapping, i.e., for vertices v and w in pattern graph with $p(v) < p(w)$ we must have $p(I(v)) < p(I(w))$. If we have isolated vertices, we additionally have to check that the sequence of vertices between v and w is a substring of the sequence of vertices between $I(v)$ and $I(w)$. This additional checking can be easily incorporated into the algorithm.

7   Maximal Common Subgraph Problem

The subgraph isomorphism algorithm is very fast for graphs corresponding to TOPS diagrams. This permits finding maximal common subgraphs by repeated extension and checking for subgraph isomorphism.

In order to find the maximal common subgraph for a given set of graphs we basically use an exhaustive search. Starting a the simple (one vertex) pattern graph, we check for subgraph isomorphism against all graphs in a given set and in the case of success attempt to extend the already matched pattern graph in all possible ways. Some restrictions on the number of different types of edges and vertices can be deduced from the given set of target graphs and are used by the algorithm. Apart from that, the previous successful match may be used to deduce information about extensions which are more likely to be successful in the next match. In general this does not prune the search space but may help to discover large common subgraphs earlier. There is also a greater probability that the largest common subgraph is found within a given time limit, even when the search has not been completed.

The advantage of this approach is that we obtain an algorithm with time complexity that is linear with respect to the number of graphs in a given set. Since there are likely to be more restrictions on the pattern for larger sets, often the most difficult cases arise for sets containing only one graph—however in this case we can simply return the given graph as the maximal common subgraph. Other methods that are known (for example as described in [8]) may be more efficient for sets containing a small number (basically just two) of graphs, but in general cannot be used to find the exact answer to the problem for larger sets.

Experiments suggest that this approach is still practical for TOPS diagrams. As mentioned above in the results section, all motifs in the Atlas for the CATH level H have been found by using repeated pattern matching and extension in 2 hours on an ordinary PC workstation. However it seems that the size of TOPS diagrams is quite close to the limit up to which such a maximal common subgraph algorithm can be successfully used, thus our solution may be quite problem specific. At the same time we expect that the subgraph isomorphism algorithm may be adapted also for considerably larger structures and may be useful for the other problems in bioinformatics.

Acknowledgments

Juris Vīksna was supported by a Wellcome Trust International Research Award.

References

1. Berman, H.M., Westbrook, J., Feng., Z., Gilliland, G., Bhat, T.N., Weissig, H., Shindyalov, I.N., Bourne, P.E.: The Protein Data Bank. Nucleic Acids Research 28 (2000) 235-242.
2. Bron, C., Kerbosch, J.: Algorithm 457: Finding all cliques of an undirected graph. Communications of ACM 16 (1973) 575-577.
3. Evans, P.A.: Finding common subsequences with arcs and pseudoknots. Proceedings of Combinatorial Pattern Matching 1999, LNCS 1645 (1999) 270-280.

4. Flores, T.P.J., Moss, D.M., Thornton, J.M.: An algorithm for automatically generating protein topology cartoons. Protein Engineering 7 (1994) 31-37.
5. Gilbert, D., Westhead, D.R., Nagano, N., Thornton, J.M.: Motif-based searching in tops protein topology databases. Bioinformatics 15 (1999) 317-326.
6. Hofmann, K., Bucher, P., Falquet, L., Bairoch, A.: The PROSITE database, its status in 1999. Nucleic Acids Research 27 (1999) 215-219.
7. Holm, L., Park, J.: DaliLite workbench for protein structure comparison. Bioinformatics 16 (2000) 566-567.
8. Koch, I., Lengauer, T., Wanke, E.: An algorithm for finding maximal common subtopologies in a set of protein structures. Journal of Computational Biology 3 (1996) 289-306.
9. McGregor, J.J.: Relational consistency algorithms and their application in finding subgraph and graph isomorphisms. Information Science 19 (1979) 229-250.
10. Orengo, C.A.: CORA—topological fingerprints for protein structural families. Protein Science 8 (1999) 699-715.
11. Orengo, C.A., Michie, A.D., Jones, S., Swindelis, M.B.: CATH—a hierarchic classification of protein domain structures. Structure 5 (1997) 1093-1108.
12. Ullmann, J.R.: An algorithm for subgraph isomorphism. Journal of the ACM 23 (1976) 31-42.
13. Westhead, D.R., Hatton, D.C., Thornton, J.M.: An atlas of protein topology cartoons available on the World Wide Web. Trends in Biochemical Sciences 23 (1998) 35-36.
14. Westhead, D.R., Slidel, T.W.F., Flores, T.P.J., Thornton, J.M.: Protein structural topology: automated analysis and diagrammatic representation. Protein Science 8 (1999) 897-904.
15. Zhang, K., Wang, L., Ma, B.: Computing similarity between RNA structures. Proceedings of Combinatorial Pattern Matching 1999, LNCS 1645 (1999) 281-293.

Computing Linking Numbers of a Filtration

Herbert Edelsbrunner[1] and Afra Zomorodian[2]

[1] Department of Computer Science
Duke University, Durham, NC 27708, USA
and Raindrop Geomagic, Research Triangle Park, NC 27709, USA
edels@cs.duke.edu
[2] Department of Computer Science
University of Illinois, Urbana, IL 60801, USA
zomorodi@uiuc.edu

Abstract. We develop fast algorithms for computing the linking number of a simplicial complex within a filtration. We give experimental results in applying our work toward the detection of non-trivial tangling in biomolecules, modeled as alpha complexes.

1 Introduction

In this paper, we develop fast algorithms for computing the linking numbers of simplicial complexes. Our work is within a framework of applying computational topology methods to the fields of biology and chemistry. Our goal is to develop useful tools by researchers in computational structural biology.

Motivation and Approach. In the 1980's, it was shown that the DNA, the molecular structure of the genetic code of all living organisms, can become knotted during replication [1]. This finding initiated interest in knot theory among biologists and chemists for the detection, synthesis, and analysis of knotted molecules [8]. The impetus for this research is that molecules with non-trivial topological attributes often display exotic chemistry. Taylor recently discovered a figure-of-eight knot in the structure of a plant protein by examining 3,440 proteins using a computer program [19]. Moreover, chemical self-assembly units have been used to create *catenanes*, chains of interlocking molecular rings, and *rotaxanes*, cyclic molecules threaded by linear molecules. Researchers are building nanoscale chemical switches and logic gates with these structures [2, 3]. Eventually, chemical computer memory systems could be built from these building blocks.

Catenanes and rotaxanes are examples of non-trivial structural tanglings. Our work is on detecting such interlocking structures in molecules through a combinatorial method, based on algebraic topology. We model biomolecules as a sequence of alpha complexes [7]. The basic assumption of this representation is that an alpha-complex sequence captures the topological features of a molecule. This sequence is also a filtration of the Delaunay triangulation, a well-studied combinatorial object, enabling the development of fast algorithms.

O. Gascuel and B.M.E. Moret (Eds.): WABI 2001, LNCS 2149, pp. 112–127, 2001.

The focus of this paper is the linking number. Intuitively, this invariant detects if components of a complex are linked and cannot be separated. We hope to eventually incorporate our algorithm into publicly available software as a tool for detecting existence of interlocked molecular rings.

Given a filtration, the main contributions of this paper are:

(i) the extension of the definition of the linking number to graphs, using a canonical basis,

(ii) an algorithm for enumerating and generating all cycles and their spanning surfaces within a filtration,

(iii) data structures for efficient enumeration of co-existing pairs of cycles in different components,

(iv) an algorithm for computing the linking number of a pair of cycles,

(v) and the implementation of the algorithms and experimentation on real data sets.

Algorithm (iv) is based on spanning surfaces of cycles, giving us an approximation to the linking number in the case of non-orientable or self-intersecting surfaces. Such cases do not arise often in practice, as shown in Section 6. However, we note in Section 2 that the linking number of a pair may be also computed by alternate algorithms. Regardless of the approach taken, pairs of potentially linked cycles must be first detected and enumerated. We provide the algorithms and data structures of such enumeration in (i-iii).

Prior Work. Important knot problems were shown to be decidable by Haken in his seminal work on normal surfaces [10]. This approach, as reformulated by Jaco and others [13], forms the basis of many current knot detection algorithms. Haas *et al.* recently showed that these algorithms take exponential time in the number of crossings in a knot diagram [12]. They also placed both the UNKNOTTING PROBLEM and the SPLITTING PROBLEM in NP, the latter being the focus of our paper. Generally, other approaches to knot problems have unknown complexity bounds, and are assumed to take at least exponential time. As such, the state of the art in knot detection only allows for very small data sets. We refer to Adams [1] background in knot theory.

Three-dimensional alpha shapes and complexes may be found in Edelsbrunner and Mücke [7]. We modify the persistent homology algorithm to compute cycles and surfaces [6]. We refer to Munkres [15] for background in homology theory that is accessible to non-specialists.

Outline. The remainder of this paper is organized as follows. We review linking numbers for collections of closed curves, and extend this notion to graphs in $\mathbb{R}^3$ in Section 2. We describe our model for molecules in Section 3. Extending the persistence algorithm, we design basic algorithms in Section 4 and use them to develop an algorithm for computing linking numbers in Section 5. We show results of some initial experiments in Section 6, concluding the paper in Section 7.

2   Linking Number

In this section, we define links and discuss two equivalent definitions of the linking number. While the first definition provides intuition, the second definition is the basis of our computational approach.

Links. A *knot* is an embedding of a circle in three-dimensional Euclidean space, $k :$ $\mathbb{S}^1 \to \mathbb{R}^3$. Two knots are *equivalent* if there is an ambient isotopy that maps the first to the second. That is, we may deform the first to the second by a continuous motion that does not cause self-intersections. A *link l* is a collection of knots with disjoint images. A link is *separable (splittable)* if it can be continuously deformed so that one or more components can be separated from other components by a plane that itself does not intersect any of the components. We often visualize a link l by a *link diagram*, which is the projection of a link onto a plane such that the over- and under-crossings of knots are presented clearly, We give an example in Figure 1(a). For a formal definition, see [12].

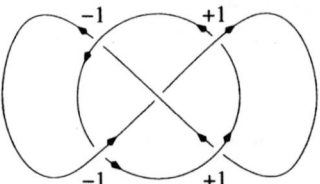

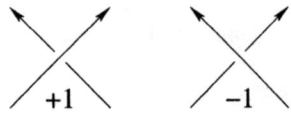

(a) A Link Diagram for the Whitehead Link.

(b) Crossing Label Convention.

Fig. 1. The Whitehead link (a) is labeled according to the convention (b) that the crossing label is $+1$ if the rotation of the overpass by 90 degrees counter-clockwise aligns its direction with the underpass, and -1 otherwise.

Linking Number. A *knot (link) invariant* is a function that assigns equivalent objects to equivalent knots (links.) Seifert first defined an integer link invariant, the linking number, in 1935 to detect link separability [18]. Given a link diagram for a link l, we choose orientations for each knot in l. We then assign integer labels to each crossing between any pair of knots k, k', following the convention in Figure 1(b). Let $\lambda(k, k')$ of the pair of knots to be one half the sum of these labels. A standard argument using Reidermeister moves shows that λ is an invariant for equivalent pairs of knots up to sign [1]. The *linking number* $\lambda(l)$ of a link l is

$$\lambda(l) = \sum_{k \neq k' \in l} |\lambda(k, k')|.$$

We note that $\lambda(l)$ is independent of knot orientations. Also, the linking number does not completely recognize linking. The Whitehead link in Figure 1(a), for example, has linking number zero, but is not separable.

Surfaces. The linking number may be equivalently defined by other methods, including one based on surfaces [17]. A *spanning surface* for a knot k is an embedded surface with boundary k. An orientable spanning surface is a *Seifert surface*. Because it is orientable, we may label its two sides as positive and negative. We show examples of spanning surfaces for the Hopf link and Möbius strip in Figure 2. Given a pair of oriented knots

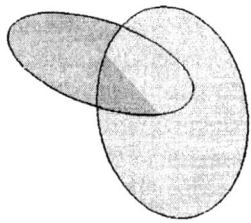
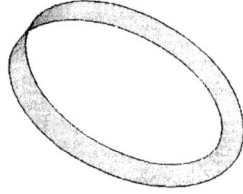

Fig. 2. The Hopf link and Seifert surfaces of its two unknots are shown on the left. Clearly, $\lambda = 1$. This link is the 200th complex for data set H in Section 6. The spanning surface produced for the cycle on the right is a Möbius strip and therefore non-orientable.

k, k', and a Seifert surface s for k, we label s by using the orientation of k. We then adjust k' via a homotopy h until it meets s in a finite number of points. Following along k' according to its orientation, we add $+1$ whenever k' passes from the negative to the positive side, and -1 whenever k' passes from the positive to the negative side. The following lemma asserts that this sum is independent of our the choice of h and s, and it is, in fact, the linking number.

SEIFERT SURFACE LEMMA. $\lambda(k, k')$ is the sum of the signed intersections between k' and any Seifert surface for k.

The proof is by a standard Seifert surface construction [17]. If the spanning surface is non-orientable, we can still count how many times we pass through the surface, giving us the following weaker result.

SPANNING SURFACE LEMMA. $\lambda(k, k')$ (mod 2) is the parity of the number of times k' passes through any spanning surface for k.

Graphs. We need to extend the linking number to graphs, in order to use the above lemma for computing linking numbers for simplicial complexes. Let $G = (V, E)$, $E \subseteq \binom{V}{2}$ be a simple undirected graph in $\mathbb{R}^3$ with c components $G^1, \ldots, G^c$. Let $z_1, \ldots, z_m$ be a fixed basis for the cycles in G, where $m = |E| - |V| + c$. We then define the linking number between two components of G to be $\lambda(G^i, G^j) = |\lambda(z_p, z_q)|$ for all cycles z_p, z_q in G^i, G^j, respectively. The linking number of G is then defined by combining the total interaction between pairs of components:

$$\lambda(G) = \sum_{i \neq j} \lambda(G^i, G^j).$$

The linking number is computed only between pairs of components following Seifert's original definition. Linked cycles within the same component may be easily unlinked by a homotopy. Figure 3 shows that the linking number for graphs is dependent on the chosen basis. While it may seem that we want $\lambda(G) = 1$ in the figure, there is no clear

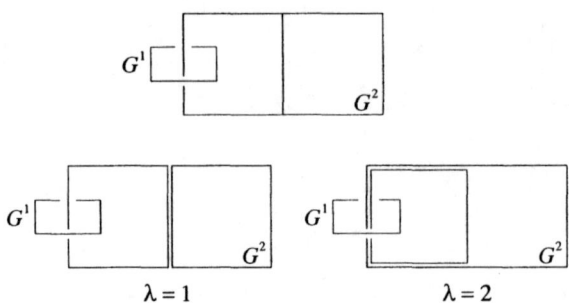

Fig. 3. We get different $\lambda(G)$ for graph G (top) depending on our choice of basis for G^2: two small cycles (left) or one large and one small cycle (right.)

answer in general. We will define a canonical basis in Section 4 using the persistent homology algorithm to compute $\lambda(G)$ for simplicial complexes.

3   Alpha Complexes

Our approach to analyzing a topological space is to assume a filtration for such a space. A filtration may be viewed as a history of a growing space that is undergoing geometric and topological changes. While filtrations may be obtained by various methods, only meaningful filtrations give meaningful linking numbers. As such, we use alpha complex filtrations to model molecules. The alpha complex captures the connectivity of a molecule that is represented by a union of spheres. This model may be viewed as the dual of the space filling model for molecules [14].

Dual Complex. A *spherical ball* $\hat{u} = (u, U^2) \in \mathbb{R}^3 \times \mathbb{R}$ is defined by its center u and square radius U^2. If $U^2 < 0$, the radius is imaginary and so is the ball. The *weighted distance* of a point x from a ball $\hat{u}$ is $\pi_{\hat{u}}(x) = \|x - u\|^2 - U^2$. Note that a point $x \in \mathbb{R}^3$ belongs to the ball iff $\pi_{\hat{u}}(x) \leq 0$, and it belongs to the bounding sphere iff $\pi_{\hat{u}}(x) = 0$. Let S be a finite set of balls. The *Voronoi region* of $\hat{u} \in S$ is the set of points for which $\hat{u}$ minimizes the weighted distance,

$$V_{\hat{u}} = \{x \in \mathbb{R}^3 \mid \pi_{\hat{u}}(x) \leq \pi_{\hat{v}}(x), \forall \hat{v} \in S\}.$$

The Voronoi regions decompose the union of balls into convex cells of the form $\hat{u} \cap V_{\hat{u}}$, as illustrated in Figure 4. Any two regions are either disjoint or they overlap along a shared portion of their boundary. We assume general position, where at most four

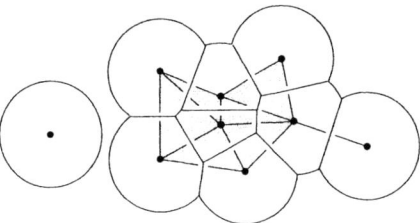

Fig. 4. Union of nine disks, convex decomposition using Voronoi regions, and dual complex.

Voronoi regions can have a non-empty common intersection. Let $T \subseteq S$ have the property that its Voronoi regions have a non-empty common intersection, and consider the convex hull of the corresponding centers, $\sigma_T = \text{conv} \{u \mid \hat{u} \in T\}$. General position implies that σ_T is a d-dimensional simplex, where $d = \text{card} \, T - 1$. The *dual complex* of S is the collection of simplices constructed in this manner,

$$K = \{\sigma_T \mid T \subseteq S, \bigcap_{\hat{u} \in T} (\hat{u} \cap V_{\hat{u}}) \neq \emptyset\}.$$

Any two simplices in K are either disjoint or they intersect in a common face which is a simplex of smaller dimension. Furthermore, if $\sigma \in K$, then all faces of σ are simplices in K. A set of simplices with these two properties is a *simplicial complex* [15]. A *subcomplex* is a subset $L \subseteq K$ that is itself a simplicial complex.

Alpha Complex. A *filtration ordering* is an ordering of a set of simplices such that each prefix of the ordering is a subcomplex. The sequence of subcomplexes defined by taking successively larger prefixes is the corresponding *filtration*. For dual complexes of a collection of balls, we generate an ordering and a filtration by literally growing the balls. For every real number $\alpha^2 \in \mathbb{R}$, we increase the square radius of a ball $\hat{u}$ by α^2, giving us $\hat{u}(\alpha) = (u, U^2 + \alpha^2)$. We denote the collection of expanded balls $\hat{u}(\alpha)$ as $S(\alpha)$. If $U^2 = 0$, then α is the radius of $\hat{u}(\alpha)$. If $\alpha^2 < 0$, then α is imaginary, and so is the ball $\hat{u}(\alpha)$. The α-complex $K(\alpha)$ of S is the dual complex of $S(\alpha)$ [7]. For example, $K(-\infty) = \emptyset$, $K(0) = K$, and $K(\infty) = D$ is the dual of the Voronoi diagram, also known as the Delaunay triangulation of S. For each simplex $\sigma \in D$, there is a unique *birth time* $\alpha^2(\sigma)$ defined such that $\sigma \in K(\alpha)$ iff $\alpha^2 \geq \alpha^2(\sigma)$. We order the simplices such that $\alpha^2(\sigma) < \alpha^2(\tau)$ implies σ precedes τ in the ordering. More than one simplex may be born at a time and such cases may arise even if S is in general position. In the case of a tie, it is convenient to order lower-dimensional simplices before higher-dimensional ones, breaking remaining ties arbitrarily. We call the resulting sequence the *age ordering* of the Delaunay triangulation.

Modeling Molecules. To model molecules by alpha complexes, we use representations of molecules as unions of balls. Each ball is an atom, as defined by its position in space and its van der Waals radius. These atoms become the spherical balls we need to define our complexes. Our representation gives us a filtration of alpha complexes for each molecule. We compute a linking number for each complex in a filtration of

m complexes. Let $[m]$ denote the set $\{1, 2, \ldots, m\}$. Then, the linking number may be viewed as a *signature function* $\lambda : [m] \to \mathbb{Z}$ that maps each index $i \in [m]$ to an integer $\lambda(i) \in \mathbb{Z}$. For other signature functions for filtrations of alpha complexes, see [5, 7].

4 Basis and Surfaces

To compute the linking numbers for an alpha complex, we need to recognize cycles, establish a basis for the set of cycles, and find spanning surfaces for the basis cycles. We do so by extending an algorithm we developed for computing persistent homology [6]. We dispense with defining persistence and concentrate on the algorithm and its extension.

Homology. We use homology to define cycles in a complex. Homology partitions cycles into equivalence classes using the boundary class of bounding cycles as the null element of a quotient group in each dimension. We use $\mathbb{Z}_2$ homology, so the group operation, which we call *addition*, is symmetric difference. Addition allows us to combine sets of simplices in a way that eliminates shared boundaries, as shown in Figure 5. Intuitively, non-bounding 1-cycles correspond to the graph notion of a cycle. We need to

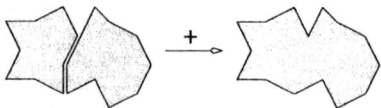

Fig. 5. Symmetric difference in dimensions one and two. We add two 1-cycles to get a new 1-cycle. We add the surfaces the cycles bound to get a spanning surface for the new 1-cycle.

define a basis for the first homology group of the complex which contains all 1-cycles, and choose representatives for each homology class. We use these representatives to compute linking numbers for the complex.

A simplex of dimension d in a filtration either creates a d-cycle or destroys a $(d-1)$-cycle by turning it into a boundary. We mark simplices as *positive* or *negative*, according to this action [5]. In particular, edges in a filtration which connect components are marked as negative. The set of all negative edges gives us a spanning tree of the complex, as shown in Figure 6. We use this spanning tree to define our canonical basis. Every time a positive edge σ_i is added to the complex, it creates a new cycle. We choose the unique cycle that contains σ_i and no other positive edge as a new basis cycle. We call this cycle a *canonical cycle*, and the collection of canonical cycles, the *canonical basis*. We use this basis for computation.

Persistence. The persistence algorithm matches positive and negative simplices to find life-times of homological cycles in a filtration. The algorithm does so by following a representative cycle z for each class. Initially, z is the boundary of a negative simplex σ_j, as z must lie in the homology class σ_j destroys. The algorithm then successively

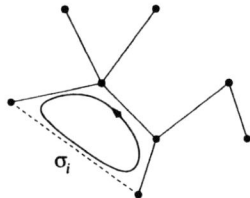

Fig. 6. Solid negative edges combine to form a spanning tree. The dashed positive edge σ_i creates a canonical cycle.

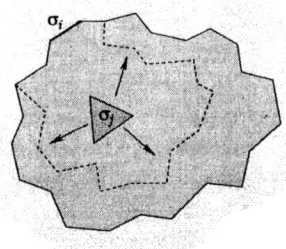

Fig. 7. Starting from the boundary of the negative triangle σ_j, the persistence algorithm finds a matching positive edge σ_i by finding the dashed 1-cycle. We modify this 1-cycle further to find the solid canonical 1-cycle and a spanning surface.

adds class-preserving boundary cycles to z until it finds the matching positive simplex σ_i, as shown in Figure 7. We call the half-open interval $[i, j)$ the *persistence interval* of both the homology class and its canonical representative. During this interval, the homology class exists as a class of homologous non-boundings cycles in the filtration. As such, the class may only affect the linking numbers of complexes $K_i, \ldots, K_{j-1}$ in the filtration. We use this insight in the next section to design an algorithm for computing linking numbers.

Computing Canonical Cycles. The persistence algorithm halts when it finds the matching positive simplex σ_i for a negative simplex σ_j, often generating a cycle z with multiple positive edges and multiple components. We need to convert z into a canonical cycle by eliminating all positive edges in z except for σ_i. We call this process *canonization*. To canonize a cycle, we add cycles associated with unnecessary positive edges to z successively, until z is composed of σ_i and negative edges, as shown in Figure 7. Canonization amounts to replacing one homology basis element with a linear combination of other elements in order to reach the unique canonical basis we defined earlier. A cycle undergoing canonization changes homology classes, but the rank of the basis never changes.

Computing Spanning Surfaces. For each canonical cycle, we need a spanning surface in order to compute linking numbers. We may compute these by maintaining surfaces while computing the cycles. Recall that initially, a cycle representative is the boundary of a negative simplex σ_j. We use σ_j as the initial spanning surface for z. Every time we add a cycle y to z in the persistence algorithm, we also add the surface y bounds to the z's surface. We continue this process through canonization to produce both canonical cycles and their spanning surfaces. Here, we are using a crucial property of our filtrations: the final complex is always the Delaunay complex of the set of weighted points and does not contain any 1-cycles. Therefore, all 1-cycles are eventually turned to boundaries and have spanning surfaces.

If the generated spanning surface is Seifert, we may apply the SEIFERT SURFACE LEMMA to compute the linking numbers. In some cases, however, the spanning surface is not Seifert, as in Figure 2(b). In these cases, we may either compute the linking number modulo 2 by applying the SPANNING SURFACE LEMMA, or compute the linking number by alternative methods.

5 Algorithm

In this section, we use the basis and spanning surfaces computed for 1-cycles to find linking numbers for all complexes in a filtration. Since we focus on 1-cycles only, we will refer to them simply as cycles.

Overview. We assume a filtration $K_1, K_2, \ldots, K_m$ as input, which we alternately view as a single complex undergoing growth. As simplices are added, the complex undergoes topological changes which affect the linking number: new components are created and merged together, and new non-bounding cycles are created and eventually destroyed. We use a basic insight from the last section: a basis cycle z with persistence interval $[i, j)$ may only affect the linking numbers of complexes $K_i, K_{i+1}, \ldots, K_{j-1}$ in the filtration, Consequently, we only need to consider basis cycles z' that exist during some subinterval $[u, v) \subseteq [i, j)$ in a different component than z's. We call the pair z, z' a *potentially-linked (p-linked) pair* of basis cycles, and the interval $[u, v)$ the *p-linking interval*.

Focusing on p-linked pairs, we get an algorithm with three phases. In the first phase, we compute all p-linked pairs of cycles. In the second phase, as shown in Figure 8, we compute the linking numbers of such pairs. In the third and final phase, we aggregate these contributions to find the linking number signature for the filtration.

```
for each p-linked pair z_p, z_q with interval [u, v) do
    Compute λ = |λ(z_p, z_q)| ;
    Output (λ, [u, v))
endfor.
```

Fig. 8. Linking Number Algorithm.

Two cycles z_p, z_q with persistence intervals $[i_p, j_p), [i_q, j_q)$ co-exist during $[r, s) = [i_p, j_p) \cap [i_q, j_q)$. We need to know if these cycles also belong to different components during some sub-interval $[u, v) \subseteq [r, s)$. Let $t_{p,q}$ be the minimum index in the filtration when z_p and z_q are in the same component. Then, $[u, v) = [r, s) \cap [0, t_{p,q})$. If $[u, v) \neq \emptyset$, z_p, z_q are p-linked during that interval. In the remainder of this section, we will first develop a data structure for computing $t_{p,q}$ for any pair of cycles z_p, z_q. Then, we use this data structure to efficiently enumerate all pairs of p-linked cycles. Finally, we give an algorithm for computing $\lambda(z_p, z_q)$ for a p-linked pair of cycles z_p, z_q.

Component History. To compute $t_{p,q}$, we need to have a history of the changes to the set of components in a filtration. There are two types of simplices that can change this set. Vertices create components and are therefore all positive. Negative edges connect components. We construct a binary tree called *component tree* recording these changes using a union-find data structure [4]. The leaves of the component tree are the vertices of the filtration. When a negative edge connects two components, we create an internal node and connect it to the nodes representing these components, as shown in Figure 9. The

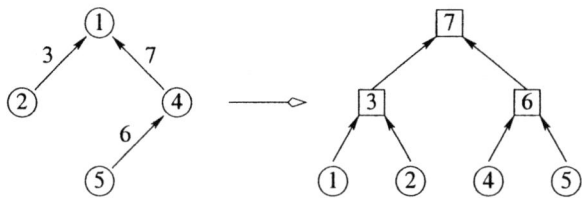

Fig. 9. The union-find data structure (left) has vertices as nodes and negative edges as edges. The component tree (right) has vertices as leaves and negative edges as internal nodes.

component tree has size $O(n)$ for n vertices, and we construct it in time $O(nA^{-1}(n))$, where $A^{-1}(n)$ is the inverse of the Ackermann's function which exhibits insanely slow growth. Having constructed the component tree, we find the time two vertices w, x are in the same component by finding their lowest common ancestor (lca) in this tree. We utilize Harel and Tarjan's optimal method to find lca's with $O(n)$ preprocessing time and $O(1)$ query time [11]. Their method uses bit operations. If such operations are not allowed, we may use van Leeuwen's method with the same preprocessing time and $O(\log \log n)$ query time [20].

Enumeration. Having constructed the component tree, we use a modified union-find data structure to enumerate all pairs of p-linked cycles. We augment the data structure to allow for quick listing of all existing canonical cycles in each component in K_i. Our augmentation takes two forms: we put the roots of the disjoint trees, representing components, into a circular doubly-linked list. We also store all existing cycles in each component in a doubly-linked list at the root node of the component, as shown in Figure 10. When components merge, the root x_1 of one component becomes the parent of the root x_2 of the other component. We concatenate the lists stored at the x_1, x_2, store the resulting list at x_1, and eliminate x_2 from the circular list in $O(1)$ time. When cycle z_p is created at time i, we first find z_p's component in time $O(A^{-1}(n))$. Then, we

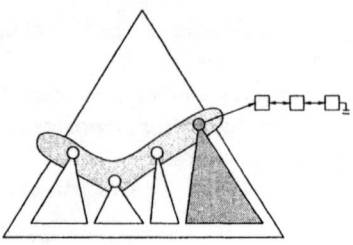

Fig. 10. The augmented union-find data structure places root nodes in the shaded circular doubly-linked list. Each root node stores all active canonical cycles in that component in a doubly-linked list, as shown for the darker component.

store z_p at the root of the component and keep a pointer to z_p with simplex σ_j, which destroys z_p. This implies that we may delete z_p from the data structure at time j with constant cost.

Our algorithm to enumerate p-linked cycles is incremental. We add and delete cycles using the above operations from the union-find forest, as the cycles are created and deleted in the filtration. When a cycle z_p is created at time i, we output all p-linked pairs in which z_p participates. We start at the root which now stores z_p and walk around the circular list of roots. At each root x, we query the component tree we constructed in the last subsection to find the time t when the component of x merges with that of z_p. Note that $t = t_{p,q}$ for all cycles z_q stored at x. Consequently, we can compute the p-linking interval for each pair z_p, z_q to determine if the pair is p-linked. If the filtration contains P p-linked pairs, our algorithm takes time $O(mA^{-1}(n) + P)$, as there are at most m cycles in the filtration.

Orientation. In the previous section, we showed how one may compute spanning surfaces s_p, s_q for cycles z_p, z_q, respectively. To compute the linking number using our lemma, we need to orient either the pair s_p, z_q or z_p, s_q. Orienting a cycle is trivial: we orient one edge and walk around to orient the cycle. If either surface has no self-intersections, we may easily attempt to orient it by choosing an orientation for an arbitrary triangle on the surface, and spreading that orientation throughout. The procedure either orients the surface or classifies it as non-orientable. We currently do not have an algorithm for orienting surfaces with self-intersections. The main difficulty is distinguishing between two cases for a self-intersection: a surface touching itself and passing through itself.

Computing λ. We now show how to compute $\lambda(z_p, z_q)$ for a pair of p-linked cycles z_p, z_q, completing the description of our algorithm in Figure 8. We assume that we have oriented s_p, z_q for the remainder of this subsection.

Let the *star* of a vertex u St u be the set of simplices containing u as a vertex. We subdivide the complex via a *barycentric subdivision* by connecting the centroid of each triangle to its vertices and midpoints of its edges, subdividing the simplices accordingly. This subdivision guarantees that no edge uv will have both ends on a Seifert surface unless it is entirely contained in that surface. We note that this approach mimics the construction of regular neighborhoods for complexes [9].

For a vertex $u \in s_p$, the edge property guaranteed by subdivision enables us to mark each edge $uv \in \mathrm{St}\, u, v \notin s_p$ as positive or negative, depending on the location of v with respect to s_p. After marking edges, we walk once around z_q, starting at a vertex not on s_p. If such a vertex does not exist, then $\lambda(z_p, z_q) = 0$. Otherwise, we create a string $S_{p,q}$ of $+$ and $-$ characters by noting the marking of edges during our walk. $S_{p,q}$ has even length as we start and end our walk on a vertex not on s_p, and each intersection of z_q with s_p produces a pair of characters, as shown in Figure 11 on the left. If $S_{p,q}$ is the

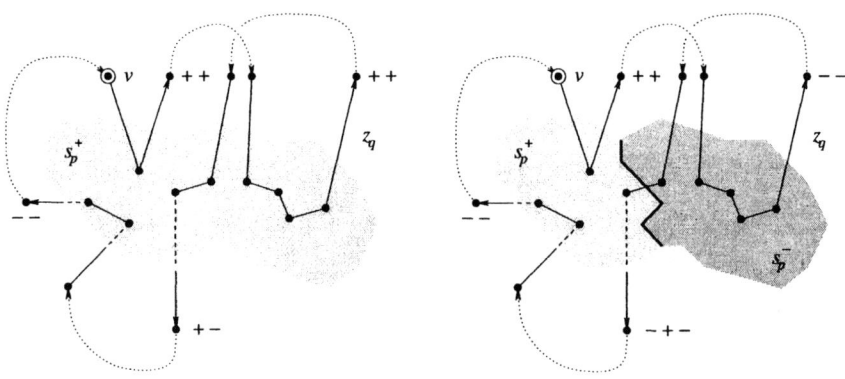

Fig. 11. On the left, starting at v, we walk on z_q according to its orientation. Segments of z_q that intersect s_p are shown, along with their contribution to $S_{p,q} = $ " $+ + + + + - - -$ ". We get $\lambda(z_p, z_q) = -1$. On the right, the bold flip curve is the border of s_p^+ and s_p^-, the portions of s_p that are oriented differently. $S_{p,q} = $ " $+ + - - - + - - -$ ", so counting all $+$'s, we get $\lambda(z_p, z_q) \bmod 2 = 3 \bmod 2 = 1$.

empty string, z_q never intersects s_p and $\lambda(z_p, z_q) = 0$. Otherwise, z_q passes through s_p for pairs $+-$ and $-+$, corresponding to z_q piercing the positive or negative side of s_p, respectively. Scanning $S_{p,q}$ from left to right in pairs, we add $+1$ for each occurrence of $-+$, -1 for each $+-$, and 0, for each $++$ or $--$. Applying the SEIFERT SURFACE LEMMA in Section 2, we see that this sum is $\lambda(z_p, z_q)$.

Computing λ mod 2. If neither of the spanning surfaces s_p, s_q of the two cycles z_1, z_2 is Seifert, we may still compute $\lambda(z_1, z_2) \bmod 2$ by a modified algorithm, provided one surface, say s_p, has no self-intersections. We choose an orientation on s_p locally, and extend it until all the stars of the original vertices are oriented. are oriented. This orientation will not be consistent globally, resulting in pair of adjacent vertices in s_p with opposite orientations. We call the implicit boundary between vertices with opposite orientations a *flip curve*, as shown in bold in Figure 11 to the right. When a cycle segment crosses the flip curve, orientation changes. Therefore, in addition to noting marked edges, we add a + to the string $S_{p,q}$ every time we cross a flip line. To compute $\lambda(z_p, z_q)$ mod 2, we only count +'s in $S_{p,q}$ and take the parity as our answer.

If s_p is orientable, there are no flip curves on it. The contribution of cycle segments to the string is the same as before: $+-$ or $-+$ for segments that pass through s_p, and $++$ and $--$ for segments that do not. By counting +'s, only segments that pass through s_p change the parity of the sum for λ. Therefore, the algorithm computes λ mod 2 correctly for orientable surfaces. For the orientable surface on the right in Figure 11, for instance, we get $\lambda(z_p, z_q)$ mod 2 = 5 mod 2 = 1, which is equivalent to the parity of the answer computed by the previous algorithm.

Remark. We are currently examining the question of orienting surfaces with self-intersections. Using our current methods, we may obtain a lower bound signature for λ by computing a mixed sum: we compute λ and λ mod 2 whenever we can to obtain the approximation. We may also develop other methods, including those based on the projection definition of the linking number in Section 2.

6   Experiments

In this section, we present some experimental timing results and statistics which we used to guide our algorithm development. We also provide visualizations of basis cycles in a filtration. All timings were done on a Micron PC with a 266 MHz Pentium II processor and 128 MB RAM running Solaris 8.

Implementation. We have implemented all the algorithms in the paper, except for the algorithm for computing λ mod 2. Our implementation differs from our exposition in three ways. The implemented component tree is a standard union-find data structure with the union by rank heuristic, but no path compression [4]. Although this structure has a $O(n \log n)$ construction time and a $O(\log n)$ query time, it is simple to implement and extremely fast in practice. We also use a heuristic to reduce the number of p-linked cycles. We store bounding boxes at the roots of the augmented union-find data structure. Before enumerating p-linked cycles, we check to see if the bounding box of the new cycle intersects with that of the stored cycles. If not, the cycles cannot be linked, so we obviate their enumeration. Finally, we only simulate the barycentric subdivision.

Data. We have experimented with a variety of data sets and show the results for six representative sets in this section. The first data set contains points regularly sampled along two linked circles. The resulting filtration contains a complex which is a Hopf link, as shown in Figure 2. The other data sets represent molecular structures with weighted points. In each case, we first compute the weighted Delaunay triangulation and the age ordering of that triangulation. The data points become vertices or 0-simplices. Table 1 gives the sizes of the data sets, their Delaunay triangulations, and age orderings. We show renderings of specific complexes in the filtration for data set K in Figure 12.

Basis. Table 2 summarizes the basis generation process. We distinguish the two steps of our algorithm: initial basis generation and canonization. We give the number of basis

Table 1. H defines a Hopf link. G is Gramicidin A, a small protein. M is a protein monomer. Z is a portion of a periodic zeolite structure. K is a human cyclin-dependent kinase. D is a DNA tile.

	# simplices of dimension d				total
	0	1	2	3	
H	100	1,752	3,240	1,587	6,679
G	318	2,322	3,978	1,973	8,591
M	1,001	7,537	13,018	6,481	28,037
Z	1,296	11,401	20,098	9,992	42,787
K	2,370	17,976	31,135	15,528	67,009
D	7,774	60,675	105,710	52,808	226,967

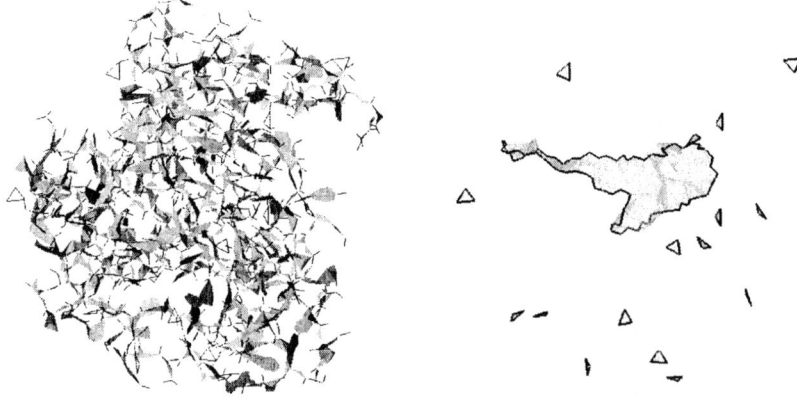

Fig. 12. Complex K_{8168} of K has two components and seventeen cycles. The spanning surfaces are rendered transparently.

cycles for the entire filtration, which is equal to the number of positive edges. We also show the effect of canonization on the size of the cycles and their spanning surfaces in Table 2. Note that canonization increases the size of cycles by one or two orders of magnitude. This is partially the reason we try to avoid performing the link detection if possible.

Links. In Table 3, we show that our component tree and augmented trees are very fast in practice to generate p-linked pairs. We also show that our bounding box heuristic for reducing the number of p-linked pairs increases the computation time negligibly. The heuristic is quite successful, moreover, in reducing the number of pairs we have to check for linkage, eliminating 99.8% of the candidates for dataset Z. The differences in total time of computation reflect the basic structure of the datasets. Dataset D has a large computation time, for instance, as the average size of the p-linked surfaces is approximately 264.16 triangles, compared to about 1.88 triangles for dataset K, and about 1.73 triangles for dataset M.

Table 2. On the left, we give the time to generate and canonize basis cycles, as well as their number. On the right, we give the average length of cycles and size of surfaces, before and after canonization.

	time in seconds			# cycles
	generate	canonize	total	
H	0.08	0.04	0.12	1,653
G	0.08	0.03	0.11	2,005
M	0.28	0.20	0.48	6,537
Z	0.46	0.46	0.92	10,106
K	0.72	1.01	1.73	15,607
D	2.63	2.94	5.57	52,902

	avg cycle length		avg surface size	
	before	after	before	after
H	3.06	51.03	1.06	63.04
G	3.26	13.02	1.38	52.28
M	3.29	34.18	1.33	71.18
Z	4.71	25.33	3.26	117.81
K	3.48	67.87	1.62	166.70
D	3.46	39.94	1.81	158.99

Table 3. Time to construct the component tree, and the computation time and number of p-linked pairs (alg), p-linked pairs with intersecting bounding boxes (heur), and links.

	tree	time in seconds			# pairs		
		alg	heur	links	alg	heur	links
H	0.01	0.00	0.00	0.01	1	1	1
G	0.00	0.01	0.02	0.02	112	0	0
M	0.03	0.06	0.06	0.23	16,503	14,968	0
Z	0.04	0.07	0.07	0.13	169,594	308	0
K	0.06	0.13	0.16	0.36	12,454	11,365	0
D	0.27	0.56	0.82	8.22	98,522	4,448	0

Discussion. Our initial experiments demonstrate the feasibility of the algorithms for fast computation of linking. The experiments fail to detect any links in the protein data, however. This is to be expected, as a protein consists of a single component, the primary structure of a protein being a single polypeptide chain of amino acids. Links, on the other hand, exist in different components by definition. We may relax this definition easily, however, to allow for links occurring in the same component. We have implementations of algorithms corresponding to this relaxed definition. Our future plans include looking for links in proteins from the Protein Data Bank [16]. Such links could occur naturally as a result of disulphide bonds between different residues in a protein.

7   Conclusion

In this paper, we develop algorithms for finding the linking numbers of a filtration. We give algorithms for computing bases of 1-cycles and their spanning surfaces in simplicial complexes, and enumerating co-existing cycles in different components. In addition, we present an algorithm for computing the linking number of a pair of cycles using the surface formulation. Our implementations show that the algorithms are fast and feasible in practice. By modeling molecules as filtrations of alpha complexes, we can detect potential non-trivial tangling within molecules. Our work is within a framework for applying topological methods for understanding molecular structures.

Acknowledgments

We thank David Letscher for discussions during the early stages of this work. We also thank Daniel Huson for the zeolite dataset Z, and Thomas La Bean for the DNA tile data set D. This work was supported in part by the ARO under grant DAAG55-98-1-0177; The first author is also supported by NSF under grants CCR-97-12088, EIA-99-72879, and CCR-00-86013.

References

1. Colin C. Adams. *The Knot Book: An Elementary Introduction to the Mathematical Theory of Knots*. W. H. Freeman and Company, New York, NY, 1994.
2. Richard A. Bissell, Emilio Córdova, Angel E. Kaifer, and J. Fraser Stoddart. A checmically and electrochemically switchable molecular shuttle. *Nature*, 369:133–137, 1994.
3. C. P. Collier, E. W. Wong, Belohradský, F. M. Raymo, J. F. Stoddart, P. J. Kuekes, R. S. Williams, and J. R. Heath. Electronically configurable moleculear-based logic gates. *Science*, 285:391–394, 1999.
4. Thomas H. Cormen, Charles E. Leiserson, and Ronald L. Rivest. *Introduction to Algorithms*. The MIT Press, Cambridge, MA, 1994.
5. C. J.Ã. Delfinado and H. Edelsbrunner. An incremental algorithm for Betti numbers of simplicial complexes on the 3-sphere. *Comput. Aided Geom. Design*, 12:771–784, 1995.
6. H. Edelsbrunner, D. Letscher, and A. Zomorodian. Topological persistence and simplification. In *Proc. 41st Ann. IEEE Sympos. Found. Comput. Sci.*, pages 454–463, 2000.
7. H. Edelsbrunner and E.P̃. Mücke. Three-dimensional alpha shapes. *ACM Trans. Graphics*, 13:43–72, 1994.
8. Erica Flapan. *When Topology Meets Chemistry : A Topological Look at Molecular Chirality*. Cambridge University Press, New York, NY, 2000.
9. P. J. Giblin. *Graphs, Surfaces, and Homology*. Chapman and Hall, New York, NY, second edition, 1981.
10. Wolfgang Haken. Theorie der Normalflächen. *Acta Math.*, 105:245–375, 1961.
11. Dov Harel and Robert Endre Tarjan. Fast algorithms for finding nearest common ancestors. *SIAM J. Comput.*, 13:338–355, 1984.
12. Joel Hass, Jeffrey C. Lagarias, and Nicholas Pippenger. The computational complexity of knot and link problems. *J. ACM*, 46:185–211, 1999.
13. William Jaco and Jeffrey L. Tollefson. Algorithms for the complete decomposition of a closed 3-manifold. *Illinois J. Math.*, 39:358–406, 1995.
14. Andrew R. Leach. *Molecular Modeling, Principles and Applications*. Pearson Education Limited, Harlow, England, 1996.
15. J.R̃. Munkres. *Elements of Algebraic Topology*. Addison-Wesley, Redwood City, California, 1984.
16. RCSB. Protein data bank. http://www.rcsb.org/pdb/.
17. Dale Rolfsen. *Knots and Links*. Publish or Perish, Inc., Houston, Texas, 1990.
18. H. Seifert. Über das Geschlecht von Knoten. *Math. Annalen*, 110:571–592, 1935.
19. William R. Taylor. A deeply knotted protein structure and how it might fold. *Nature*, 406:916–919, 2000.
20. Jan van Leeuwen. Finding lowest common ancestors in less than logarithmic time. Unpublished report.

Side Chain-Positioning as an Integer Programming Problem

Olivia Eriksson[1], Yishao Zhou[2], and Arne Elofsson[3]

[1] Stockholm Bioinformatics Center
Stockholm University, SE-106 91 Stockholm, Sweden
olivia@sbc.su.se
[2] Department of Mathematics
Stockholm University, SE-106 91 Stockholm, Sweden
yishao@matematik.su.se
[3] Stockholm Bioinformatics Center
Stockholm University, SE-106 91 Stockholm, Sweden
Fax: +46-8-158057, Tel: +46-8-16 1553 arne@sbc.su.se

Abstract. An important aspect of homology modeling and protein design algorithms is the correct positioning of protein side chains on a fixed backbone. Homology modeling methods are necessary to complement large scale structural genomics projects. Recently it has been shown that in automatic protein design it is of the uttermost importance to find the global solution to the side chain positioning problem [1]. If a suboptimal solution is found the difference in free energy between different sequences will be smaller than the error of the side chain positioning. Several different algorithms have been developed to solve this problem. The most successful methods use a discrete representation of the conformational space. Today, the best methods to solve this problem, are based on the dead end elimination theorem. Here we introduce an alternative method. The problem is formulated as a linear integer program. This programming problem can then be solved by efficient polynomial time methods, using linear programming relaxation. If the solution to the relaxed problem is integral it corresponds to the global minimum energy conformation (GMEC). In our experimental results, the solution to the relaxed problem has always been integral.

1 Introduction

Within the near future the approximate fold of most proteins will be known, thanks to structural genomics projects. The approximate fold of a protein is not enough though, to obtain a full understanding of a molecular mechanism or to be able to utilize the structure in drug design. For this a complete model of the protein is often needed. The main procedure today to obtain a complete model is by "homology modeling". The process of homology modeling often includes the positioning of amino acid side chains on a fixed backbone of a protein. Another area that has recently become important is the area of automatic "protein design". Here the goal is to obtain a sequence that folds to a given structure. Mayo and coworkers have shown that it is possible to perform automatic designs [1]. One crucial step in their procedure is to find the optimal side chain conformation.

O. Gascuel and B.M.E. Moret (Eds.): WABI 2001, LNCS 2149, pp. 128–141, 2001.
© Springer-Verlag Berlin Heidelberg 2001

There are two common features to most algorithms that try to solve this problem. The first one is to discretize the allowed conformational space into "rotamers" representing the statistically dominant side chain orientations in naturally occurring proteins [2, 3]. The rotamer approximation reduces the conformation space and makes it possible to use a discrete formulation of the problem. The second feature is the use of an energy function that can be divided into terms depending only on pairwise interactions between different parts of the protein. The total energy of a protein in a specific conformation, E_C, can therefore be described as

$$E_C = E_{backbone} + \sum_i E(i_r) + \sum_i \sum_{j>i} E(i_r j_s) \tag{1}$$

$E_{backbone}$ is the self-energy of the backbone, i.e. the interaction between all atoms in the backbone. $E(i_r)$ is the self-energy of side chain i in its rotamer conformation i_r, including its interaction with the backbone. $E(i_r j_s)$ is the interaction energy between side chain i in the rotamer conformation i_r and the side chain j in the rotamer conformation j_s. In this study we will keep the backbone of the protein fixed and only change the side chain rotamers. The term $E_{backbone}$ will therefore not contribute to any difference in energy between two protein conformations and can be ignored. The problem we want to solve can thus be defined as; *given the coordinates of the backbone and a specific rotamer library and energy function, find the set of rotamers that minimizes the energy function.* The solution space of this problem obviously increases exponentially with the number of residues included.

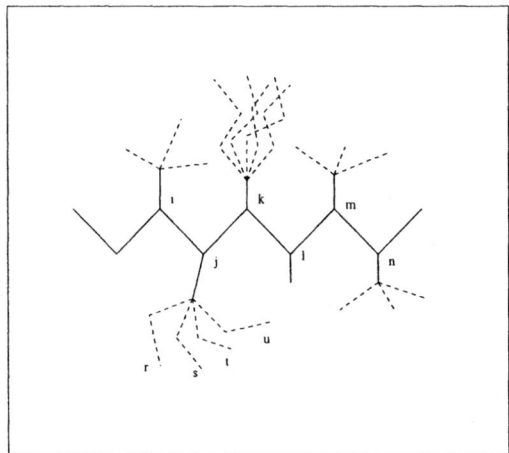

Fig. 1. Schematic Drawing of Different Rotamers in a Fantasy Protein.

1.1 Methods Used Today

There are several types of solution methods to the side chain positioning problem. Here we briefly review three of them.

Stochastic Algorithms. Several groups have developed stochastic algorithms, such as Monte Carlo simulations [4] and Genetic Algorithms [5] to solve the side chain positioning problem. The rotamer approximation together with the energy function makes it possible to view the problem as a discrete energy landscape where each point represents a specific rotamer combination and an assigned energy. To find a global minimum of the energy landscape these algorithms sample solutions semi-randomly and then move from one possible solution to another in a manner that depends both on the nature of the energy landscape and on specific rules for movement. With this approach you only need to compute the energy for the sampled conformations. Another advantage of these algorithms is that you can allow the structure of the protein to vary continuously if you want. One limitation, on the other hand, is that you are never guaranteed to have found the global minimum.

Pruning Algorithms. The most frequently used algorithms in this category are based on the dead end elimination theorem (DEE) [6–9]. DEE-based methods iteratively use rejection criteria. If a criterion is fulfilled it guarantees that a certain rotamer or combination of rotamers cannot be a part of the GMEC. This reduces the conformation space significantly and hopefully at the end a single solution remains. If DEE-based methods converge to a single solution they are guaranteed to have found the global minimum energy. During the last few years methods based on this theorem have developed considerably and can now solve most side-chain positioning problems [10]. However in protein design there is a limit beyond which this method fails to converge and other inexact methods have to be used [10]. The reason for this is that protein design requires a large rotamer library [9].

Mean-Field Algorithms. A third approach is to use mean field algorithms [11]. Here all rotamers of all side chains can be thought of as existing at the same time, but with different probabilities. Each residue is considered in turn and the probabilities for all rotamers of that side chain are updated, based on the mean field generated by the multiple side chains at neigbouring residues. The procedure is repeated until it converges. The predicted conformation of a side chain is chosen to be the rotamer with the highest probability. An advantage of self consistent mean field algorithms is that the computational time scales linearly with the number of residues. Unfortunately, there is no guarantee that the minimum of the mean-field landscape corresponds to the true GMEC.

2   The Side Chain Positioning Problem Formulated as an Integer Program

The side-chain positioning problem can be formulated as a classical *mathematical programming problem*. To this end it is necessary to introduce some notations and background. Mathematical programming concerns maximizing (or minimizing) a function of *decision variables*, which we denote by x, in the presence of *constraints*, which restrict the variables x to belong to a set F of admissible values, the *feasible set*. The function to optimize, *the objective function*, is denoted $f(x)$. A maximization problem can easily be turned into a minimization problem by changing the sign of $f(x)$. A general mathematical programming problem can then be stated as

$$\text{Minimize} \qquad f(x)$$
$$\text{subject to} \quad g_i(x) \leq 0, \ i = 1, \ldots, m$$

In that case the decision variables only can take integer values we have an *integer programming problem*. The easiest programming problems have continues variables and linear constraints and objective functions. These are called *linear programming (LP) problems*. Please refer to [12, 13] for a thorough introduction to linear programming and integer programming.

It is possible [8], to rewrite the energy function (1) so that it only contains terms depending on pairs of side chains, i.e., the energy of a conformation can be written as

$$E_C = \sum_i \sum_{j>i} E'(i_r, j_s). \tag{2}$$

This benefits our problem formulation. Let the decision variables be

$$x_{i_r, j_s} = \begin{cases} 1 \text{ if side chain } i \text{ is in rotamer } r \text{ and side chain } j \text{ is in rotamer } s \\ 0 \text{ else} \end{cases} \tag{3}$$

where $i = 1 \ldots n_{sc}$, $j = 2 \ldots n_{sc}$, $i < j$ and n_{sc} is the number of side chains. The total energy of the system (not just one conformation as before) can then be calculated as a sum, consisting of all possible variables, one for each rotamer combination, i_r, j_s, times their respective energy contribution to the total energy $e_{i_r, j_s} = E'(i_r, j_s)$, see equation (2). The energy e_{i_r, j_s} is calculated assuming both of these rotamers are included in the conformation. The *total* energy, E_{tot} is then:

$$E_{tot} = \sum_i \sum_r \sum_{j>i} \sum_s e_{i_r, j_s} x_{i_r, j_s} \tag{4}$$

where

$$\sum_r \sum_s x_{i_r, j_s} = 1 \text{ for all } i \text{ and } j, \ i < j \tag{5}$$

$$\sum_q x_{h_q, i_r} = \sum_p x_{g_p, i_r} = \sum_s x_{i_r, j_s} = \sum_t x_{i_r, k_t} \tag{6}$$

for all g, h, i, j, k and $r, \ h, g < i < j, k$

$$x_{i_r, j_s} \in \{0, 1\} \tag{7}$$

The first condition (5) together with (7) are to assure, that a certain pair of side chains, i and j, only can exist in *one* rotamer-state, i_r and j_s. If you consider a certain pair of side chains, say i and j, and think of all possible combinations of rotamer states that these two side chains could possibly take, then only *one* of these combinations can exist. This could for example be i_r and j_s. The second condition (6) states that one side chain can only exist in one rotamer state, independent of the rotamer states of other side chains. This assures that it can not exist in one state in relation to one side chain and another state in relation to another side chain.

This allows us to formulate a minimization problem as:

$$\text{Minimize:} \quad E_{tot} \qquad (8)$$

$$\text{subject to:} \ (5) \ (6) \ (7)$$

This is a linear integer programming problem. We can rewrite it in matrix form:

$$z_{IP} = \min\{cx|Ax = b, x \in \{0,1\}^n\} \qquad (9)$$

where A is an $m \times n$ matrix and b and c vectors of appropriate dimensions. In our case all components of A are $0, 1$ or -1, the components of b 0 or 1 and c consists of real values. The condition (6) can of course be implemented in different ways to obtain this form. We refer to the Appendix for a description of our approach. A small example in Table 1 shows what the integer program can look like for a small side chain positioning problem and our implementation.

In general, integer problems are hard to solve. By relaxing the integral constraints, that is, allowing $0 \leq x_{i_r,j_s} \leq 1$ instead of $x_{i_r,j_s} \in \{0,1\}$, we can turn this problem into a LP-problem to which there exist several efficient algorithms. If the solution to the LP relaxed problem is integral, $x_{i_r,j_s} \in \{0,1\}$ then it is an optimal solution to the original problem [13]. This means that if we solve the relaxation of the side chain positioning problem (8) and the solution turns out to be integral, it corresponds to the GMEC. The *linear programming relaxation* is as follows

$$z_{LP} = \min\{cx|Ax = b, 0 \leq x \leq 1\} \qquad (10)$$

The feasible solution set of the problem, $F(LP)$, defined by $Ax = b$ and $0 \leq x \leq 1$, forms a convex polyhedron, as it is an intersection of a finite number of hyper-planes. The LP-problem is well defined in that either it is infeasible, unbounded, or has an optimal solution. An optimal solution to an LP problem can always be found in an extreme point of the polyhedron (if there exists an optimal solution). For a more thorough introduction to the ideas of the linear programming methods we refer to [12].

It can be shown that there is a polynomial time method to solve LP, the proof is normally based on the *ellipsoid method* [14]. This method is, however, not appreciated in practice due to its slowness. Today, two popular methods are used. The idea of the *simplex method* is to move from one extreme point to another in such a way that the value of the objective function will always decrease or at least not increase. This is done until a minimum is reached or until the problem is found to be unbounded. Although there is an example showing that the simplex algorithm is of exponential time, it is very efficient if implemented well and in reality it is often of polynomial time [15]. The *interior point methods* is a family of algorithms that stay in the strict interior of the feasible region, such as using a barrier function. The term grew from Karmarkar's algorithm to solve a LP-problem [16]. Except for certain extreme worst cases, the interior point method runs in polynomial time.

Table 1. The A, b and c matrix, see (10) of a trivial example protein with four residues a, b, c and d, where a has three rotamers a_1, a_2, a_3, b two, c three and d two rotamers. These matrices are used as the input to the simplex algorithm. The order of the decision variables x are also shown. In these matrices the row: (ab=1) corresponds to $\sum_r \sum_s x_{a_r,b_s} = 1$, (a1b=a1c) corresponds to $\sum_s x_{a_1,b_s} = \sum_t x_{a_1,c_t}$ and (a1c=a1d) corresponds to $\sum_t x_{a_1,c_t} = \sum_u x_{a_1,d_u}$. So the rows (a1b=a1c) and (a1c=a1d) together say that rotamer a_1 either exists or not and can not do both on the same time. The rest of the rows correspond to similar constraints.

$$x = \left(x_{a_1b_1} \, x_{a_1b_2} \, x_{a_1c_1} \, x_{a_1c_2} \, x_{a_1c_3} \, x_{a_1d_1} \, x_{a_1d_2} \, x_{a_2b_1} \ldots x_{a_2d_2} \, x_{a_3b_1} \ldots x_{a_3d_2} \, x_{b_1c_1} \, x_{b_1c_2} \right.$$
$$\left. x_{b_1c_3} \, x_{b_1d_1} \, x_{b_1d_2} \, x_{b_2c_1} \ldots x_{b_2d_2} \, x_{c_1d_1} \, x_{c_1d_2} \, x_{c_2d_1} \, x_{c_2d_2} \, x_{c_3d_1} \, x_{c_3d_2} \right)$$
$$c = (1.43 \ldots -0.54)$$

A	b
1 1 0 0 0 0 0 1 1 0 0 0 0 0 1 1 0 0 0 0 0 0 0 0 0 0 0 0 0 0 0 0 0 0 0 0 0 1	(ab=1)
1 1 -1 -1 -1 0 0 0 0 0 0 0 0 0 0 0 0 0 0 0 0 0 0 0 0 0 0 0 0 0 0 0 0 0 0 0 0 0	(a1b=a1c)
0 0 1 1 1 -1 -1 0 0 0 0 0 0 0 0 0 0 0 0 0 0 0 0 0 0 0 0 0 0 0 0 0 0 0 0 0 0 0	(a1c=a1d)
0 0 0 0 0 0 0 1 1 -1 -1 -1 0 0 0 0 0 0 0 0 0 0 0 0 0 0 0 0 0 0 0 0 0 0 0 0 0 0	(a2b=a2c)
0 0 0 0 0 0 0 0 1 1 1 -1 -1 0 0 0 0 0 0 0 0 0 0 0 0 0 0 0 0 0 0 0 0 0 0 0 0 0	(a2c=a2d)
1 0 0 0 0 0 0 1 0 0 0 0 0 1 0 0 0 0 0 0 -1 -1 -1 0 0 0 0 0 0 0 0 0 0 0 0 0 0 0	(b1a=b1c)
0 0 0 0 0 0 0 0 0 0 0 0 0 0 0 0 0 1 1 1 -1 -1 -1 0 0 0 0 0 0 0 0 0 0 0 0 0 0 0	(b1c=b1d)
0 1 0 0 0 0 0 0 1 0 0 0 0 0 1 0 0 0 0 0 0 0 -1 -1 -1 0 0 0 0 0 0 0 0 0 0 0 0 0	(b2a=b2c)
0 0 1 0 0 0 0 0 0 1 0 0 0 0 0 0 1 0 0 0 0 -1 0 0 0 0 -1 0 0 0 0 0 0 0 0 0 0 0	(c1a=c1b)
0 0 0 0 0 0 0 0 0 0 0 0 0 0 0 0 0 0 0 1 0 0 0 0 1 0 0 0 0 -1 -1 0 0 0 0 0 0 0	(c1b=c1c)
0 0 0 1 0 0 0 0 0 0 1 0 0 0 0 0 0 1 0 0 0 -1 0 0 0 0 -1 0 0 0 0 0 0 0 0 0 0 0	(c2a=c2b)
0 0 0 0 0 0 0 0 0 0 0 0 0 0 0 0 0 0 0 0 1 0 0 0 0 1 0 0 0 0 0 -1 -1 0 0 0	(c2b=c2c)
0 0 0 0 1 0 0 0 0 0 0 1 0 0 0 0 0 0 1 0 0 0 -1 0 0 0 0 -1 0 0 0 0 0 0 0 0 0 0	(c3a=c3b)
0 0 0 0 0 0 0 0 0 0 0 0 0 0 0 0 0 0 0 0 0 1 0 0 0 0 1 0 0 0 0 0 0 -1 -1 0	(c3b=c3d)
0 0 0 0 0 1 0 0 0 0 0 0 1 0 0 0 0 0 0 1 0 0 0 -1 0 0 0 0 -1 0 0 0 0 0 0 0 0 0	(d1a=d1b)
0 0 0 0 0 0 0 0 0 0 0 0 0 0 0 0 0 0 0 0 0 0 1 0 0 0 0 1 0 -1 0 -1 0 -1 0 0	(d1b=d1c)
0 0 0 0 0 0 1 0 0 0 0 0 0 1 0 0 0 0 0 0 1 0 0 0 -1 0 0 0 0 -1 0 0 0 0 0 0 0	(d2a=d2b)
0 0 0 0 0 0 0 0 0 0 0 0 0 0 0 0 0 0 0 0 0 0 0 1 0 0 0 0 1 0 -1 0 -1 0 -1 0	(d2b=d2c)

To use LP-methods alone when solving an integer problem we have to prove that the solution we find is always integral. We have not been able to do so but our results indicate that this could be the case. If the the solution is fractional, branch and bound methods can be used in combination with LP to get an integer solution.

Time Complexity. Let n_{sc} be the number of side chains and n_{rot} be the average number of rotamer states for each side chain. Then the integer programming formulation of the problem (8) will give approximately $n_{sc}^2 n_{rot}$ number of constraints. For the simplex algorithm it is usually counted on that the number of iterations grows linearly with the number of constraints of the problem and the work for each iteration grows as the square of the number of constraints [15]. That is the computational time of solving the relaxed problem by the simplex method should be in the order of $O(n_{sc}^6)$. If we look at the average number of rotamers for each residue the time complexity for the simplex algorithm ought to be $O(n_{rot}^3)$. This is under the assumption that the solution to the relaxed problem is integral and therefore the final solution.

3   The Dead End Elimination Theorem

Dead end elimination methods seeks to systematically eliminate bad rotamers and combination of rotamers until a single solution remains. This is done by the iterative use of different criteria of elimination, which make it possible to exclude rotamers from being a part of the GMEC.

It is necessary for the DEE methods that the energy description is pairwise as described in equation (1). Let us consider one residue, say residue i and let $\{C'\}$ be the set of all possible conformations of the rest of the protein, that is, all possible combinations of rotamers of all side chains *except* side chain i. Now consider two rotamers of residue i, namely i_r and i_t. If every protein conformation with i_r is higher in energy than the corresponding conformation with i_t for all possible configurations of non-i residues then,

$$E(i_r) + \sum_{j,j \neq i} E(i_r j_s) > E(i_t) + \sum_{j,j \neq i} E(i_t j_s) \quad \textbf{all } C' \tag{11}$$

and the rotamer i_r cannot be a member of the GMEC. The inequality (11) is not easy to check computationally since $\{C'\}$ can be huge. Instead one can get a more practical, but weaker, condition by just comparing that case where the left side of (11) is *highest* in energy with the case where the right side is *lowest* in energy. This can be done by the following inequality known as the DEE-theorem introduced by Desmat *et al.* [6]

$$E(i_r) + \sum_{j,j \neq i} \min_s E(i_r j_s) > E(i_t) + \sum_{j,j \neq i} \max_s E(i_t j_s) \quad i \neq j. \tag{12}$$

The theorem states that the rotamer i_r is a dead ending, thus precluding i_r to be a member of the GMEC, if the inequality (12) holds true for any rotamer $i_t \neq i_r$ of the side chain i . For a proof of this theorem we refer to [6]. In words inequality (12) says that rotamer i_r must be a dead ending if the energy of the system taken at i_r:s most favorable interactions with all the other residues are bigger than the energy of another rotamer (i_t) at its worst interactions. The best and worst interactions are given by "choosing" that rotamer which gives the lowest, respectively the highest, value for each interaction term. Desmat also described how the criteria could be extended to the elimination of pairs of rotamers inconsistent with the GMEC.

A more powerful version of these criteria was introduced by Goldstein [7]. It subtracts the right-hand side from the left in equation (12) before applying the min operator.

$$E(i_r) - E(i_t) + \sum_{j,j \neq i} \min_s (E(i_r j_s) - E(i_t j_s)) > 0 \tag{13}$$

This criterion says that i_r is a dead ending if we can always lower the energy by taking rotamer i_t instead of i_r while keeping the other rotamers fixed. Goldstein's criterion can also be extended to pairs of rotamers inconsistent with the GMEC. These criteria are used iteratively and one after each other until no more rotamers can be eliminated. If the method converges to a single solution, this is guaranteed to be the GMEC. In computational time calculation using the rotamer-pairs is significantly more expensive than with single rotamers.

Other improvements of the DEE-method have been done [9, 17]. They have made it possible to eliminate more rotamers and to accelerate the process of elimination. Today DEE based methods can handle most side chain positioning problems [10]. One still reach a limit though, in design problems where this method fails to converge in a reasonable amount of time [10]. This means that the conformational space can not be reduced to a single solution. In this study we have only utilized Goldstein's criterion for single residues (13) .

4   Methods

The energy function used in this study consists of van der Waals forces between atoms that are more than three bonds away from each other. Van der Waals parameters were taken from CHARMM22 [18] parameter set (par_all22_prot). In this study we have focused on the algorithmic part of the side chain positioning problem, i. e. finding the global minimum energy conformation of the model. The next step would be to use a more appropriate energy function. We used a backbone-independent rotamer library of R. Dunbrack [3] (bbind99.Aug.lib), that contains a maximum of 81 rotamers per residue. Further, all hydrogen atoms were ignored. All calculations have been performed using the backbone coordinates of the lambda repressor (PDB-code: 1r69). The energy contributions of each rotamer pair to the total energy were calculated in advance and stored in a vector c, see (10). The two matrices A and b were constructed. An example of these matrices for a small problem can be seen in Table 1. These matrices were then used as the input to the linear programming algorithms. First a simplex algorithm, lp_solve_3.1 [19], was used. Secondly the problem was solved using a mixed integer programming (mip) algorithm from the CPLEX package [20]. This algorithm is designed to solve problems with both integral and real variables. It makes use of the fact that we have an integer problem, and finds structures in the input that could be used to reduce the problem. After a relaxation of the integer constraints it solves the obtained LP-program by the simplex method. If the solution is not integral a branch and bound procedure takes place. We have also tried other algorithms from the CPLEX package such as primal and dual; simplex, hybrid network and hybrid barrier solvers. The mip solver gave the best results.

For comparison we have also performed exhaustive search for up to 10 side chains and implemented a single-residue based DEE-algorithm, using Goldstein's criterion (13). We also implemented a combination of the DEE-theorem and the linear programming (CPLEX-mip). Here we used condition (13) iteratively until no more rotamers could be eliminated and then applied linear programming on the rest of the solution space to find the GMEC.

All algorithms were tested on identical systems, and we assured that the correct solution was found. In Table 2 the size of the conformational space of some of the problems can be seen. As several different programs were used for the calculations, it is not trivial to compare the absolute computational time needed. It is, however, possible to compare their complexity.

Table 2. Some of the problems used in the study; n_{sc} is the number of free side chains, $\bar{n}_{rot}$ is the average number of rotamers per side chain. The fraction of rotamers left after the DEE-search and the total number of conformations in the problems are also shown. The protein used is the lambda repressor (PDB-code:1r69).

n_{sc}	$\bar{n}_{rot}$	Number of conformations	% Remaining rot. after DEE
4	15	$10^{3.8}$	0
5	27	$10^{6.2}$	0
6	24	$10^{7.1}$	0
8	15.7	$10^{8.1}$	4.2
10	16.2	$10^{10.5}$	3.3
12	21	$10^{13.4}$	5.0
13	19.6	$10^{13.8}$	5.0
15	27.8	$10^{17.6}$	3.0
20	28.8	$10^{22.9}$	9.9
27	25.9	$10^{30.5}$	14.3
33	27.5	$10^{38.2}$	17.0
35	28.5	$10^{41.0}$	16.0
43	26.5	$10^{48.1}$	15.4

5   Results and Discussion

Today there are mainly three types of methods used for the side chain positioning problem, stochastic methods, mean-field algorithms and DEE-theorem based methods. Stochastic methods and mean-field algorithms always find a solution, however this is not guaranteed to be optimal. If a DEE method converges to one solution, it is guaranteed to be the global optimal solution. However, for larger problems DEE methods do not always converge. Therefore, the remaining solutions have to be tested by e.g. exhaustive search. Here we introduce a novel method using linear programming. If the solution from the LP-method is integral it corresponds to the GMEC.

Integral Solutions. In our experiments, see below, the solutions were controlled for integrality and so far we have never found a fractional solution. The *mip* method from the CPLEX package is made for integer programs. It uses a simplex solver, followed by branch-and-bound with simplex if the solution is not integral, see Section 4. In our experiments the branch-and-bound part of the algorithm was never used. We have also used primal and dual; simplex, hybrid network and hybrid barrier solvers from the CPLEX package. We received integral results from all these solvers. To examine if the energy function has any effect on the integrality of the solutions we have tried different nonsense energy functions on a test case with the simplex algorithm. All the solutions found were integral.

Experimental Time Complexity. First the number of free side chains was increased, n_{sc}, to examine the time behavior of the linear programming methods. The computational time for the *mip* method and the simplex method (from lp_solve_3.1) are shown

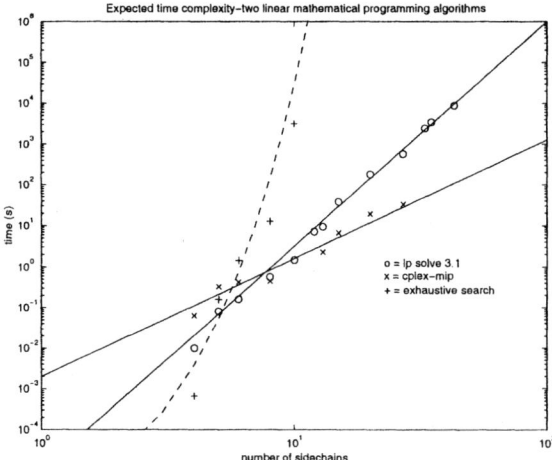

Fig. 2. Experimental studies of the complexity of the two linear programming algorithms versus the number of side chains. The circles represent the simplex method and the crosses the mip method. The complexity is approximately $O(n_{sc}^5)$ and $O(n_{sc}^3)$ respectively. For comparison a curve representing the exhaustive search is also included.

in Figure 2. It can be seen that the time scales approximately as $O(n_{sc}^3)$ for the mip method and as $O(n_{sc}^5)$ for the simplex method. This agrees quite well with the estimation of the complexity (see Section 2), where the computational time was calculated to scale as $O(n_{sc}^6)$ for the simplex method. Furthermore, it shows the superiority of the mip-method. The reason that the mip-method has a better complexity is due to the preprocessor, which finds structures in the input that can reduce the problem size and increase the speed. This also means that there probably exists better ways to implement the conditions of (8).

In protein design the average number of rotamers for each side chain, n_{rot}, is often very large. Therefore, it is interesting to study the complexity of the LP-algorithms versus n_{rot}. Our estimation of the time complexity for the simplex method on the side chain positioning problem is $O(n_{rot}^3)$, see Section 2. This agrees quite well with our experimental study, see Figure 3, where the complexity is approximately $O(n_{rot}^{2.2})$ both for the simplex and mip algorithms.

Comparison with the DEE-Method. The DEE-method is a pruning method that consecutively eliminate parts of the conformation space. First, simple criteria with a low complexity are used and then slower methods are applied until convergence. For large design problems DEE-methods do not converge in a reasonably amount of time [10]. This means that the time complexity for large system could be exponential and not polynomial. All this makes it difficult to calculate an overall time complexity of DEE-methods. In a study by Pierce et al [17], the time complexity of DEE was estimated in terms of the cost of the nested loops required to implement each approach. For the different criteria their estimation was between $O(n_{sc}^2 n_{rot}^2)$ and $O(n_{sc}^3 n_{rot}^5)$.

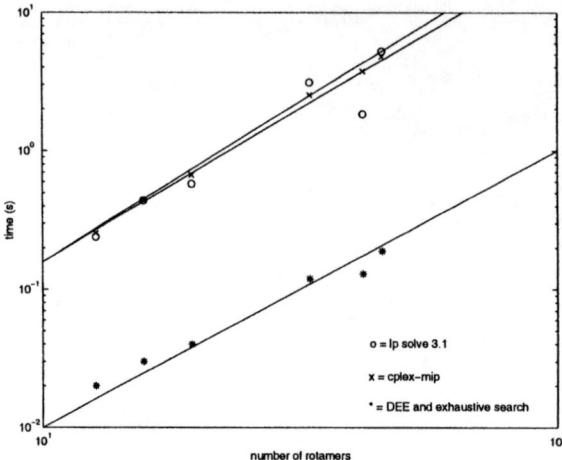

Fig. 3. Experimental studies of the complexity of the two linear programming algorithms used in this study versus the number of rotamers. The number of side chains is 8. The circles represent the simplex method, the crosses the mip method and the stars a DEE-algorithm with a final exhaustive search, see methods. For a problem of this size the DEE-algorithm almost converges. Therefore, the computational time of the exhaustive search is negligible. The complexity is approximately $O(n_{rot}^{2.2})$ (mip), $O(n_{rot}^{2.3})$ (simplex) and $O(n_{rot}^{2.0})$ (DEE).

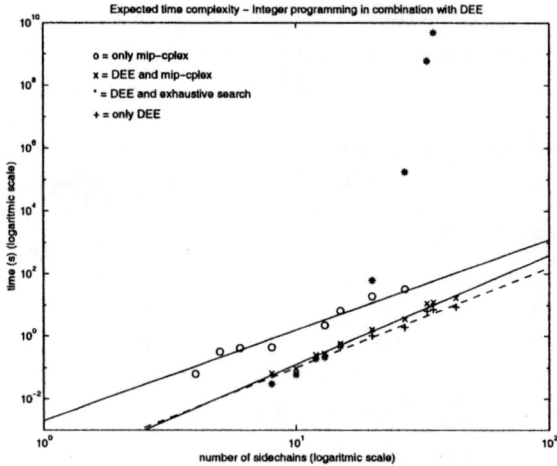

Fig. 4. Experimental studies of the complexity of the mip linear programming algorithm and the Dead End Elimination algorithm, versus the number of side chains. The circles represent the linear programming mip method alone, the crosses a combination of DEE and LP, the stars the DEE with a final exhaustive search and the '+'-signs the DEE-algorithm alone. The complexity for mip and the combined methods is approximately $O(n_{sc}^{3})$ and $O(n_{sc}^{3.5})$ respectively. The complexity of the DEE part of the computation is $O(n_{sc}^{3.2})$, the dashed line.

To perform a comparison between DEE and LP a part of the DEE method, Goldstein's criterion for single residues was implemented. However, we did not implement other criteria. The comparison of the two methods can therefore only be seen as an indication of their relative performance.

In our implementation, where the DEE did not converge to one solution, the remaining conformational space was searched exhaustively. When the problem contained more than approximately 8 residues the DEE algorithm did not converge. In Figure 4, it can be seen that for larger systems our DEE implementation does not show a polynomial complexity. With a more complete DEE-method it ought to be possible to limit the remaining conformational space so that a single solution is obtained for much larger systems. However, the complexity for the DEE algorithm would then be larger.

What might be more interesting is to compare the DEE implementation with the mip-linear programming method. If one only consider the complexity of the DEE-algorithm before the exhaustive search is started, the complexity is approximately equal to the LP-method, see Figure 4. However, while the LP-method has found the GMEC, the DEE-algorithm has not converge to a single solution for most problems, Table 2. In Figure 4, the time complexity of a combination of this DEE-algorithm and the LP-algorithm mip is also shown. Here, the time for the combined method is less than for mip alone, but the complexity is a little bit better for mip.

We have also made a study of the DEE-algorithm (Goldstein's singles criterion) with an increasing number of rotamers, see Figure 3. Here the problem was small enough (8 residues) for the DEE-method to eliminate almost all rotamers. The complexity of the DEE-algorithm was $O(n_{rot}^2)$, i.e. almost identical with the LP-methods. However, the DEE-algorithm was faster.

6 Conclusions

We have introduced a novel solution method to the problem of optimal side chain positioning on a fixed backbone. It has earlier been shown that finding the optimal solution to this problem is essential for the success of automatic protein designs [1]. The state of the art methods include several rounds using different versions of the Dead End Elimination theorem, with an increasing time complexity. This method do not necessarily converge to a single solution but the conformational space is reduced significantly and the remaining solutions can be searched.

By using linear programming (LP) we are guaranteed to find an optimal solution in polynomial time. If this solution is integral it corresponds to the global minimum energy conformation. This far in our studies the solutions have always been integral.

Linear programming is a well studied area of research and many algorithms for fast solutions are available. We obtained the best results using the mip-method from the CPLEX package. The time complexity for the mip-method to find the GMEC was $O(n_{sc}^3 n_{rot}^{2.2})$, while our DEE implementation (Goldstein's singles criterion), had a similar complexity of $O(n_{sc}^3 n_{rot}^2)$. However, for the mip-method the GMEC was found while for the DEE-method a fraction of the conformation space remained to be searched. More advanced DEE implementations converge to a single solution for larger problems, but they use more time consuming criteria, the worst with an estimated complexity of $O(n_{sc}^3 n_{rot}^5)$ [17]. As the complexity for the mip-method has a smaller dependency on the number of rotamers, the use of LP-algorithms might be best for problems with many rotamers, as in protein design. One reason for the effectiveness of the mip-algorithm is most likely due to the preprocessing of the problem. Therefore, a reformulation of the input matrices (10) to the simplex algorithm, perhaps as a network [21], might improve the complexity even further.

Acknowledgments

We thank Jens Lagergren for the idea of using linear programming and helpful discussions. This project was supported in part by Swedish Natural Science Research Council (NFR) and the Strategic Research Foundation (SSF).

References

1. Dahiyat, B.I. and Mayo, S.L: De novo protein design: fully automated sequence selection. Science **278** (1997) 82–87
2. Ponder, J.W. and Richards, F.M.: Tertiary Templates for Proteins: Use of Packing Criteria in the Enumeration of Allowed Sequences for Different Structural Classes J. Mol. Biol. **193** (1987) 775–791
3. Dunbrack, R. and Karplus, M.: Backbone-dependent rotamer library for proteins - application to sidechain prediction. J. Mol. Biol. **230** (1993) 543–574
4. Metropolis, N., Metropolis, A.W., Rosenbluth, M.N. and Teller, A.H.: Equation of state computing on fast computing machines. J. Chem. Phys. **21** (1953) 1087–1092
5. LeGrand, S.M. and Mers Jr, K.M. The genetic algorithm and the conformational search of polypeptides and proteins. J. Mol. Simulation **13** (1994) 299–320
6. Desmat, J., De Maeyer, M., Hazes, B. and Lasters, I.: The dead end elimination theorem and its use in side-chain positioning. Nature **356** (1992) 539–542
7. Goldstein, R.F.: Efficient rotamer elimination applied to protein side-chains and related spin glasses. Biophysical J. **66** (1994) 1335–1340
8. Lasters, I., De Maeyer, M. and Desmet J.: Enhanced dead-end elimination in the search for the global minimum energy conformation of a collection of protein side chains. Protein Eng. **8** (1995) 815–822
9. Gordon, D.B. and Mayo, S.L.: Radical performance enhancements for combinatorial optimization algorithms based on the dead-end elimination theorem. J. Comp. Chem. **19** (1998) 1505–1514
10. Voigt, C.A., Gordon, D.B. and Mayo, S.L.: Trading accuracy for speed: A quantitative comparison of search algorithms in protein sequence design. J. Mol. Biol. **299** (2000) 789–803
11. Koehl, P. and Levitt, M.: De Novo Protein Design. I. In search of stability and specificity. J. Mol. Biol. **293** (1999) 1161–1181
12. Lueneberger, D.G.: Linear and nonlinear programming. Addison-Wesley publishing company (1973)
13. Nemhauser, G.L. and Wolsey, L.A.: Integer and Combinatorial Optimization. Wiley-Interscience series in discrete mathematics and optimization (1988)
14. Korte, B. and Vygen, J.: Combinatorial optimization. Theory and algorithms Springer-Verlag (1991)
15. Lindeberg, P.O.: Opimeringslara: en introduction. Department of Optimization and Systems Theory,Royal institute of Technology, Stockholm (1990)

16. Karmarkar, N.: A New Polynomial Time Algorithm for Linear Programming. Combinatorica **4** (1984) 375–395
17. Pierce, N.A., Spriet, J.A., Desmet, J. and Mayo, S.L.: Conformational Splitting:A More Powerful criterion for Dead-End Elimination. J. Comp. Chem.**21** (2000) 999–1009
18. Brooks, B.R., Bruccoleri, R.E., Olafson, B.D., States, D.J., Swaminathan, S. and Karplus, M.: CHARMM: A program for Macromolecular Energy, Minimization and Dynamics Calculations. Journal of Computational Chemistry **4** (1983) 187–217
19. Berkelaar, M: lp_solve_3.1: Program with the Simplex algorithm.
    ftp://ftp.es.ele.tue.nl/pub/lp_solve (1996)
20. http://www.cplex.com
21. Rockafellar, R.T.: Network Flows and Monotropic Optimization. Wiley Interscience (1984)

A   Implementation of the Integer Program

To transform the integer program describing the side chain positioning problem (8) into the following form

$$z_{IP} = \min\{cx | Ax = b, x \in \{0,1\}^n\} \tag{14}$$

we have formulated the condition (6) as

$$\sum_s x_{i_r,j_s} - \sum_t x_{i_r,(j+1)_t} = 0, \ i = 1 \ldots (n_{sc} - 2) , j = 2 \ldots (n_{sc} - 1), \ i < j$$

$$\sum_r x_{i_r,j_s} - \sum_t x_{(i+1)_t,j_s} = 0, \ i = 1 \ldots (n_{sc} - 2) , j = 3 \ldots n_{sc}, \ (i+1) < j$$

$$\sum_r x_{i_r,j_s} - \sum_t x_{j_s,(i+2)_t} = 0, \ i = 1 \ldots (n_{sc} - 2) , j = i + 1.$$

A Chemical-Distance-Based Test
for Positive Darwinian Selection

Tal Pupko[1], Roded Sharan[2], Masami Hasegawa[1], Ron Shamir[1], and Dan Graur[3]

[1] The Institute of Statistical Mathematics
4-6-7 Minami Azabu, Minato ku, Tokyo, Japan
{tal,hasegawa}@ism.ac.jp
[2] School of Computer Science
Tel-Aviv University, Tel-Aviv 69978, Israel
{roded,rshamir}@post.tau.ac.il
[3] Department of Zoology, Faculty of Life Sciences
Tel-Aviv University, Tel-Aviv, Israel
graur@post.tau.ac.il

Abstract. There are very few instances in which positive Darwinian selection has been convincingly demonstrated at the molecular level. In this study, we present a novel test for detecting positive selection at the amino-acid level. In this test, amino-acid replacements are characterized in terms of chemical distances, i.e., degrees of dissimilarity between the exchanged residues in a protein. The test identifies statistically significant deviations of the mean observed chemical distance from its expectation, either along a phylogenetic lineage or across a subtree. The mean observed distance is calculated as the average chemical distance over all possible ancestral sequence reconstructions, weighted by their likelihood. Our method substantially improves over previous approaches by taking into account the stochastic process, tree phylogeny, among site rate variation, and alternative ancestral reconstructions. We provide a linear time algorithm for applying this test to all branches and all subtrees of a given phylogenetic tree. We validate this approach by applying it to two well-studied datasets, the MHC class I glycoproteins serving as a positive control, and the house-keeping gene carbonic anhydrase I serving as a negative control.

1 Introduction

The neutral theory of molecular evolution maintains that the great majority of evolutionary changes at the molecular level are caused not by Darwinian selection acting on advantageous mutants, but by random fixation of selectively neutral or nearly neutral mutants [12]. There are very few cases in which positive Darwinian selection was convincingly demonstrated at the molecular level [10, 22, 34, 30, 23]. These cases are vital to understanding the link between sequence variability and adaptive evolution. Indeed, it has been estimated that positive selection has occurred in only 0.5% of all protein-coding genes [2].

The most widely used method for detecting positive Darwinian selection is based on comparing synonymous and nonsynonymous substitution rates between nucleotide

O. Gascuel and B.M.E. Moret (Eds.): WABI 2001, LNCS 2149, pp. 142–155, 2001.
© Springer-Verlag Berlin Heidelberg 2001

sequences [17]. Synonymous substitutions are assumed to be selectively neutral. If only purifying selection operates, then the rate of synonymous substitution should be higher than the rate of nonsynonymous substitution. In the few cases where the opposite pattern was observed, positive selection was invoked as the likely explanation (see, e.g., [33, 14]). One critical shortcoming of this method is that it requires estimating numbers of synonymous substitutions. Because of saturation, such estimation is virtually impossible when the sequences under study are evolutionarily distant. The estimation is problematic even if close species are concerned. For example, saturation of substitutions in the third position is evident even when comparing cytochrome b sequences among species within the same mammalian order [5].

Another method for detecting positive selection is searching for parallel and convergent replacements. It is postulated that such molecular changes in different parts of a phylogenetic tree can only be explained by the same selective pressure being exerted on different taxa that became exposed to the same conditions [23, 32]. This method is limited to the few cases in which the same type of positive Darwinian selection occurs in two or more unrelated lineages.

A third method of detecting positive selection is based on comparing conservative and radical nonsynonymous differences [9, 7]: Nonsynonymous sites are divided into conservative sites and radical sites based on physiochemical properties of the amino-acid side chain, such as volume, hydrophobicity, charge or polarity. Radical and conservative sites and radical and conservative replacements are separately counted, and the number of radical replacements per radical site is compared to the number of conservative replacements per conservative site. If the former ratio is significantly higher than the latter, then positive Darwinian selection is invoked. By using this method, positive selection was inferred for the antigen binding cleft of class I major-histocompatibility-complex (MHC) glycoproteins [9] and rat olfactory proteins [8]. This method for detecting positive selection has the advantage that distant protein sequences can be compared even when synonymous substitutions are saturated. Another virtue of this method is its flexibility with respect to the sequence characteristic tested. For example, if we suspect that polar replacements might be advantageous, a test can be applied with radical replacements defined as those occurring between amino-acids with polar and non-polar residues only. However, this method also has many shortcomings. First, no correction for multiple substitutions is applicable [7]. Second, each codon in a pair of aligned amino-acid is used twice: Once for estimating the number of radical and conservative sites, and once for estimating the number of radical and conservative replacements. Third, the method treats replacements between different amino-acids as equally probable. Fourth, the method ignores branch lengths, implicitly assuming independence of the replacement probabilities between the amino acids and the evolutionary distance between the sequences under study. Finally, the phylogenetic signal is ignored, i.e., the test is applied to pairwise sequence comparisons rather than testing hypotheses on a phylogenetic tree.

The test for positive selection proposed in this study overcomes the shortcomings of the radical-conservative test. Our test incorporates a probabilistic framework for dealing with radical vs. conservative replacements. It applies a novel method for averaging over ancestral sequence assignments, weighted by their likelihood, thus eliminating bias which might result from assuming a specific ancestral sequence reconstruction. The rationale underlying our proposed test is that the evolutionary acquisition of a new function requires a significant change of the biochemical properties of the amino-acid sequence [7]. To quantify this biochemical difference between two amino-acid sequences, we define a chemical distances measure based on, e.g., Grantham's matrix [4]. Our test identifies large deviations of the mean observed chemical distance from the expected distance along a branch or across a subtree in a phylogenetic tree. If the observed chemical distance between two sequences significantly exceeds the chance expectation, then it is unlikely that this is the result of random genetic drift, and positive Darwinian selection should be invoked.

Based on the assumed stochastic process, the tree topology and its branch lengths, we calculate both the mean observed chemical distance and its underlying distribution for the branch or subtree in question. The mean observed chemical distance is calculated as the average chemical distance over all ancestral sequence reconstructions, weighted by their likelihood, thus, eliminating possible bias in a calculation based on a particular ancestral sequence reconstruction. The underlying distribution of this random variable is calculated using the JTT stochastic model [11], the tree topology and branch lengths, taking into account among site rate variation. We provide a linear time algorithm to perform this test for all branches and subtrees of a phylogenetic tree with n leaves.

In order to validate our approach, we applied it to two control datasets: Class I major-histocompatibility-complex (MHC) glycoproteins, and carbonic anhydrase I. These datasets were chosen since they were already used as standard positive control (MHC) and negative control (carbonic anhydrase) for positive selection [24]. For the MHC class I dataset, as reported in [9], we observe positive selection which favors charge replacements only when applying the test to the subsequences of the binding cleft ($P < 0.01$). In addition we observe positive selection which favors polarity replacements when using Grantham's polarity indices [4] ($P < 0.01$). When applying the test to the carbonic anhydrase dataset, no positive selection is observed.

The paper is organized as follows: Section 2 contains the notations and terminology used in the paper. Section 3 presents the new test for positive Darwinian selection. Section 4 describes the application of this test to the two control datasets. Finally, Section 5 contains a summary and a discussion of our approach.

2   Preliminaries

Let A be the set of 20 amino-acids. We assume that sequence evolution follows the JTT probabilistic reversible model [11]. For amino-acid sequences this model is described by a 20×20 matrix M, indicating the relative replacement rates of amino-acids, and a vector $(P_A, \ldots, P_Y)$ of amino-acid frequencies. For each branch of length t and amino-acids i and j, the $i \to j$ replacement probability, denoted by $P_{ij}(t)$, can be calculated from the eigenvalue decomposition of M [13]. (In practice, an approximation to $P_{ij}(t)$ is used to speedup the computation [19].) We denote by $f_{ij}(t) = P_i \cdot P_{ij}(t) = P_j \cdot P_{ji}(t)$ the probability of observing i and j in the same position in two aligned sequences of evolutionary distance t.

Let s be an amino-acid sequence. The amino-acid at position i in s is denoted by s_i. For two amino-acids $a, b \in \mathcal{A}$, we denote their *chemical distance* by $d(i, j)$. We assume we have a table of chemical distances between every pair of amino-acids. One such distance is Grantham's chemical distance [4]. (Other similar distance measures appear in [27, 16].) This chemical distance measures the difference between two amino-acids in terms of their volume, polarity and composition of the side chain. The choice of which distance measure to use, reflects the type of test we wish to perform. For example, Grantham's distance is appropriate when testing whether the replacements between the sequences under question are more radical with respect to a range of physiochemical properties (volume, charge and composition of the side chain). For testing whether polarity differences between sequences are higher than the random expectation, two distance measures are applicable: The first measure is based on dividing the set of amino-acids into 2 categories: Polar (C,D,E,H,K,N,Q,R,S,T,W,Y) and non-polar (the rest). The polarity distance between two amino-acids is then defined as 1 if one is polar and the other is not, and 0 otherwise [9]. The second polarity distance is defined as the absolute difference between the polarity indices of the two amino-acids, and yields real values [4]. For testing charge differences 3 categories of amino-acids are defined: Positive (H,K,R), negative (D,E) and neutral (all other). The charge distance between two amino-acids is defined as 1 if they belong to two different categories, and 0 if they belong to the same category [9].

We define the *average chemical distance* between two sequences s^1 and s^2 of length N as the average of the chemical distances between pairs of amino-acids occupying the same position in a gapless alignment of s^1 and s^2:

$$D(s^1, s^2) = \frac{1}{N} \sum_{i=1}^{N} d(s_i^1, s_i^2)$$

Let $\mathcal{T}$ be an unrooted phylogenetic tree. For a node v, we denote by $N(v)$ the set of nodes adjacent to v. For an edge $(u, v) \in \mathcal{T}$ we denote by $t(u, v)$ the length of the branch connecting u and v.

3   A Test for Positive Darwinian Selection

In this section we describe a new test for detecting positive Darwinian selection. The input to the test is a set of gap-free aligned sequences and a phylogenetic tree for them. We first present a version of our test for a pair of known sequences. We then extend this method to test positive selection on specific branches of a phylogenetic tree under study. Finally we generalize the test to subtrees (clades) and incorporate among site rate variation.

3.1   Testing Two Known Sequences

Let s^1 and s^2 be two amino-acid sequences of length N and evolutionary distance t. The underlying distribution of $D(s^1, s^2)$ is inferred as follows. The expectation of the chemical distance at position i is:

$$E(d(s_i^1, s_i^2)) = \sum_{a,b \in \mathcal{A}} d(a,b) f_{ab}(t)$$

Assuming that the distribution of the chemical distance in each position is identical, we obtain

$$E(D(s^1, s^2)) = \frac{1}{N} \sum_{i=1}^{N} E(d(s_i^1, s_i^2)) = E(d(s_1^1, s_1^2))$$

The variance of the chemical distance at position i is:

$$V(d(s_i^1, s_i^2)) = E(d(s_i^1, s_i^2)^2) - E(d(s_i^1, s_i^2))^2 = \sum_{a,b \in \mathcal{A}} d(a,b)^2 f_{ab}(t) - E(d(s_i^1, s_i^2))^2$$

and assuming further that sequence positions are independent, we obtain

$$V(D(s^1, s^2)) = \frac{V(d(s_1^1, s_1^2))}{N}$$

For practical values of N, $D(s^1, s^2)$ is approximately normally distributed with expectation $E(D(s^1, s^2)$ and standard deviation $\sqrt{V(D(s^1, s^2))}$. This allows us to compute for each observed chemical distance d, the probability that it occurs by chance, i.e., its p-value. If the observed chemical distance is found above the 0.99 percentile of the normal distribution, we conclude that replacements in these two sequences significantly deviate from the expectation, and suggest positive selection to explain this phenomenon.

3.2 Testing a Tree Lineage

Here we first describe a general method to apply pairwise tests to a phylogenetic tree. Suppose that we wish to test a statistical hypothesis on a specific branch of the phylogenetic tree. Also suppose that we have a procedure to test our hypothesis on a pair of known sequences, like the procedure described above. In order to test our hypothesis on a specific branch, we could first infer the corresponding ancestral sequences (using, e.g., maximum likelihood estimation [20]) and then check our hypothesis. Inferring ancestral sequences and then using these sequences as observations was done in e.g., [31]. This approach, which treats estimated reconstructions as observations may lead to erroneous conclusions due to bias in the reconstruction. A more robust approach is to average over all possible reconstructions, weighted by their likelihood. By averaging over all possible ancestral assignments, we extend our test to hypothesis testing on a phylogenetic tree, without possible bias that results from reconstructing particular sequences at internal tree nodes.

We describe in the following how to apply our test to a specific branch connecting nodes x and y in a tree $\mathcal{T}$. Since we assume that different positions evolve independently we restrict the subsequent description to a single site.

Each branch $(u, v) \in \mathcal{T}$ partitions the tree into two subtrees. Let $L(u, v, a)$ denote the likelihood of the subtree which includes v, given that v is assigned the amino-acid a. $L(u, v, a)$ can be computed by the following recursion equation:

$$L(u, v, a) = \prod_{w \in (N(v) \setminus \{u\})} \left\{ \sum_{b \in \mathcal{A}} P_{ab}(t(v, w)) \cdot L(v, w, b) \right\}$$

For a leaf v at the base of the recursion we have $L(u, v, a) = 1$, assuming amino-acid a in v, and $L(u, v, a) = 0$ otherwise.

The likelihood of $\mathcal{T}$ is thus:

$$P_{\mathcal{T}} = \sum_{a,b \in \mathcal{A}} f_{ab}(t(u, v)) \cdot L(u, v, b) \cdot L(v, u, a)$$

where (u, v) is any branch of $\mathcal{T}$.

Suppose that the data at the leaves of $\mathcal{T}$ is $\boldsymbol{w} = (w_1, \ldots, w_n)$. The *mean observed chemical distance* for a given branch $(x, y) \in \mathcal{T}$ can be calculated as follows:

$$D(x, y) = \sum_{a,b \in \mathcal{A}} Pr(x = a, y = b | \boldsymbol{w}) \cdot d(a, b)$$

$$= \frac{1}{P_{\mathcal{T}}} \sum_{a,b \in \mathcal{A}} \left\{ d(a, b) \cdot f_{ab}(t(x, y)) \cdot L(x, y, b) \cdot L(y, x, a) \right\}$$

It remains to compute the null distribution of this statistic. The expectation of $D(x, y)$ (with respect to all possible leaf-assignments) is as follows:

$$E(D(x, y)) = \sum_{\boldsymbol{z} \in \mathcal{A}^n} Pr(\boldsymbol{z}) \sum_{a,b \in \mathcal{A}} Pr(x = a, y = b | \boldsymbol{z}) \cdot d(a, b)$$

$$= \sum_{a,b \in \mathcal{A}} d(a, b) \sum_{\boldsymbol{z} \in \mathcal{A}^n} Pr(\boldsymbol{z}) \cdot Pr(x = a, y = b | \boldsymbol{z})$$

$$= \sum_{a,b \in \mathcal{A}} d(a, b) \cdot f_{ab}(t(x, y))$$

We conclude that $E(D(x, y))$ is the same as in the known-sequences case. For the variance of $D(x, y)$ we have no explicit formula. Instead, we evaluate $V(D(x, y))$ using parametric bootstrap [25]. Specifically, we draw at random many assignments of amino-acids to the leaves of $\mathcal{T}$ and compute $D(x, y)$ for each of them, thereby evaluating its variance. An assignment to the leaves of $\mathcal{T}$ is obtained as follows: We first root $\mathcal{T}$ at an arbitrary node r. We then draw at random an amino-acid for r according to the amino-acid frequencies. We next draw amino-acids for each child of r according to the appropriate replacement probabilities of our model, and continue in this manner till we reach the leaves.

Finally, since $D(x, y)$ is approximately normally distributed, we can compute a p-value for the test, which is simply $Pr(Z \geq \frac{D(x,y) - E(D(x,y))}{\sqrt{V(D(x,y))}})$ where $Z \sim Normal(0, 1)$. Note, that if the test is applied to several (or all) branches of the tree, then the significance level of the test should be corrected in accordance with the number of tests performed, e.g., using Bonferroni's correction which multiplies the p-value by the number of branches tested.

The algorithm for testing the branches of a phylogenetic tree $\mathcal{T}$ is summarized in Figure 1. For each branch $(x, y) \in \mathcal{T}$ the algorithm outputs the p-value of the test for that branch. In the actual implementation we used $M = 100$.

PositiveSelectionTest($\mathcal{T}$):
Root $\mathcal{T}$ at an arbitrary node r.
Draw M assignments to the leaves of $\mathcal{T}$ using parametric bootstrap.
Traverse $\mathcal{T}$ bottom-up, computing along the way for every $(u, v) \in \mathcal{T}, a \in \mathcal{A}$
   the value of $L(u, v, a)$, where u is the parent of v.
Traverse $\mathcal{T}$ top-down, computing along the way for every $(u, v) \in \mathcal{T}, a \in \mathcal{A}$
   the value of $L(v, u, a)$, where u is the parent of v.
For every $(x, y) \in \mathcal{T}$ **do**:
   Calculate $D(x, y)$ and $E(D(x, y))$.
   Evaluate $V(D(x, y))$.
   Output the p-value for the branch (x, y).

Fig. 1. An Algorithm for Testing the Branches of a Phylogenetic Tree $\mathcal{T}$.

Theorem 1. *For a given phylogenetic tree $\mathcal{T}$ with n leaves, the algorithm tests all branches of $\mathcal{T}$ in $O(n)$ time.*

Proof. Given $L(u, v, a)$ for every $(u, v) \in \mathcal{T}$ and every $a \in \mathcal{A}$, it is straightforward to compute $D(u, v)$ for all $(u, v) \in \mathcal{T}$ in linear time. The computation of $E(D(u, v))$ and $V(D(u, v))$ is clearly linear. The complexity follows.

3.3 Testing a Subtree

In this section we present an extension of our method to test subtrees of a given phylogenetic tree $\mathcal{T}$. This is motivated by the consideration that if a clade of contemporary sequences has undergone positive Darwinian selection, we cannot necessarily assume that this selection occurred solely along the branch leading to that clade. A reasonable scenario is that the selection was continuous and occurred along several or all branches of the subtree corresponding to this clade. In such a case, the test we have just described may not detect any significant positive selection along any specific branch. Hence, we are interested at testing for positive selection across subtrees as well.

For a subtree $\mathcal{T}'$ of $\mathcal{T}$, we define the mean observed chemical distance $D(\mathcal{T}')$ as the average observed distance along its branches (i.e., the sum of the observed distance for each branch divided by the number of branches in $\mathcal{T}'$). Clearly, the expectation of $D(\mathcal{T}')$ is equal to the average expectation of the branches of $\mathcal{T}'$. The variance of $D(\mathcal{T}')$ can be evaluated using parametric bootstrap. We then use the normal approximation to compute a p-value for this test. We conclude:

Theorem 2. *For a given phylogenetic tree $\mathcal{T}$ with n leaves, the complexity of testing all its subtrees is $O(n)$.*

3.4 Introducing among Site Rate Variation

The rate of evolution is not constant among amino-acid sites [28]. Consider two sequences of length N. Suppose that there are on average l replacements per site between these sequences. This means that we expect lN replacements altogether. How many replacements should we expect at each particular site? Naive models assume that the variation of mutation rate among sites is zero, i.e., that all sites have the same replacement probability. Models that take this Among Site Rate Variation (ASRV) into account assume that at the j-th position the average number of replacement is $lr[j]$, where each $r = r[j]$ is a rate parameter drawn from some probability distribution. Since the mean rate over all sites is l, the mean of r is equal to 1. Yang suggested the gamma distribution with parameters α and β as the distribution for r, and since the mean of the gamma distribution α/β, must be equal to 1, $\alpha = \beta$ [28], that is:

$$f(r; \alpha, \beta) = \frac{\alpha^\alpha}{\Gamma(\alpha)} e^{-\alpha r} r^{\alpha-1}$$

Maximum likelihood models incorporating ASRV are statistically superior to those assuming among site rate homogeneity [28]. They also help avoiding the severe underestimation of long branch lengths that can occur with the homogeneous models [15].

In this study we use the discrete gamma model with k categories whose means are $r_1, \ldots, r_k$ to approximate the continuous gamma distribution [29]. The categories are selected so that the probabilities of r falling into each category are equal. We thus assume that $Pr(r = r_i) = 1/k$.

The incorporation of the discrete gamma model in our test is straightforward. For each rate category i we calculate both the expected and observed chemical distance, given that the rate is r_i. This is equivalent to making the computation in the homogeneous case, where all branch lengths are multiplied by the factor r_i. The observed and expected chemical distance for each branch are then averaged over all rate categories.

4 Biological Results

In order to validate our approach, we applied it to two control datasets: Class I major-histocompatibility-complex (MHC) glycoproteins, and carbonic anhydrase I. We have chosen to analyze these datasets since they were already used as standard positive control (MHC) and negative control (carbonic anhydrase) for positive selection tests [24].

The datasets contain aligned sequences (all sequences are of the same length, and the best alignment is gapless). Phylogenetic trees were constructed using the MOLPHY software [1], with the neighbor-joining method [21] for MHC class I, and with the maximum likelihood method for carbonic anhydrase I. The reason for the use of two tree construction methods is that in the MHC case we are dealing with 42 sequences and, therefore, an exhaustive maximum likelihood approach is impractical. Branch lengths for each topology were estimated using the maximum likelihood method [3] with the JTT stochastic model [11], assuming that the rate is discrete gamma distributed among sites with 4 rate categories.

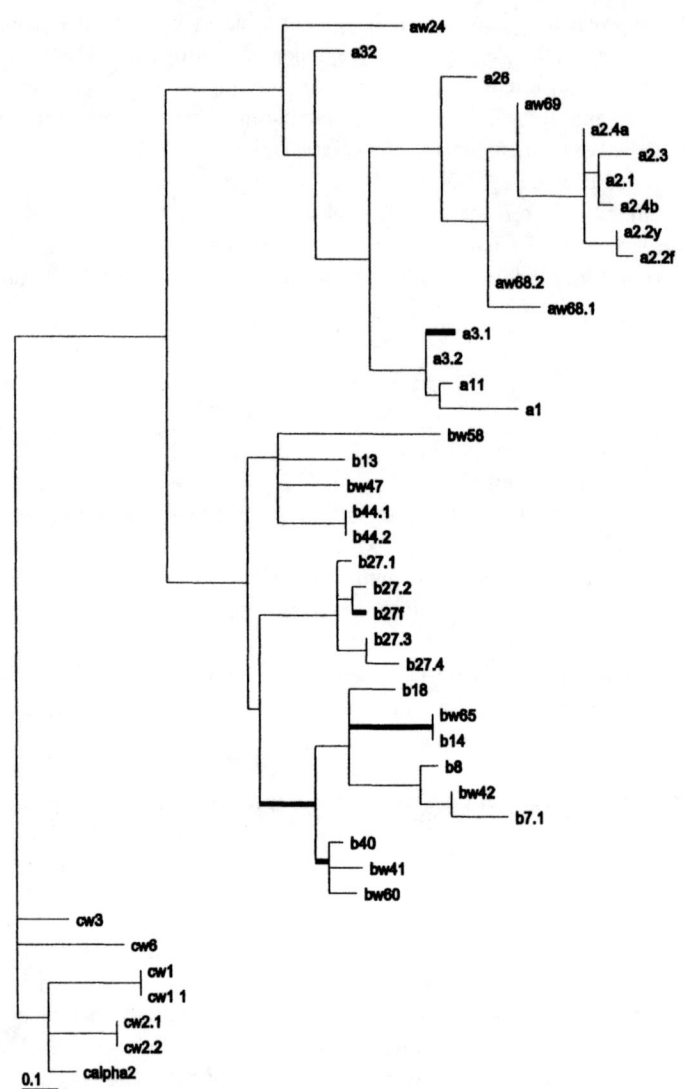

Fig. 2. A Phylogenetic Tree for MHC Class I Sequences. Species labels are as in [9]. The tree topology was estimated by using whole sequences. Branch lengths were estimated for the cleft subsequences only. Each branch was subjected to the positive selection test on the cleft subsequences. Branches in bold-face indicate *p*-value< 0.01.

4.1 MHC Class I

The primary immunological function of MHC class I glycoproteins is to bind and "present" antigenic peptides on the surface of cells, for recognition by antigen-specific T cell receptors. MHC class I glycoproteins are expressed on the surface of most cells and are recognized by CD8-positive cytotoxic T cells, an essential step for initiating the elimination of virally infected cells by T-cell mediated lysis. These molecules are very polymorphic, and it was claimed that this polymorphism is the result of positive Darwinian selection that operates on the antigen-binding cleft [9]. Using pairwise comparisons of sequences, it was shown that the proportion of nonsynonymous differences in the antigen-binding cleft that cause charge changes was significantly higher than the proportion that conserve charge. This suggests that peptide binding is at the basis of the positive selection acting on these loci [7].

Following [9] we analyzed 42 human MHC class I sequences from three allelic groups: HLA-A, -B, and -C loci. Most of these sequences are not available in Genbank, and were copied from Parham *et al.* [18]. The length of each MHC class I sequence is 274 amino acids. The binding site is a subsequence of 29 residues [18]. The phylogenetic tree for MHC class I sequences is given in Figure 2. The α parameter found for this tree was 0.24. When our clade-based test was applied to the whole tree, no indication for positive selection was found. The respective z-scores are shown in Table 1.

Table 1. A list of z-scores for each of the tests performed on the MHC class I dataset. The first row contains scores with respect to whole sequences. The second row contains results with respect to the binding cleft subsequences, with branch lengths as for the whole sequences. The third row contains results with respect to the binding cleft subsequences, with branch lengths reestimated on this part of the sequence only. Significant z-scores (p-value< 0.01) appear in bold-face.

Dataset/Distance	Grantham	Charge	Polarity	
			Grantham	Hughes *et al.* [9]
Whole	-1.30	0.01	-1.25	1.10
Cleft	**9.38**	**9.32**	**13.23**	**5.79**
Cleft & cleft-based lengths	1.08	**3.14**	**2.78**	0.01

When we applied our test to the binding site only, positive selection was found with very high confidence ($P < 0.001$). The respective z-scores are shown in Table 1. However, it might be argued, that when only the binding site part of the sequence is analyzed, the branch lengths estimated for the whole sequences are irrelevant. Since it is known that the rate of evolution in the binding site is faster relative to the rest of the sequence, the branch lengths estimated from the whole sequences are underestimated. This underestimation can result in a false positive conclusion of positive selection, since we expect in this case an excess of radical replacements. To overcome this problem, branch lengths were reestimated on the binding site part of the sequence only. Significant excess of polar and charge replacements were found also with these new estimates ($P < 0.01$). The corresponding z-scores are shown in Table 1. We note, that using the 0-1 polarity distance of [9], we found no evidence for positive selection. On the other hand, when we used Grantham's polarity indices [4], significant deviations from the random expectations were observed (see Table 1). The latter distance measure is clearly more accurate since it is not restricted to 0-1 values. We conclude that there is a significant excess in both charge and polar replacements, and not only in charge replacements, as reported in [9].

Finally, we tested specific branches in the tree to find those branches which contribute the most to the excess of charge replacements. Branches whose corresponding p-value was found to be smaller than 0.01 appear in bold-face in Figure 2. We note, that since we have no prior knowledge of which branches are expected to show excess of charge replacements, these p-values should be scaled according to the number of branches tested. Nevertheless, these high scoring branches lie all in the subtrees corresponding to the A and B alleles, matching the findings of Hughes et al. who report positive selection for these alleles only [9].

4.2 Carbonic Anhydrase I

This dataset comprises of 6 sequences of the carbonic anhydrase I house-keeping gene, for which there is no evidence of positive selection [24]. The carbonic anhydrase I sequences were the same as in [24], except that amino-acid sequences were used instead of nucleotide sequences. Sequence accession numbers are: JN0835 (*Pan troglodytes*), JN0836 (*Gorilla gorilla*), P00915 (*Homo sapiens*), P35217 (*Macaca nemestrina*), P48282 (*Ovis aries*) and P13634 (*Mus musculus*). The maximum likelihood estimate of the α parameter for this dataset was 0.52.

When analyzing carbonic anhydrase I sequences, no evidence for positive selection was found. This was true, irrespective of the distance measures we used: Grantham (z-score= 0.01), Grantham's polarity (z-score=-1.04), Hughes et al. polarity (z-score= -0.49), and charge (z-score= -1.73).

5 Discussion

Natural selection may act to favor amino acid replacements that change certain properties of amino acids [7]. Here we propose a method to test for such selection. Our method takes into account the stochastic model of amino-acid replacements, among site rate variation and the phylogenetic relationship among the sequences under study. The method is based on identifying large deviations of the mean observed chemical distance between two proteins from the expected distance. Our test can be applied to a specific branch of a phylogenetic tree, to a clade in the tree or, alternatively, over all branches of the phylogenetic tree. The calculation of the mean observed chemical distance is based on a novel procedure for averaging the chemical distance over all possible ancestral sequence reconstructions weighted by their likelihood. This results in an unbiased estimate of the chemical distance along a branch of a phylogenetic tree. The underlying distribution of this random variable is calculated using the JTT model, taking into account among site rate variation. We give a linear time algorithm to perform this test for all branches and subtrees of a given phylogenetic tree.

Two variants of the test are presented: The first is a statistical test of a single branch in a phylogenetic tree. Positive selection along a tree lineage can be the result of a specific adaptation of one taxon to some special environment. In this case, the branch in question is known a priori, and the branch-specific test should be used. Alternatively, if the selection constraints are continuous, as for example, the selection that promotes diversity among alleles of the MHC class I, the test should be applied to all the sequences under the assumed selection pressure - a clade-based test.

We validated our method on two datasets: Carbonic anhydrase I sequences served as a negative control, and the cleft of MHC class I sequences as a positive control. MHC class I sequences were previously shown to be under positive selection pressure, acting to favor amino-acid replacements that are radical with respect to charge.

There are, however, some limitations to our method. The method relies heavily on an assumed stochastic model of evolution. If this model underestimates branch lengths, one might get false positive results. It is for this reason that it is important to estimate branch lengths under realistic models, taking into account among site rate variation. Furthermore, if the test is applied to specific parts of the protein, such as an alpha helix, a replacement matrix that is specific for this part might be preferable over the more general JTT model used in this study (see [26]). One might claim that if excess of, say, polar replacements is found, it should not be interpreted as indicative of positive selection, but rather, as an indication that a more sequence-specific amino-acid replacement model is required. In MHC class I glycoproteins, however, other lines of evidence [9, 24] suggest positive Darwinian selection.

In the future, we plan to make the test more robust by accommodating uncertainties in branch lengths and topology. This can be achieved by Markov-Chain Monte-Carlo methods [6]. The sensitivity of our test to different assumptions regarding the stochastic process and the phylogenetic tree will be better understood when more datasets are analyzed.

Acknowledgments

We thank Hirohisa Kishino and Yoav Benjamini for their suggestions regarding the statistical analysis. The first author was supported by a JSPS fellowship. The second author was supported by an Eshkol fellowship from the Ministry of Science, Israel. The fourth author was supported in part by the Israel Science Foundation (grant number 565/99). This study was also supported by the Magnet Da'at Consortium of the Israel Ministry of Industry and Trade and a grant from Tel Aviv Univeristy (689/96).

References

1. J. Adachi and M. Hasegawa. Molphy: programs for molecular phylogenetics based on maximum likelihood, version 2.3. Technical report, Institute of Statistical Mathematics, Tokyo, Japan, 1996.
2. T. Endo, K. Ikeo, and T. Gojobori. Large-scale search for genes on which positive selection may operate. *Mol. Biol. Evol.*, 13:685–690, 1996.

3. J. Felsenstein. Evolutionary trees from DNA sequences: a maximum likelihood approach. *J. Mol. Evol.*, 17:368–376, 1981.
4. R. Grantham. Amino acid difference formula to help explain protein evolution. *Science*, 185:862–864, 1974.
5. K.M. Halanych and T.J. Robinson. Multiple substitutions affect the phylogenetic utility of cytochrome b and 12 rDNA data: Examining a rapid radiation in Leporid (Lagomorpha) evolution. *J. Mol. Evol.*, 48(3):369–379, 1999.
6. J.P. Huelsenbeck, B. Rannala, and J.P. Masly. Accommodating phylogenetic uncertainty in evolutionary studies. *Science*, 288:2349–2350, 2000.
7. A.L. Hughes. *Adaptive Evolution of Genes and Genomes*. Oxford University Press, New-York, 1999.
8. A.L. Hughes and M.K. Hughes. Adaptive evolution in the rat olfactory receptor gene family. *J. Mol. Evol.*, 36:249–254, 1993.
9. A.L. Hughes, T. Ota, and M. Nei. Positive Darwinian selection promotes charge profile diversity in the antigen-binding cleft of class I major-histocompatibility-complex molecules. *Mol. Biol. Evol*, 7:515–524, 1990.
10. A.L. Hughes and M. Yeager. Natural selection at major histocompatibility complex loci of vertebrates. *Annu. Rev. Genet.*, 32:415–435, 1998.
11. D.T. Jones, W.R. Taylor, and J.M. Thornton. The rapid generation of mutation data matrices from protein sequences. *Comput. Appl. Biosci.*, 8:275–282, 1992.
12. M. Kimura. *The neutral theory of molecular evolution*. Cambridge university press, Cambridge, 1983.
13. H. Kishino, T. Miyata, and M. Hasegawa. Maximum likelihood inference of protein phylogeny and the origin of chloroplasts. *J. Mol. Evol.*, pages 151–160, 1990.
14. Y.H. Lee, T. Ota, and V.D. Vacquier. Positive selection is a general phenomenon in the evolution of abalone sperm lysin. *Mol. Biol. Evol.*, 12:231–238, 1995.
15. P. Lio and N. Goldman. Models of molecular evolution and phylogeny. *Genome research*, 8:1233–1244, 1998.
16. T. Miyata, S. Miyazawa, and T. Yashunaga. Two types of amino acid substitutions in protein evolution. *J. Mol. Evol.*, 12:219–236, 1979.
17. M. Nei and T. Gojobori. Simple methods for estimating the numbers of synonymous and nonsynonymous nucleotide substitutions. *Mol. Biol. Evol.*, 3:418–426, 1986.
18. P. Parham, C.E. Lomen, D.A. Lawlor, J.P. Ways, N. Holmes, H.L. Coppin, R.D. Salter, A.M. Won, and P.D. Ennis. Nature of polymorphism in HLA-A, -B, and -C molecules. *Proc. Natl. Acad. Sci. USA*, 85:4005–4009, 1998.
19. T. Pupko and I. Pe'er. Maximum likelihood reconstruction of ancestral amino-acid sequences. In S. Miyano, R. Shamir, and T. Takagi, editors, *Currents in Compuational Molecular Biology*, pages 184–185. Universal Academy Press, 2000.
20. T. Pupko, I. Pe'er, R. Shamir, and D. Graur. A fast algorithm for joint reconstruction of ancestral amino acid sequences. *Mol. Biol. Evol.*, 17(6):890–896, 2000.
21. N. Saitou and M. Nei. The neighbor-joining method: a new method for reconstructing phylogenetic trees. *Mol. Biol. Evol*, 4:406–425, 1987.
22. S.A. Seibert, C.Y. Howell, M.K. Hughes, and A.L. Hughes. Natural selection on the gag, pol, and env genes of human immunodeficiency virus 1 (HIV-1). *Mol. Biol. Evol.*, 12:803–813, 1995.
23. C.B. Stewart and A.C. Wilson. Sequence convergence and functional adaptation of stomach lysozymes from foregut fermenters. *Cold Spring Harbor Symp. Quant. Biol.*, 52:891–899, 1987.
24. W.J. Swanson, Z. Yang, M.F. Wolfner, and C.F. Aquadro. Positive Darwinian selection drives the evolution of several female reproductive proteins in mammals. *Proc. Natl. Acad. Sci. USA*, 98:2509–2514, 2001.

25. D.L. Swofford, G.J. Olsen, P.J. Waddell, and D.M. Hillis. Phylogenetic inference. In D.M. Hillis, C. Moritz, and B.K. Mable, editors, *Molecular systematics, 2nd Ed.*, pages 407–514. Sinauer Associates, Sunderland, MA, 1995.

26. J.L. Thorne, N. Goldman, and D.T. Jones. Combining protein evolution and secondary structure. *Mol. Biol. Evol.*, 13:666–673, 1996.

27. H.C. Wang, J. Dopazo, and J.M. Carazo. Self-organizing tree growing network for classifying amino acids. *Bioinformatics*, 14:376–377, 1998.

28. Z. Yang. Maximum-likelihood estimation of phylogeny from DNA sequences when substitution rates differ over sites. *Mol. Biol. Evol.*, 10:1396–1401, 1993.

29. Z. Yang. Maximum likelihood phylogenetic estimation from DNA sequences with variable rates over sites: approximate methods. *J. Mol. Evol.*, 39:306–314, 1994.

30. Z. Yang. Maximum likelihood estimation on large phylogenies and analysis of adaptive evolution in human influenza virus A. *J. Mol. Evol.*, 51:423–432, 2000.

31. Z. Yang, S. Kumar, and M. Nei. A new method of inference of ancestral nucleotide and amino acid sequences. *Genetics*, 141:1641–1650, 1995.

32. J. Zhang and S. Kumar. Detection of convergent and parallel evolution at the amino acid sequence level. *Mol. Biol. Evol.*, 14:527–536, 1997.

33. J. Zhang, S. Kumar, and M. Nei. Small-sample tests of episodic adaptive evolution: a case study of primate lysozymes. *Mol. Biol. Evol.*, 14:1335–1338, 1997.

34. J. Zhang, H.F. Rosenberg, and M. Nei. Positive Darwinian selection after gene duplication in primate ribonuclease genes. *Proc. Natl. Acad. Sci. USA*, 95:3708–3713, 1998.

Finding a Maximum Compatible Tree for a Bounded Number of Trees with Bounded Degree Is Solvable in Polynomial Time

Ganeshkumar Ganapathysaravanabavan and Tandy Warnow

Department of Computer Sciences
University of Texas, Austin, TX 78712
{gsgk,tandy}@cs.utexas.edu

Abstract. In this paper, we consider the problem of computing a maximum compatible tree for k rooted trees when the maximum degree of all trees is bounded. We show that the problem can be solved in $O(2^{2kd}n^k)$ time, where n is the number of taxa in each tree, and every node in every tree has at most d children. Hence, a maximum compatible tree for k unrooted trees can be found in in $O(2^{2kd}n^{k+1})$ time.

1 Introduction

A "phylogeny" (or evolutionary tree) represents the evolutionary history of a set of species S by a rooted (typically binary) tree in which the leaves are labelled with elements from S, and unlabelled internal nodes.

For a variety of reasons, a typical outcome of a phylogenetic analysis of a dataset can consist of many different unrooted trees, and each tree represents an equally believable estimate of the true tree. Making sense of the set of these trees is then a challenging prospect.

There are two basic approaches that have been used for this problem. The first approach (and the most popular) represents the set of trees by a single tree on the full dataset (i.e. the "consensus tree"). Consensus tree techniques such as "strict consensus" and "majority consensus" are the most popular, and have the advantage that they are polynomial time. However, consensus tree methods have the disadvantage that they tend to create consensus trees that lack resolution (and this lack of resolution can be significant), meaning that the trees can contain high degree nodes. An alternate approach seeks a subset of the taxa on which all the trees agree (the "maximum agreement subset" (MAST) approach), which means that when restricted to this subset of the taxa, the trees are identical. By contrast with the consensus tree approach, the agreement subset approach is computationally intensive (unless at least one of the trees has low degree); furthermore, the size of the maximum agreement subset can be very small [7].

In this paper we consider a different approach to this problem, in which we seek the largest subset of taxa on which all the trees are "compatible" (meaning that when restricted to this subset of taxa, the trees share a common refinement). The problem is suited for the case where the set of trees contains unresolved trees (such as can happen

O. Gascuel and B.M.E. Moret (Eds.): WABI 2001, LNCS 2149, pp. 156–163, 2001.

when returning a set of consensus trees, one for each phylogenetic island obtained during a search for the maximum parsimony or maximum likelihood tree). In such cases, it may return a larger subset of taxa than the maximum agreement subset approach, as illustrated in Figure 1. This "maximum compatible set" (MCS) problem is NP-hard for

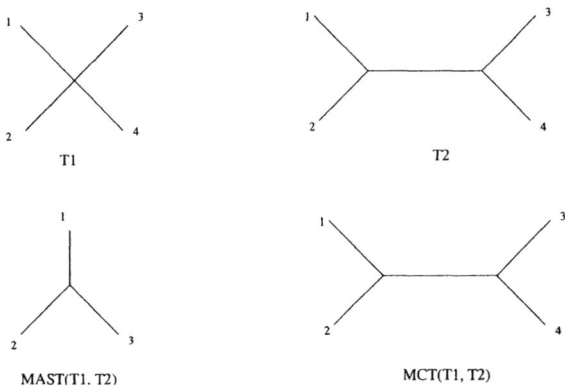

Fig. 1. The MCT of T1 and T2 has more leaves than the MAST.

6 or more trees [5]. We will denote by MCT the tree constructed on the MCS, and call it the maximum compatible tree. Occasionally, when the context is clear, we will use MAST to also denote the maximum agreement subtree.

Our result for the MCS problem is an algorithm for the two-tree MCS problem which runs in time $O(2^{4d}n^3)$ time, where $n = |S|$ and d is the maximum degree of the two unrooted trees. (The algorithm we present is a dynamic programming algorithm and is an extension of the earlier dynamic programming algorithm for the two-tree MAST problem in [6].) Thus, we show that the two-tree MCS problem is fixed-parameter tractable. We extend this algorithm to the k-tree MCS problem in which all trees have bounded degree, obtaining a $O(2^{2kd}n^{k+1})$ algorithm.

The organization of the paper is as follows. In Section 2 we give some basic definitions (we will introduce and explain other terminology as needed). We then present the dynamic programming algorithm for two-tree MAST, and discuss the challenges in extending the algorithm to the two-tree MCS problem. In Section 3 we present the dynamic programming algorithm for the two-tree MCS problem, the proof of correctness, and running time analysis. We then show how we extend this algorithm to the k tree case. In Section 4 we discuss further work in the area.

2   Computing a Maximum Agreement Subtree of Two Trees

In this section, we first present some definitions and then we describe the dynamic programming algorithm for computing the maximum agreement subtree of two trees, as given in [6]. The algorithm actually computes the cardinality of the maximum agreement subset but can be easily extended to give the maximum agreement subset (or equivalently the maximum agreement subtree).

Definition 1. *Given a leaf-labelled tree T on a set S, the restriction of T to a set $X \subseteq S$ is obtained by removing all leaves in $(S - X)$ and then suppressing all internal nodes of degree two. This restriction will be denoted by $T|X$.*

Definition 2. *Given a set $\Sigma = \{T_1, T_2, \ldots, T_k\}$ of leaf-labelled trees on the set S, a maximum agreement subset for the set of trees Σ is a set X of maximum cardinality such that the trees $T_{i|X}$ are all isomorphic. (The isomorphism must map leaves with the same label to each other).*

Definition 3. *Given a set $\Sigma = \{T_1, T_2, \ldots, T_k\}$ of leaf-labelled trees on the set S, a maximum compatible subset for the set of trees Σ is a set X of maximum cardinality such that the trees $T_i|X$ all share a common refinement.*

We will first assume that the two trees T and T' are both rooted (we will later discuss how to extend the algorithm for unrooted trees). The trees can have any degree, but both are leaf-labelled by the same set S of taxa.

Let v be a node in T, and denote by T_v the subtree of T rooted at v. Similarly denote by T'_w the subtree of T' rooted at a node w in T'. Denote by $\text{MAST}(T_v, T'_w)$ the number of leaves in a maximum agreement subset of T_v and T'_w. The dynamic programming algorithm operates by computing $\text{MAST}(T_v, T'_w)$ for all pairs of nodes (v, w) in $V(T) \times V(T')$, "bottom-up".

We now describe the basic idea of the dynamic programming algorithm. First, the value $\text{MAST}(T_v, T'_w)$ is easy to compute when either v or w are leaves. Now consider the computation of $\text{MAST}(T_v, T'_w)$ where both v and w are not leaves, and let X be a maximum agreement subset of T_v and T'_w. The least common ancestor of X in T_v may be v, or it may be a node below v. Similarly, the least common ancestor of X in T'_w may be w or it may be a node below w. We have four cases to consider:

1. If the least common ancestor of X is below v in T and similarly below w in T', then $|X| = \text{MAST}(v_i, w_j)$ for some pair (v_i, w_j) of nodes where v_i is a child of v and w_j is a child of w.
2. If the least common ancestor of X is below v in T but equal to w in T', then $|X| = \text{MAST}(v_i, w)$ for some child v_i of v.
3. If the least common ancestor of X is below w in T' but equal to v in T, then $|X| = \text{MAST}(v, w_j)$ for some child w_j of w.
4. If the least common ancestor of X is v in T and w in T', then X is the disjoint union of $X_1, X_2, \ldots, X_p$ where each X_i is a maximum agreement subset for some pair of subtrees T_{v_a} and T'_{w_b} (in which v_a is a child of v and w_b is a child of w). Furthermore, for each i and j, the pairs of subtrees associated to X_i and X_j are disjoint. Hence X is the maximum value of a matching in the weighted complete bipartite graph $G(A, B, E)$ in which $A = \{v_1, v_2, \ldots, v_j\}$ (for the children of v), $B = \{w_1, w_2, \ldots, w_p\}$ (for the children of w), and the weight of edge (v_i, w_j) is $\text{MAST}(v_i, w_j)$.

This discussion suggests a straightforward dynamic programming algorithm which involves computing $O(n^2)$ subproblems, each of which involves computing the maximum of a number of $O(d)$ values (where d is the maximum degree of any node in both T and T'). Each of these values in turn is easy to compute, though the maximum weighted bipartite matching of an $O(d)$ vertex graph takes $O(d^{2.5} \log d)$ time [4]. The running time analysis of the MAST algorithm given in [6] shows it is $O(n^{4.5} \log n)$ if d is not bounded, but $O(n^2)$ if d is bounded.

3   Algorithms for the MCS Problem

3.1   Relation between MCS and MAST

We begin by observing the following:

Lemma 1. *Let T_1 and T_2 be two unrooted leaf-labelled trees. Let X be an MCS of T and T'. Then there exists a binary tree T_1' refining T_1 and a binary tree T_2' refining T_2 such that X is a MAST of T_1' and T_2'.*

Proof. Since X is a compatible subset of taxa for the pair of trees T_1 and T_2, there is a common refinement T^* of $T_1|X$ and $T_2|X$. Hence we can refine T_1, obtaining T_1', and refine T_2, obtaining T_2', so that T_1' restricted to X yields T^*, and similarly T_2' restricted to X also yields T^*. Then X is an agreement subset of T_1' and T_2'.

This observation is illustrated in Figures 2 and 3. This observation suggests an obvious

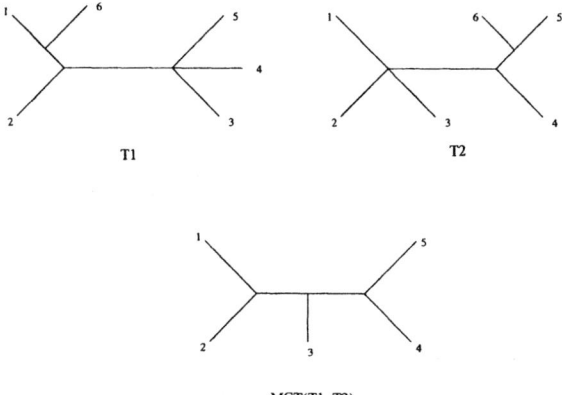

MCT(T1. T2)

Fig. 2. The MCT of Trees T1 and T2.

algorithm for computing an MCS of two trees: for each way of refining the two trees into a binary tree, compute a MAST. However, this algorithm is computationally expensive, since the number of binary refinements of an n leaf tree with maximum degree d can be $O(4^{nd})$. Hence this brute force algorithm will not be acceptably fast.

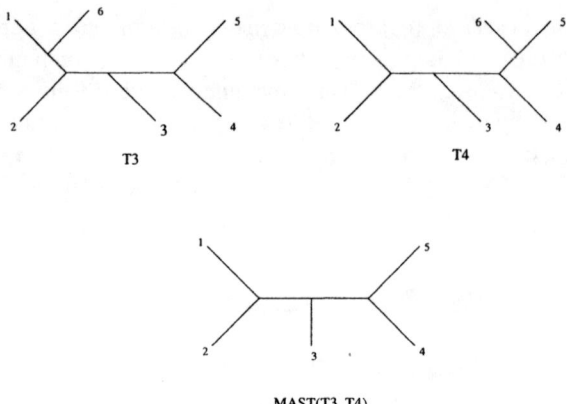

Fig. 3. The MAST of Trees T3 and T4.

3.2 Computing an MCS of Two Rooted Trees

We now describe the dynamic programming algorithm for the maximum compatible set (MCS) of two rooted trees, both with degree bounded by d (by this we mean that every node has at most d children). As for the MAST problem, here too the algorithm computes the cardinality of an MCS, but can be easily extended to compute an MCS itself. This algorithm can easily be extended to produce a dynamic programming algorithm for computing an MCS of two unrooted trees, by computing an MCS of each of the n pairs of rooted trees (obtained by rooting the unrooted trees at each of the n leaves). Furthermore, we also show how to extend the algorithm to handle k rooted or unrooted trees.

The basic set of problems we need to compute must include the computation of an MCS of subtrees T_v and T'_w, for every pair of nodes v and w. (T_v denotes the subtree of T rooted at v, and similarly T'_w denotes the subtree of T' rooted at w.) We will also need to include the computation of MCS's of other pairs of trees, but begin our discussion with these MCS calculations.

Let T and T' be two rooted trees, and let v and w denote nodes in T and T' respectively. Let the children of v be $v_1, v_2, \ldots, v_p$ and the children of w be $w_1, w_2, \ldots, w_q$. Let X be the set of leaves involved in an MCS T^* of T_v and T'_w. Note that $T|X$ and $T'|X$ will only include those children of v and w which have some element(s) of X below them. Let A be the children of v included in $T|X$ and B be the children of w included in $T'|X$. (Note that X defines the sets A and B.)

Note also that any MCS of T and T' actually defines an agreement subset of some binary refinement of T and some binary refinement of T' (Lemma 1). Hence, T^* defines a binary refinement at the node v if $|A| > 1$, and a binary refinement at the node w if $|B| > 1$. In these cases, T^* defines a partition of the nodes in A into two sets, and a partition of the nodes in B into two sets.

There are four cases to consider:

1. $|A| = |B| = 1$, i.e $A = \{v_i\}$ and $B = \{w_j\}$ for some i and j. In this case, any MCS of T_v and T'_w is an MCS of T_{v_i} and T'_{w_j}.
2. $|A| = 1$ and $|B| > 1$, i.e, any MCS of T_v and T'_w is an MCS of T_{v_i} and T'_w for some i.
3. $|A| > 1$ and $|B| = 1$, i.e, any MCS of T_v and T'_w is an MCS of T_v and T'_{w_j} for some j.
4. $|A| > 1$ and $|B| > 1$.

The analysis of the fourth case is somewhat complicated, and is the reason that we need additional subproblems. Recall that T^* defines a bipartition of A into $(A', A - A')$ and B into $(B', B - B')$. Further, recall that T^* is a binary tree with two subtrees off the root; we call these subtrees T_1 and T_2. It can then be observed that T_1 is an MCS of the subtree of T_v obtained by restricting to the nodes below $A - A'$ and the subtree of T'_w obtained by restricting to the nodes below $B - B'$. Similarly, T_2 is an MCS of the subtree of T_v obtained by restricting to the nodes below A' and the subtree of T'_w obtained by restricting to the nodes below B'. Hence we need to define additional subproblems as follows. For each $A' \subset A$ define the tree $T(v, A')$ to be subtree of T_v obtained by deleting all the children of v (and their descendents) not in A'. Similarly define the tree $T'(w, B')$ to be the subtree of T'_w obtained by deleting all the children of w (and their descendents) not in B'. The construction of tree $T(v, A')$ is illustrated in Figure 4. Now define $MCS(v, A', w, B')$ to be the size of an MCS of $T(v, A')$ and

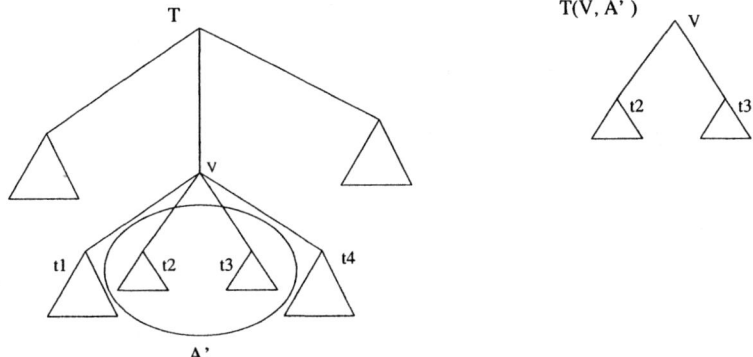

Fig. 4. The Tree T, the Set A' and the Tree T(v, A').

$T'(w, B')$[1]. From the above discussion it follows that:

$MCS(v, A', w, B')$ is the maximum of:

- $max\{MCS(v, A', w_j, Children(w_j)) : w_j$ a child of $w\}$,
- $max\{MCS(v_i, Children(v_i), w, B') : v_i$ a child of $v\}$,

---

[1] The size of a tree is taken to mean the number of leaves in the tree.

- $max\{MCS(v, A'', w, B'') + MCS(v, A' - A'', w, B' - B'') : A''$ and B'' are non-empty proper subsets of A' and B', respectively. $\}$.

The computation of these subproblems follows the obvious partial ordering, in which $MCS(v, A, w, B)$ must follow $MCS(v', A', w', B')$ if both of the following conditions holds:

- v lies above v' or $[v = v'$ and $A' \subseteq A]$.
- w lies above w' or $[w = w'$ and $B' \subseteq B]$.

The base cases, in which v is a leaf or w is a leaf, are easy to compute, and equal 1 or 0. For example, if v is a leaf then necessarily $A = \emptyset$, and so $MCS(v, \emptyset, w, B) = 1$ if $v \in T'(w, B)$, and 0 otherwise.

Running time analysis There are $O(2^d n)$ trees $T(v, A)$, and hence $O(2^{2d} n^2)$ subproblems. The computation of $MCS(v, A, w, B)$ involves computing the maximum of $2d + 2^{2d}$ values, and hence takes $O(2^{2d})$ time. Hence the running time is $O(2^{4d} n^2)$.

3.3   Algorithm for the MCS Problem of k Rooted Trees with Bounded Degree

We now show how to extend the analysis to k rooted trees. In this case, the subproblems are $2k$-tuples of the form $MCS(v_1, A_1, v_2, A_2, \ldots, v_k, A_k)$ where v_i is a node in T_i and $A_i \subset Children(v_i)$. Hence there are $O(2^{kd} n^k)$ subproblems. Computing each subproblem involves taking the maximum of $O(kd + 2^{kd})$ values. Hence the running time for the algorithm is $O(2^{2kd} n^k)$ time.

3.4   Extension to Unrooted Trees

For each of the algorithms described, extending the algorithm to handle unrooted trees involves rooting each of the trees at each of the n leaves (for each leaf all the trees are rooted at that leaf), and then finding the best rooting. This is based on the observation that given a leaf l, the size of an MCS of the trees rooted at l is the maximum size of a compatible set for the unrooted trees that includes l (while rooting the trees at the leaf l, l itself must be excluded from the trees. Hence the size of a maximum compatible set for the unrooted trees that includes l will be actually one more than the size of an MCS of the trees rooted at l). Since there are n leaves, this multiplies the running time by n.

Theorem 1. *We can compute an MCS of k unrooted trees in which each tree has degree at most $d + 1$ in $O(2^{2kd} n^{k+1})$ time.*

4   Future Work

To conclude we point out that many questions about the MCS problem remain unsolved. We know that MCS is NP-hard for 6 trees with unbounded degree, but we do not know the minimum number of trees for which MCS becomes hard. In particular, we do not know if MCS is NP-hard or polynomial for two trees. It also remains to be seen if there are any approximation algorithms for the problem, or exact algorithms when only some of the trees have bounded degree.

Acknowledgments

Tandy Warnow's work was supported in part by the David and Lucile Packard Foundation.

References

1. A. Amir and D. Kesselman. Maximum agreement subtree in a set of evolutionary trees—metrics and efficient algorithms. *Proceedings of the 35th IEEE FOCS, Santa Fe, 1994.*
2. M. Farach Colton, T.M. Przytycka, and M.Thorup, On the Agreement of Many Trees. *Information Processing Letters* 55 (1995), 297–301.
3. C.R. Finden and A.D. Gordon. Obtaining common pruned trees, *Journal of Classification* 2, 255–276 (1985).
4. H.N. Gabow and R.E. Tarjan. Faster scaling algorithms for network problems. *SIAM J. Comput.* 18 (5), 1013–1036 (1989).
5. A.M. Hamel and M.A. Steel. Finding a maximum compatible tree is NP-hard for sequences and trees. *Research Report No. 114, Department of Mathematics and Statistics, University of Canterbury*, Christchurch, New Zealand, 1994.
6. M. Steel and T. Warnow. Kaikoura tree theorems: computing the maximum agreement subtree. *Information Processing Letters* 48, 77–82 (1993).
7. D. Swofford, personal communication.

Experiments in Computing Sequences of Reversals

Anne Bergeron and François Strasbourg

LACIM, Université du Québec à Montréal
C.P. 8888 Succ. Centre-Ville, Montréal, Québec, Canada, H3C 3P8
bergeron.anne@uqam.ca

Abstract. This paper discusses a bit-vector implementation of an algorithm that computes an optimal sequence of reversals that sorts a signed permutation. The main characteristics of the implementation are its simplicity, both in terms of data structures and operations, and the fact that it exploits the parallelism of bitwise logical operations.

1 Introduction

For several good reasons, the problem of sorting signed permutations has received a lot of attention in recent years. One of the attractive features of this problem is its simple and precise formulation: Given a permutation π of integers between 1 and n, some of which may have a minus sign,

$$\pi = (\pi_1 \ \pi_2 \ldots \pi_n)$$

find $d(\pi)$, the minimum number of *reversals* that transform π into the identity permutation:

$$(+1 \ +2 \ \ldots +n).$$

The reversal operation reverses the order of a block of consecutive elements in π, while changing their signs.

Another good reason to study this problem is comparative genomics. The genome of a species can be thought of as a set of ordered sequences of genes—the ordering devices being the chromosomes—, each gene having an orientation given by its location on the DNA double strand. Different species often share similar genes that were inherited from common ancestors. However, these genes have been shuffled by mutations that modified the content of chromosomes, the order of genes within a particular chromosome, and/or the orientation of a gene. Comparing two sets of similar genes appearing along a chromosome in two different species yields two (signed) permutations. It is widely accepted that the reversal distance between these two permutations, that is, the minimum number of reversals that transform one into the other, faithfully reflects the evolutionary distance between the two species.

The last, and probably the best feature of the sorting problem from an algorithmic point of view, is the dramatic increase in efficiency its solution exhibited in a few years. From a problem of unknown complexity in the early nineties, a time when approximate solutions were plenty [6], polynomial solutions of constantly decreasing degree were

O. Gascuel and B.M.E. Moret (Eds.): WABI 2001, LNCS 2149, pp. 164–174, 2001.

successively found: $\mathcal{O}(n^4)$ [5], $\mathcal{O}(n^2\alpha(n))$ [3], $\mathcal{O}(n^2)$ [7], and finally $\mathcal{O}(n)$ [1], for the distance computation alone. A computer scientist's delight, knowing that the unsigned version was proved to be NP-hard [4].

This high level of scientific activity inevitably generated undesirable side-effects. Complex constructions, useful in the initial investigations, were later proven unnecessary. Terminology is not yet standard. For example, the *overlap graphs* of [7] are different from the ones in [1]. Most importantly, the complexity measures tend to mix two different problems, as pointed out in [1]: the computation of the *number* of necessary reversals, and the reconstruction of one possible sequence of reversals that realizes this number.

The first problem has an efficient and simple linear solution [1] that can hardly be improved on. In this paper, we address the second problem with elementary tools, further developing the ideas of [2]. We show that, with any problem of biologically relevant size, it is possible to implement efficient algorithms using the simplest data structures and operations: In this case, bit-vectors and standard logical and arithmetic operations.

The next section presents a brief introduction to the current theory. It is followed by a discussion of the implementation, and results on simulated biological data.

2 Sorting by Reversals

The basic construction used for computing $d(\pi)$, the *reversal distance* of a signed permutation π, is the *cycle graph* [1] associated with π. Each positive element x in the permutation π is replaced by the sequence $2x - 1\ 2x$, and each negative element $-x$ by the sequence $2x\ 2x - 1$. Integers 0 and $2n + 1$ are added as first and last elements. For example, the permutation

$$\pi = (\ -2\ -1\ +4\ +3\ +5\ +8\ +7\ +6\ )$$

becomes

$$\pi' = (\ 0\ 4\ 3\ 2\ 1\ 7\ 8\ 5\ 6\ 9\ 10\ 15\ 16\ 13\ 14\ 11\ 12\ 17\ ).$$

The elements of π' are the vertices of the cycle graph. Edges join every other pair of consecutive elements of π', starting with 0, and every other pair of consecutive integers, starting with $(0, 1)$. The first group of edges, the horizontal ones, is often referred to as *black edges*, and the second as *arcs* or *gray edges*.

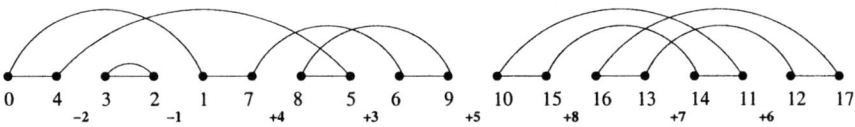

Fig. 1. The Cycle Graph of $\pi = (\ -2\ -1\ +4\ +3\ +5\ +8\ +7\ +6\ )$.

---

[1] A cycle graph in which all cycles of length two have been removed is often called a *breakpoint graph*.

Every connected component of the cycle graph is a cycle, which is a consequence of the fact that each vertex has exactly two incident edges. The graph of Figure 1 has 4 cycles.

The *support* of an arc is the interval of elements of π' between, and including, its endpoints. Two arcs *overlap* if their support intersect, without proper containment. An arc is *oriented* if its support contains an odd number of elements, otherwise it is *unoriented*. Note that an arc is oriented if and only if its endpoints belong to elements with different signs in the original permutation.

The *arc overlap graph* is the graph whose vertices are arcs of the cycle graph, and whose edges join overlapping arcs[2]. The overlap graph corresponding to the cycle graph of Figure 1 is illustrated in Figure 2, in which each vertex is labeled by an arc $(2i, 2i+1)$. Oriented vertices—those for which the corresponding arc is oriented—are marked by black dots. Orientation extends to connected component in the sense that a connected component with at least one oriented vertex is oriented. It is easy to show that a vertex is oriented if and only if its degree is odd.

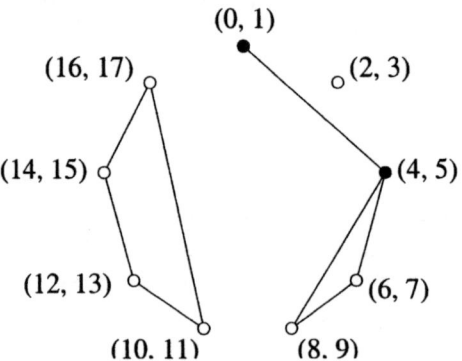

Fig. 2. The Arc Overlap Graph of p.

A *hurdle* is an unoriented component of the arc overlapping graph whose vertices are consecutive on the circle, except for isolated points [7], [2]. The graph of Figure 2 has one hurdle spanning the vertices $(10, 11)$ to $(16, 17)$.

Hannenhalli and Pevzner have shown [5] that the reversal distance of an unsigned permutation of length n is given by the formula:

$$d(\pi) = n + 1 - c + h + f$$

where c is the number of cycles in the cycle graph, h is the number of hurdles, and f is a correction factor equals to 1 when there are at least 3 hurdles satisfying a particular condition. With the above formula, it is easy to compute the reversal distance of the permutation of Figure 1 and Figure 2 as $9 - 4 + 1 = 6$.

---

[2] The *cycle overlap graph* is obtained in a similar way, by defining two cycles to overlap if they have at least two overlapping arcs.

Computing the distance $d(\pi)$ can be done in linear time [1]. However, this number alone gives no clue to the reconstruction of a possible sequence of reversals that realizes $d(\pi)$.

Reconstructing a possible sequence of reversals raises two different problems. The first one is how to deal with unoriented components. This problem is solved by carefully choosing a sequence of reversals that merges these components while creating at least one oriented vertex in each component [5], [3], [7]. The selection of these reversals can be done efficiently while computing $d(\pi)$.

The second step, called the *oriented sort*, requires to choose, among several candidates, a *safe* reversal, that is a reversal that decreases the reversal distance. Such a reversal always exists, but can be hard to find. For example, choosing the obvious reversal of the first two elements in the permutation:

$$(\ \underline{-2 \ -1} \ +4 \ +3 \ )$$

for which $d = 3$, yields the permutation

$$(\ +1 \ +2 \ +4 \ +3 \ )$$

that still has $d = 3$, since one hurdle was created by the reversal. On the other hand, the original permutation can be sorted by the sequence:

$$(\ \underline{-2 \ -1} \ +4 \ +3 \ )$$

$$(\ \underline{-4 \ +1} \ +2 \ +3 \ )$$

$$(\ \underline{-4 \ -3 \ -2 \ -1} \ )$$

$$(\ +1 \ +2 \ +3 \ +4 \ )$$

In general, to any oriented vertex of the overlap graph, there is an associated reversal that creates consecutive elements in the permutation, and the search for safe reversals can be restricted to reversals associated to oriented vertices. Several criteria have been proposed to select safe reversals. In [5], the selection of a safe reversal is the most expensive iteration of the algorithm; [3] reduces the complexity of the search by considering only $\mathcal{O}(\log n)$ candidates; and [7] gives a characterization of a safe reversal in terms of cliques. In [2], we showed that there is a much simpler way to identify a safe reversal:

Theorem 1. *The reversal that maximizes the number of oriented vertices is safe.*

In the following sections we discuss a bit-vector implementation of the oriented sort that uses this property.

3 Implementing the Oriented Sort with Bit-Vectors

The idea underlying the sorting algorithm is that the reversal corresponding to an oriented vertex v modifies the overlap graph as follows: it isolates v, complements the subgraph of vertices adjacent to v, and reverses their orientation. Therefore, the net change in the number of oriented vertices depends only on the orientation of vertices adjacent to v.

The *score* of a reversal associated to vertex v is defined by the difference between the number of its unoriented neighbors U_v, and the number of its oriented neighbors O_v. The score of a reversal is a "local" property of the graph, and this locality suggests the possibility of a parallel algorithm to keep the scores and to compute the effects of a reversal.

For a signed permutation of length n, we will denote by bold letters the characteristic bit-vectors of subsets of the $n + 1$ arcs $(0, 1)$ to $(2n, 2n + 1)$. We will use only three operations on these vectors: the exclusive-or operator $\oplus$; the conjunction $\wedge$; and the negation $\neg$.

3.1   The Data Structure

Given an overlap graph, we construct a bit-matrix in which each line v_i is the set of adjacent vertices to arc $(2i, 2i+1)$. For example, the bit-matrix associated to the overlap graph of permutation $(\ 0 +3 +1 +6 +5 -2 +4 +7\ )$:

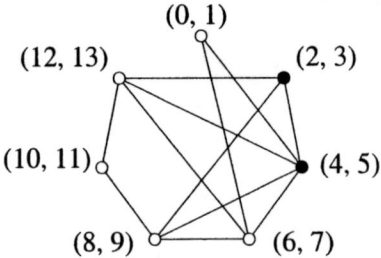

is the following:

	v_0	v_1	v_2	v_3	v_4	v_5	v_6
v_0	0	0	1	1	0	0	0
v_1	0	0	1	0	1	0	1
v_2	1	1	0	1	1	0	1
v_3	1	0	1	0	1	0	1
v_4	0	1	1	1	0	1	0
v_5	0	0	0	0	1	0	1
v_6	0	1	1	1	0	1	0
p	0	1	1	0	0	0	0
s	0	1	3	2	0	2	0

The last two lines contain, respectively, the parity, or orientation, of the vertex, and the score of the associated reversal. We will discuss efficient ways to initialize the structure and to adjust scores in Sections 3.2 and 3.3.

Given the vectors p and s, selecting the oriented reversal with maximal score is elementary. In the above example, vertex 2 would be the selected candidate.

The interesting part is how a reversal affects the structure. These effects are summarized in the following algorithm, which recalculates the bit-matrix v, the parity vector p, and the score vector s, following the reversal associated to vertex i, whose set of adjacent vertices is denoted by v_i.

$$s \leftarrow s + v_i$$
$$v_{ii} \leftarrow 1$$

For each vertex j adjacent to i

If j is oriented
$$s \leftarrow s + v_j$$
$$v_{jj} \leftarrow 1$$
$$v_j \leftarrow v_j \oplus v_i$$
$$s \leftarrow s + v_j$$

Else
$$s \leftarrow s - v_j$$
$$v_{jj} \leftarrow 1$$
$$v_j \leftarrow v_j \oplus v_i$$
$$s \leftarrow s - v_j$$

$$p \leftarrow p \oplus v_i$$

The logic behind the algorithm is the following. Since vertex i will become unoriented and isolated, each vertex adjacent to i will automatically gain a point of score. Next, if j is a vertex adjacent to i, vertices adjacent to j after the reversal are either existing vertices that were not adjacent to i, or vertices that were adjacent to i but not to j. The exceptions to this rule are i and j themselves, and this problem is solved by setting the diagonal bits to 1 before computing the direct sum.

If j is oriented, each of its former adjacent vertices will gain one point of score, since j will become unoriented, and each of its new adjacent vertices will gain one point of score. Note that a vertex that *stays* connected to w will gain a total of two points. For unoriented vertices, the gains are converted to losses.

The amount of work done to process a reversal corresponding to vertex i, in terms of vector operations, is thus proportional to the number of adjacent vertices to vertex i.

3.2   Representing the Scores

The additions and subtractions to adjust the score vector are the usual arithmetic operations performed component-wise. In order to have a truly bit-vector implementation, we represented the score vector as a $\lceil \log(n) \rceil \times n$ bit-matrix, each column containing the binary representation of a score. With this representation, component-wise addition of a bit-vector v to s can be realized with the following:

For k from 1 to $\lceil \log(n) \rceil$
$$t \leftarrow v$$
$$v \leftarrow v \wedge s_k$$
$$s_k \leftarrow t \oplus s_k$$

Subtraction is implemented in a similar way. A side benefit of this structure is that the selection of the next reversal can be also done in parallel, by "sifting" the score matrix through the parity vector. The set c of candidates contains initially all the oriented vertices. Going from the higher bit of scores to the lower, if at least one of the candidates has bit i set to 1, we eliminate all candidates for which bit i is 0.

$$
\begin{aligned}
&c \leftarrow p \\
&i \leftarrow \lceil \log(n) \rceil \\
&\text{While } i \geq 0 \text{ do} \\
&\qquad \text{While } (c \wedge s_i) = 0 \\
&\qquad\qquad i \leftarrow i - 1 \\
&\qquad \text{If } i \geq 0 \\
&\qquad\qquad c \leftarrow c \wedge s_i \\
&\qquad\qquad i \leftarrow i - 1
\end{aligned}
$$

At the end of the loop, c is the set of oriented vertices of maximal score.

3.3   Initializing the Data Structure

We saw, in Section 2, that the overlap graph of a signed permutation $\pi = (\pi_1 \ \pi_2 \dots \pi_n)$ contains $n + 1$ vertices corresponding to the arcs joining $2i$ and $2i + 1$ in the equivalent unsigned permutation. In this section, we will construct a representation of the overlap graph without explicitly referring to the unsigned permutation, thus removing one more step between the actual algorithm, and the original formulation of the problem.

The construction is based on the following simple lemma. Let I be a set of intervals with distinct endpoints in an ordered set S. Let $i = (b_i, e_i)$ and $j = (b_j, e_j)$ be intervals in I. Define the sets l_i and r_i as follows:

$$
l_i = \{j \in I \mid b_j < e_i < e_j\}
$$
$$
r_i = \{j \in I \mid b_j < b_i < e_j\}
$$

We have the following:

Lemma 1. *The set v_i of intervals that overlap i in I is given by:* $v_i = l_i \oplus r_i$.

Starting with a signed permutation $\pi = (\pi_1 \ \pi_2 \dots \pi_n)$, we first read the elements from left to right. Let a represent the set of arcs—i,e., vertices—for which exactly one endpoint has been read. Initially, a is the set $\{0\}$, corresponding to the arc $(0, 1)$. When element π_i is read, we have to process two arcs: $(2\pi_i - 2, 2\pi_i - 1)$ and $(2\pi_i, 2\pi_i + 1)$. In increasing order, if π_i is positive, and decreasing order, otherwise. Processing an arc $(2j, 2j + 1)$ is done by the following instructions:

$$
\begin{aligned}
&\text{If } a_j = 0 \\
&\quad \text{Then } a_j \leftarrow 1 \quad (\text{* First endpoint of arc } (2j, 2j + 1) \text{ *}) \\
&\quad \text{Else} \qquad\qquad (\text{* Second endpoint *}) \\
&\qquad a_j \leftarrow 0 \\
&\qquad v_j \leftarrow a \quad (\text{* } a \text{ is the set } l_j \text{ *})
\end{aligned}
$$

We then repeat the process in the reverse order, reading the permutation from right to left, initializing a to the set $\{n\}$, and changing the last instruction to $v_j \leftarrow v_j \oplus a$.

4 Analysis and Performances

The formal analysis of the algorithm of Section 3 raises interesting questions. For example, what is an elementary operation? Except for a few control statements, the only operations used by the algorithm are very efficient bit-wise logical operators on words of size w—typically 32 or 64, depending on implementation. The most expensive instructions in the main loop are additions and subtractions, such as

$$s \leftarrow s + v_j,$$

where s is a bit matrix of size $n \log(n)$, and v_j is a bit vector of size n. Such an operation requires a total of $(2n \log(n))/w$ elementary operations with the loop described in Section 3.2. Hopefully, $\log(n)$ is much smaller than w, and, in the range of biologically meaningful values, n is often a small multiple of w. In the actual implementation, the loop is controlled by the value of $\log$(maximal score) which tends to be much less than $\log(n)$. We thus have a, very generous, $\mathcal{O}(n)$ estimate for the instructions in the main loop.

The overall work done by the algorithm depends on the total number v of vertices adjacent to vertices of maximal score. We can easily bound it by n^2, noting that the number d of reversals needed to sort the permutation is bounded by n, and the degree of a vertex is also bounded by n. We thus get an $\mathcal{O}(n^3)$ estimate for the algorithm, assuming that $\log n < w$. However, experimental results suggest that v is better estimated by $\mathcal{O}(n)$, at least for values of n up to 4096, which seems largely sufficient for most biological applications.

The value of v is hard to control experimentally, but if we write the quantity governing the running time as $d v_m n$, in which v_m is the mean number of adjacent vertices, then both d and n can be fixed in an independent way.

4.1 Experimental Setup

In the following experiments, we generated sets of simulated genomes of various lengths n, by applying k random reversals to the identity permutation. The parameter k is often called the *evolution rate*. The—almost nonexistent—permutations that contained an oriented component were rejected from the compilations.

The code is written in C, and is quite "bare". For example, loops controlled by bit-vectors are implemented with right shifts and tests of bit values. The tests were conducted on an 800MHz Pentium 3.

4.2 Speed and Range

The first observation is that the implementation is very fast for typical biological problems. With $n = 128$, and $k = 32$, we can compute 10,000 sequences of reversals in 11s. On the other end of the spectrum, we applied the algorithm to values up to $n = 4096$. In this case, the computation of reversal sequences for $k = 512$ took a mean time of 3.86s for 100 random permutations.

4.3   Effects of the Variation of n

In order to study the effect of the variation of n on the running time, we choose four different evolution rates $k = 32, 128, 256,$ and 512. For each value of k, we generated sets of 100 permutations of length ranging from 256 to 4,096. Figure 3 displays the results for the mean sorting time for $k = 256$ and $k = 512$—values for smaller ks were too low for a significant analysis.

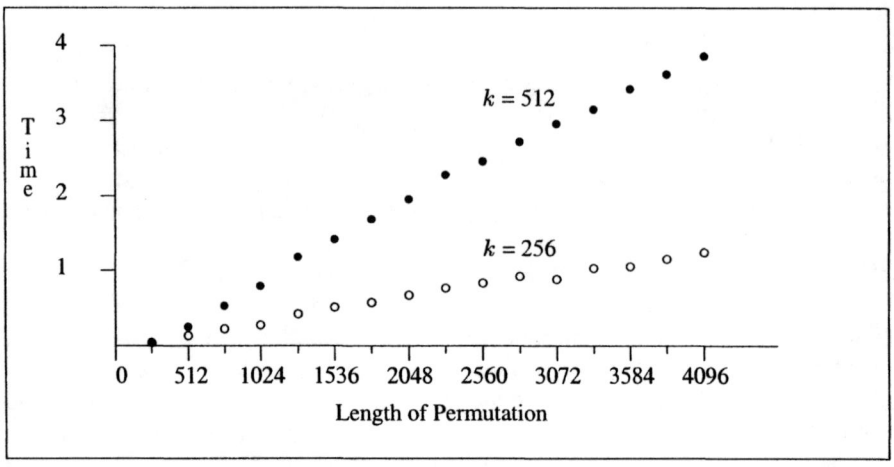

Fig. 3. Mean Time to Sort a Permutation of Length n (in sec.).

In this range of values, the behavior of the algorithm is clearly linear. Recall from the analysis that the estimated running time is governed by the quantity $dv_m n$. With k constant and n sufficiently large, then d is also constant. It seems that for large n, the shape of the overlap graph depends only on d, which would be certainly be true in the limit case of a continuous interval.

4.4   Effects of the Variation of d

In this series of experiments, we studied the effect of the variation of d, the number of reversals, for a fixed $n = 1,024$. We generated sets of 500 permutations with evolution rates varying from $k = 64$ to $k = 1,024$, with equal increments.

For each set, we computed the mean number of reversals. Figure 4 presents the pairs of mean values (d, t). The fact that the points are closer together in the right part of the graph is called *saturation*: when k grows close to n, the value of d tends to stabilize.

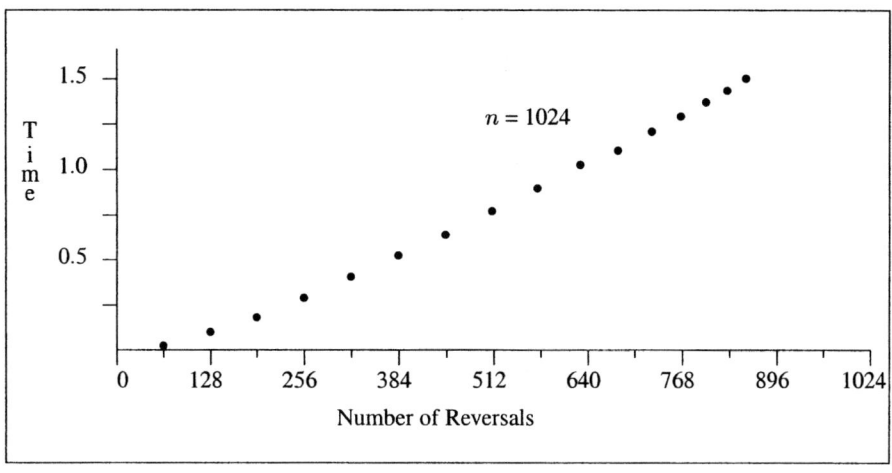

Fig. 4. Mean Time vs. Number of Reversals.

At least for the studied range of values, the performance of the algorithm on the value of d, for a fixed n, seems to be much less than $\mathcal{O}(d^2)$, which is a bit surprising, given that what is measured here is dv_m. Factoring out d from the data yields the curve of Figure 5, which appears to be $\mathcal{O}(\log^2(d))$.

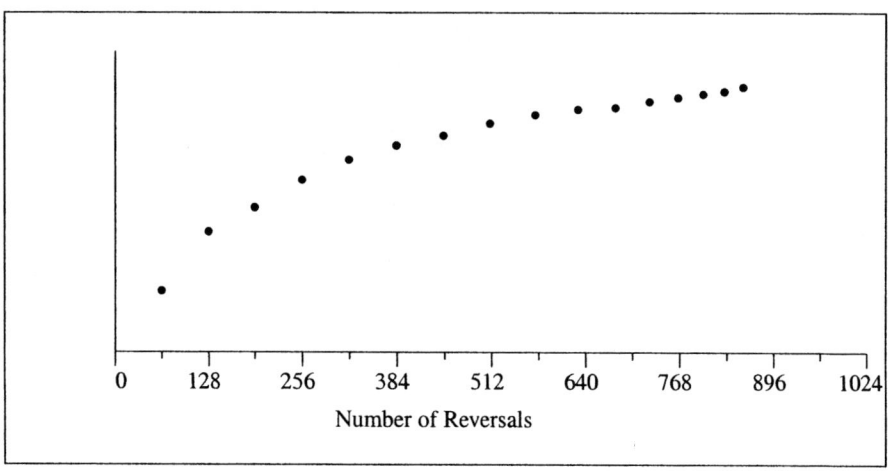

Fig. 5. (Mean Time)/(Number of Reversals) vs. Number of Reversals.

5   Conclusions

Our goal is eventually to be able to study combinatorial properties of optimal sequences of reversals, and our algorithm has the power and performance to serve as a basic tool in such studies. But we think that one of the most desirable feature of the algorithm is its simplicity: the basic ideas can be implemented using only arrays and vector operations.

The experimental running times are still a bit of a puzzle. The theoretical $\mathcal{O}(n^3)$ seems greatly overestimated for mean performances. The role of the parameter v_m is still under study.

In one experiment, we generated two sets of 50 000 permutations of length $n = 256$, with evolution rate $k = 64$. In the first group, we restricted the length of the random reversals to less than $(1/3)n$, and in the second group, to more than $(1/3)n$, under the hypothesis that shorter reversals would produce a less dense overlap graph. Indeed, the "short" group was sorted in 85% of the time of the "long" group.

References

1. David Bader, Bernard Moret, Mi Yan, *A Linear-Time Algorithm for Computing Inversion Distance Between Signed Permutations with an Experiemntal Study*. Proc. 7th Workshop on Algs. and Data Structs. WADS01, Providence (2001), to appear in Lecture Notes in Computer Science, Springer Verlag.
2. Anne Bergeron, *A Very Elementary Presentation of the Hannenhalli-Pevzner Theory*. CPM 2001, Jerusalem (2001), to appear in Lecture Notes in Computer Science, Springer Verlag.
3. Piotr Berman and Sridhar Hannenhalli, *Fast Sorting by Reversal*. CPM 1996, LNCS 1075: 168–185 (1996).
4. Alberto Caprara, *Sorting by reversals is difficult*. RECOMB 1997, ACM Press: 75–83 (1997).
5. Sridhar Hannenhalli and Pavel Pevzner, *Transforming Cabbage into Turnip: Polynomial Algorithm for Sorting Signed Permutations by Reversals*. JACM 46(1): 1–27 (1999).
6. John Kececioglu and David Sankoff, *Efficient bounds for oriented chromosome-inversion distance*. CPM 1994, LNCS 807: 307–325 (1994).
7. Haim Kaplan, Ron Shamir, Robert Tarjan, *A Faster and Simpler Algorithm for Sorting Signed Permutations by Reversals*. SIAM J. Comput. 29(3): 880–892 (1999).
8. Pavel Pevzner, Computational Molecular Biology, MIT Press, Cambridge, Mass., 314 pp. (2000).
9. David Sankoff, *Edit Distances for Genome Comparisons Based on Non-Local Operations*. CPM 1992, LNCS 644: 121–135 (1992).

Exact-IEBP: A New Technique for Estimating Evolutionary Distances between Whole Genomes

Li-San Wang

Department of Computer Sciences
University of Texas, Austin, TX 78712
lisan@cs.utexas.edu

Abstract. Evolution operates on whole genomes by operations that change the order and strandedness of genes within the genomes. This type of data presents new opportunities for discoveries about deep evolutionary rearrangement events, provided that sufficiently accurate methods can be developed to reconstruct evolutionary trees in these models [3, 11, 13, 18]. A necessary component of any such method is the ability to accurately estimate the *true evolutionary distance* between two genomes, which is the number of rearrangement events that took place in the evolutionary history between them. We improve the technique (IEBP) in [21] with a new method, *Exact-IEBP*, for estimating the true evolutionary distance between two signed genomes. Our simulation study shows Exact-IEBP is a better estimation of true evolutionary distances. Furthermore, Exact-IEBP produces more accurate trees than IEBP when used with the popular distance-based method, neighbor joining [16].

1   Introduction

Genome Rearrangement Evolution. The genomes of some organisms have a single chromosome or contain single chromosome organelles (such as mitochondria [4, 14] or chloroplasts [13, 15]) whose evolution is largely independent of the evolution of the nuclear genome for these organisms. Many single-chromosome organisms and organelles have circular chromosomes. Gene maps and whole genome sequencing projects can provide us with information about the ordering and strandedness of the genes, so the chromosome is represented by an ordering (linear or circular) of signed genes (where the sign of the gene indicates which strand it is located on). The evolutionary process on the chromosome can thus be seen as a transformation of signed orderings of genes. The process includes inversions, transpositions, and inverted transpositions, which we will define later.

True Evolutionary Distances. Let T be the true tree on which a set of genomes has evolved. Every edge e in T is associated with a number k_e, the actual number of rearrangements along edge e. The *true evolutionary distance* (*t.e.d.*) between two leaves G_i and G_j in T is $k_{ij} = \sum_{e \in P_{ij}} k_e$, where P_{ij} is the simple path on T between G_i and G_j. If we can estimate all k_{ij} sufficiently accurately, we can reconstruct the tree T using very simple methods, and in particular, using the neighbor joining method (NJ) [1, 16]. Estimates of pairwise distances that are close to the true evolutionary distances will in

O. Gascuel and B.M.E. Moret (Eds.): WABI 2001, LNCS 2149, pp. 175–188, 2001.
© Springer-Verlag Berlin Heidelberg 2001

general be more useful for evolutionary tree reconstruction than edit distances, because edit distances *underestimate* true evolutionary distances, and this underestimation can be very significant as the number of rearrangements increases [7, 20].

There are two criteria for evaluating a t.e.d. estimator: how close the estimated distances are to the true evolutionary distance between two genomes, and how accurate the inferred trees are when a distance-based method (e.g. neighbor joining) is used in conjunction with these distances. The importance of obtaining good *t.e.d.* estimates when analyzing DNA sequences (under stochastic models of DNA sequence evolution) is understood, and well-studied [20].

Representations of Genomes. If we assign a number to the same gene in each genome, a linear genome can be represented by a signed permutation of $\{1, \ldots, n\}$ — a permutation followed by giving each number a plus or minus sign — where the sign shows which strand the gene is on. A circular genome can be represented the same way as a linear genome by breaking off the circle between two neighboring genes and choosing the clockwise or counter-clockwise direction as the positive direction. For example, the following are representations for the same circular genome: $(1, 2, 3)$, $(2, 3, 1)$, $(-1, -3, -2)$. The *canonical representation* for a circular genome is the representation where gene 1 is at the first position with positive sign. The first representation in the previous example is the canonical representation.

The Generalized Nadeau-Taylor Model. We are particularly interested in the following three types of rearrangements: inversions, transpositions, and inverted transpositions. Starting with a genome $G = (g_1, g_2, \ldots, g_n)$ an *inversion* between indices a and b, $1 \leq a < b \leq n + 1$, produces the genome with linear ordering

$$(g_1, g_2, \ldots, g_{a-1}, -g_{b-1}, \ldots, -g_a, g_b, \ldots, g_n)$$

If $b < a$, we can still apply an inversion to a circular (but not linear) genome by simply rotating the circular ordering until g_a precedes g_b in the representation — we consider all rotations of the complete circular ordering of a circular genome as equivalent. A *transposition* on the (linear or circular) genome G acts on three indices, a, b, c, with $1 \leq a < b \leq n$ and $2 \leq c \leq n + 1$, $c \notin [a, b]$, and operates by picking up the interval $g_a, g_{a+1}, \ldots, g_{b-1}$ and inserting it immediately after g_{c-1}. Thus the genome G above (with the additional assumption of $c > b$) is replaced by

$$(g_1, \ldots, g_{a-1}, g_b, g_{b+1}, \ldots, g_{c-1}, g_a, g_{a+1}, \ldots, g_{b-1}, g_c, \ldots, g_n)$$

An *inverted transposition* is the combination of a transposition and an inversion on the transposed subsequence, so that G is replaced by

$$(g_1, \ldots, g_{a-1}, g_b, g_{b+1}, \ldots, g_{c-1}, -g_{b-1}, -g_{b-2}, \ldots, -g_a, g_c, \ldots, g_n)$$

The *Generalized Nadeau-Taylor (GNT) model* [12, 21] assumes a phylogeny (i.e. a rooted binary tree leaf-labeled by species) and models inversions, transpositions, and inverted transpositions along the edges. Different inversions have equal probability, and so do different transpositions and inverted transpositions. Each model tree has two parameters: α is the probability a rearrangement event is a transposition, and β is the probability a rearrangement event is an inverted transposition. Hence, the probability for a rearrangement event to be an inversion is $1 - \alpha - \beta$. The number of events on each edge e is Poisson distributed with mean λ_e. This process produces a set of signed gene orders at the leaves of the model tree.

IEBP. The IEBP (Inverting the Expected BreakPoint distance) method [21] estimates the true evolutionary distance by approximating the expected breakpoint distance (see Section 2) under the GNT model with provable error bound. The method can be applied to any dataset of genomes with equal gene content, and for any relative probabilities of rearrangement event classes. Moreover, the method is robust when the assumptions about the model parameters are wrong.

EDE. In the EDE (Empirically Derived Estimator) method [10] we estimate the true evolutionary distance by inverting the expected inversion distance. We estimate the expected inversion distance by a nonlinear regression on simulation data. The evolutionary model in the simulation is inversion only, but NJ using EDE distance has very good accuracy even when transpositions and inverted transpositions are present.

Our New t.e.d. Estimator. In this paper we improve the result in [21] by introducing the Exact-IEBP method. The method replaces the approximation in the IEBP method by computing the expected breakpoint distance exactly. In Section 3 we show the derivation for our new method. The technique is then checked by computer simulations in Section 4. The simulation shows that the new method is the best t.e.d. estimator, and the accuracy of the NJ tree using the new method is comparable to that of the NJ tree using the EDE distances, and better than that of the NJ tree using other distances.

2   Definitions

We first define the *breakpoint distance* [3] between two genomes. Let genome $G_0 = (g_1, \ldots, g_n)$, and let G be a genome obtained by rearranging G_0. The two genes g_i and g_j are *adjacent* in genome G if g_i is immediately followed by g_j in G_0, or, equivalently, if $-g_j$ is immediately followed by $-g_i$. A breakpoint in G with respect to G_0 is defined as an ordered pair of genes (g_i, g_j) such that g_i and g_j are adjacent in G_0, but are not adjacent in G (neither (g_i, g_j) nor $(-g_j, -g_i)$ appear consecutively in that order in G). The breakpoint distance between two genomes G and G_0 is the number of breakpoints in G with respect to G_0 (or vice versa, since the breakpoint distance is symmetric). For example, let $G = (1, 2, 3, 4)$ and let $G' = (1, -3, -2, 4)$; there is a breakpoint between genes 1 and 3 in G' (w.r.t. G) but genes 2 and 3 are adjacent in G' (w.r.t. G). The breakpoint distance between two genomes is the number of breakpoints in one genome with respect to the other.

A rearrangement ρ is a permutation of the genes in the genome, followed by either negating or retaining the sign of each gene. For any genome G, let ρG be the genome obtained by applying ρ on G. Let R_I, R_T, R_V be the set of all inversions, transpositions, and inverted transpositions, respectively. We assume the evolutionary model is the GNT model with parameters α and β. Within each of the three types of rearrangement events, all events have the same probability.

Let $G_0 = (g_1, g_2, \ldots, g_n)$ be the signed genome of n genes at the beginning of the evolutionary process. For linear genomes we add the two sentinel genes 0 and $n + 1$ in the front and the end of G_0 that are never moved. For any $k \geq 1$, let $\rho_1, \rho_2, \ldots \rho_k$ be k random rearrangements and let $G_k = \rho_k \rho_{k-1} \ldots \rho_1 G_0$ (i.e. G_k is the result of applying these k rearrangements to G_0). Given any linear genome $G = (0, g'_1, g'_2, \ldots, g'_n, n+1)$, where 0 and $n+1$ are sentinel genes, we define the function $B_i(G), 0 \leq i \leq n$ by setting $B_i(G) = 0$ if genes g_i and g_{i+1} are adjacent, and $B_i(G) = 1$ if not; in other words, $B_i(G) = 1$ if and only if G has a breakpoint between g_i and g_{i+1}. When G is circular there are at most n breakpoints $B_i(G), 1 \leq i \leq n$. We denote the breakpoint distance between two genomes G and G' by $BP(G, G')$. Let $P_{i|k} = \Pr(B_i(G_k) = 1)$; then $E[BP(G_0, G_k)] = \sum_{i=0}^{n} P_{i|k}$ for linear genomes and $E[BP(G_0, G_k)] = \sum_{i=1}^{n} P_{i|k}$ for circular genomes.

3 The Exact-IEBP Method

3.1 Derivation of the Exact-IEBP Method

Signed Circular Genomes. We now assume that all genomes are given in the canonical representation. Under the GNT model for circular genomes, $P_{i|k}$ has the same distribution for all $i, 1 \leq i \leq n$. Therefore $E[BP(G_0, G_k)] = nP_{1|k}$. Let $\mathcal{G}_n^C$ be the set of all signed circular genomes, and let $W_n^C = \{\pm 2, \pm 3, \ldots, \pm n\}$. We define the function $K : \mathcal{G}_n^C \to W_n^C$ as follows: for any genome $G \in \mathcal{G}_n^C$, $K(G) = x$ if g_2 is at position $|x|$ with the same sign of x. For example, in the genome $G = (g_1, g_3, g_5, g_4, -g_2)$ we have $K(G) = -5$. Since the sign and the position of gene g_2 uniquely determine $P_{1|k}$, $\{K(G_k) : k \geq 0\}$ is a homogeneous Markov chain where the state space is W_n^C. We will use these states for indexing elements in the transition matrix and the distribution vectors. For example, if M is the transition matrix for $\{K(G_k) : k \geq 0\}$, then $M_{i,j}$ is the probability of jumping to state i from state j in one step in the Markov chain for all $i, j \in W_n^C$.

For every rearrangement $\rho \in R_I \cup R_T \cup R_V$, we construct the matrix Y_ρ as follows: for every $i, j \in W_n^C$, $(Y_\rho)_{i,j} = 1$ if ρ changes the state of gene g_2 from j to i. We then let $M_I = \frac{1}{|R_I|} \sum_{\rho \in R_I} Y_\rho$, $M_T = \frac{1}{|R_T|} \sum_{\rho \in R_T} Y_\rho$, and $M_V = \frac{1}{|R_V|} \sum_{\rho \in R_V} Y_\rho$. The transition matrix M for $\{K(G_k) : k \geq 0\}$ is therefore $M = (1 - \alpha - \beta)M_I + \alpha M_T + \beta M_V$. Let x_k be the distribution vector for $K(G_k)$, we have

$$(x_0)_2 = 1$$
$$(x_0)_i = 0, \quad i \in W_n^C, i \neq 2$$
$$x_k = M^k x_0$$
$$E[BP(G_0, G_k)] = nP_{1|k} = n(1 - (x_k)_2)$$

The result in [17] is a special case where $\alpha = \beta = 0$.

Signed Linear Genomes. When the genomes are linear, we no longer have the luxury of placing gene g_1 at some fixed position with positive sign; different breakpoints may have different distributions. We need to solve the distribution of each breakpoint individually by considering the positions and the signs of both genes involved at the same time. Let $\mathcal{G}_n^L$ be the set of all signed linear genomes, and let $W_n^L = \{(u, v) : u, v = \pm 1, \ldots, \pm n, |u| \neq |v|\}$. We define the functions $J_i : \mathcal{G}_n^L \to W_n^L, i = 1, \ldots, n - 1$, as follows: for any genome $G \in \mathcal{G}_n^L$, $J_i(G) = (x, y)$ if g_i is at position $|x|$ having the same sign of x, and g_{i+1} is at position $|y|$ having the same sign of y. Therefore

$\{J_i(G_k) : k \geq 0\}, 1 \leq i \leq n - 1$ are $n - 1$ homogeneous Markov chains where the state space is W_n^L. For example, in the genome $G = (g_1, g_2, g_4, g_5, g_6, -g_3, -g_7, g_8)$ we have $L_3(G) = (-6, 3)$ and $L_7(G) = (-7, 8)$. As before we use the states in W_n^L as indices to the transition matrix and the distribution vectors. Let $x_{i,k}$ be the distribution vector of $L_i(G_k)$. For every rearrangement $\rho \in R_I, R_T$, and R_V, Y_ρ is defined similarly as before (for circular genomes), except the dimension of the matrix is different. We then let $M_I = \frac{1}{|R_I|} \sum_{\rho \in R_I} Y_\rho$, $M_T = \frac{1}{|R_T|} \sum_{\rho \in R_T} Y_\rho$, and $M_V = \frac{1}{|R_V|} \sum_{\rho \in R_V} Y_\rho$. The transition matrix M has the same form as that for the circular genomes: $M = (1 - \alpha - \beta)M_I + \alpha M_T + \beta M_V$. Let e be the vector where $e_{(u,v)} = 1$ if $v = u + 1$, and 0 otherwise (that is, $e_s = 1$ if s is the state where the two genes are adjacent so there is no breakpoint between them). Therefore

$$(x_{i,0})_{(i,i+1)} = 1$$
$$(x_{i,0})_{(u,v)} = 0, \quad (u, v) \in W_n^L, (u, v) \neq (i, i + 1)$$
$$x_{i,k} = M^k x_{i,0}$$
$$P_{i|k} = 1 - e^T x_{i,k} = 1 - e^T M^k x_{i,0}$$

Since the two sentinel genes 0 and $n + 1$ never change their positions and signs, their states are fixed. This means the distribution of the two breakpoints B_0 and B_n depend on the state of one gene each (g_1 and g_n, respectively); we can use the method for circular genomes. Under the GNT model they have the same distribution. Then the expected breakpoint distance after k events is

$$E[BP(G_0, G_k)] = \sum_{i=0}^{n} P_{i|k} = 2P_{0|k} + \sum_{i=1}^{n-1} P_{i|k} = 2P_{0|k} + \sum_{i=1}^{n-1}(1 - e^T M^k x_{i,0})$$

$$= 2P_{0|k} + (n - 1) - e^T M^k \sum_{i=1}^{n-1} x_{i,0}$$

We now define the Exact-IEBP estimator $\widehat{k}(G, G')$ for the true evolutionary distance between two genomes G and G':

1. For all $k = 1, \ldots, r$ (where r is some integer large enough to bring a genome to random) compute $E[BP(G_0, G_k)]$ using the results above.
2. To compute $k' = \widehat{k}(G, G')(0 \leq k' \leq r)$, we
   (a) compute the breakpoint distance $b = BP(G, G')$, then
   (b) find the integer k', $0 \leq k' \leq r$ such that $|E[BP(G_0, G_{k'})] - b|$ is minimized.

3.2   The Transition Matrices for Signed Circular Genomes

We now derive closed-form formulas of the transition matrix M for the GNT model on signed circular genomes with n genes. Let $\binom{a}{b}$ denote the binomial coefficient; in addition, we let $\binom{a}{b} = 0$ if $b > a$. First consider the number of rearrangement events in each class:

1. *Inversions.* By symmetry of the circular genomes and the model, each inversion has a corresponding inversion that inverts the complementary subsequence (the solid vs. the dotted arc in Figure 1(a)); thus we only need to consider the $\binom{n}{2}$ inversions that do not invert gene g_1.
2. *Transpositions.* In Figure 1(b), given the three indices in a transposition, the genome is divided into three subsequences, and the transposition swaps two subsequences without changing the signs. Let the three subsequences be A, B, and C, where A contains gene g_1. A takes the form (A_1, g_1, A_2), where A_1 and A_2 may be empty. In the canonical representation there are only two possible unsigned permutations: (g_1, A_2, B, C, A_1) and (g_1, A_2, C, B, A_1). This means we only need to consider transpositions that swap the two subsequences not containing g_1.
3. *Inverted Transpositions.* There are $3\binom{n}{3}$ inverted transpositions. In Figure 1(c), given the three endpoints in an inverted transposition, exactly one of the three subsequences changes signs. Using the canonical representation, we interchange the two subsequences that do not contain g_1 and invert one of them (the first two genomes right of the arrow in Figure 1(c)), or we invert both subsequences without swapping (the rightmost genome in Figure 1(c)).

For all $u, v \in W_n^C$, let $\iota_n(u, v)$, $\tau_n(u, v)$ and $\nu_n(u, v)$ be the numbers of inversions, transpositions, and inverted transpositions that bring a gene in state u to state v (n is the number of genes in each genome). Then

$$M_{u,v} = (1 - \alpha - \beta)(M_I)_{u,v} + \alpha(M_T)_{u,v} + \beta(M_V)_{u,v}$$

$$= \frac{1 - \alpha - \beta}{\binom{n}{2}}\iota_n(u, v) + \frac{\alpha}{\binom{n}{3}}\tau_n(u, v) + \frac{\beta}{3\binom{n}{3}}\nu_n(u, v)$$

The following lemma gives formulas for $\iota_n(u, v)$, $\tau_n(u, v)$, and $\nu_n(u, v)$.

Lemma 1.

$$\iota_n(u, v) = \begin{cases} \min\{|u| - 1, |v| - 1, n + 1 - |u|, n + 1 - |v|\}, & \text{if } uv < 0 \\ 0, & \text{if } u \neq v, uv > 0 \\ \binom{|u|-1}{2} + \binom{n+1-|u|}{2}, & \text{if } u = v \end{cases}$$

$$\tau_n(u, v) = \begin{cases} 0, & \text{if } uv < 0 \\ (\min\{|u|, |v|\} - 1)(n + 1 - \max\{|u|, |v|\}), & \text{if } u \neq v, uv > 0 \\ \binom{n+1-|u|}{3} + \binom{|u|-1}{3}, & \text{if } u = v \end{cases}$$

$$\nu_n(u, v) = \begin{cases} (n - 2)\iota_n(u, v), & \text{if } uv < 0 \\ \tau_n(u, v), & \text{if } u \neq v, uv > 0 \\ 3\tau_n(u, v), & \text{if } u = v \end{cases}$$

Proof. The proof of (a) is omitted—this result is first shown in [17]. We now prove (b). Consider the gene with state u. Let v be the new state of that gene after the transposition with indices (a, b, c), $2 \leq a < b < c \leq n + 1$. Since transpositions do not change the

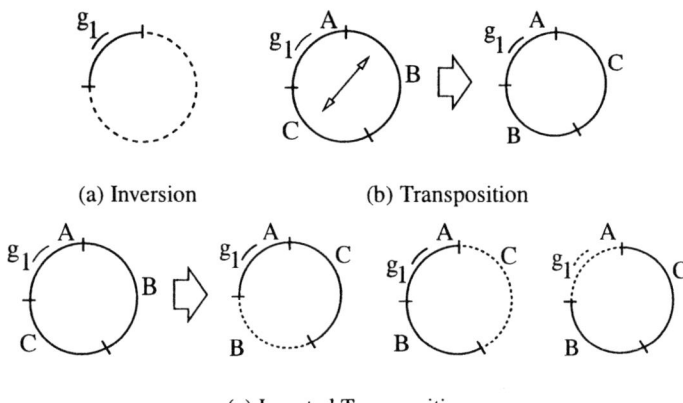

(a) Inversion   (b) Transposition

(c) Inverted Transposition

Fig. 1. The three types of rearrangement events in the GNT model on a signed circular genome. (a) We only need to consider inversions that do not invert g_1. (b) A transposition corresponds to swapping two subsequences. (c) The three types of inverted transpositions. Starting from the left genome, the three distinct results are shown here; the broken arc represents the subsequence being transposed and inverted.

sign, $\tau_n(u, v) = \tau_n(-u, -v)$, and $\tau_n(u, v) = 0$ if $uv < 0$. Therefore we only need to analyze the case where $u, v > 0$.

We first analyze the case when $u = v$. Assume that either $a \leq u < b$ or $b \leq u < c$. In the first case, from the definition in Section 1 we immediately have $v = u + (c - b)$, therefore $v - u = c - b > 0$. In the second case, we have $v = u + (a - b)$, therefore $v - u = a - b < 0$. Both cases contradict the assumption that $u = v$, and the only remaining possibilities that makes $u = v$ are when $2 \leq u = v < a$ or $c \leq u = v \leq n$. This leads to the third line in the $\tau_n(u, v)$ formula. Next, the total number of solutions (a, b, c) for the following two problems is $\tau_n(u, v)$ when $u \neq v$ and $u, v > 0$:

(i) $u < v : b = c - (v - u)$, $2 \leq a \leq u < b < c \leq n + 1, u < v \leq c$.
(ii) $u > v : b = a + (u - v)$, $2 \leq a < b \leq u < c \leq n + 1, a \leq v < u$.

In the first case $\tau_n(u, v) = (u - 1)(n + 1 - v)$, and in the second case $\tau_n(u, v) = (v - 1)(n + 1 - u)$. The second line in the $\tau_n(u, v)$ formula follows by combining the two results.

For inverted transpositions there are three distinct subclasses of rearrangement events. The result in (c) follows by applying the above method to the three cases.

3.3   Running Time Analysis

Let m be the number of genomes and the dimension of the distance matrix. Since for every pair of genomes we can compute the breakpoint distance between them in linear time, computing the breakpoint distance matrix takes $O(m^2 n)$ time. Consider the value r, the number of inversions needed to produce a genome that is close to random; we can use this as an upper bound of k. In [21] we showed by the simulation and the IEBP formula that it is reasonable to set $r = \gamma n$ for some constant γ larger than 1. We used $\gamma = 2.5$ for 120 genes in our experiment.

Constructing the transition matrix M for circular genomes takes $O(n^2)$ time by Lemma 1. We believe results similar to Lemma 1 can be obtained for linear genomes, though it is still an open problem. Instead, we use the construction in Section 3.1 for linear genomes. For each rearrangement ρ, constructing the Y_ρ matrix takes $O(n^4)$ time. Since there are $O(n^2)$ inversions and $O(n^3)$ transpositions and inverted transpositions, constructing the transition matrix M takes $O(n^7)$ time. The running time for computing x_k in Exact-IEBP for $k = 1, \ldots, r$ is $O(rn^2) = O(n^3)$ for circular genomes and $O(rn^4) = O(n^5)$ for linear genomes by r matrix-vector multiplications. Since the breakpoint distance is always an integer between 0 and n, we can construct the array $\hat{k}(b)$ that converts the breakpoint distance b to the corresponding Exact-IEBP distance in $O(n^2)$ time. Transforming the breakpoint distance matrix into the Exact-IEBP distance matrix takes $O(m^2)$ additional array lookups.

We summarize the discussion as follows:

Theorem 1. *Given a set of m genomes on n genes, we can estimate the pairwise true evolutionary distance in*

1. $O(m^2 n + n^3)$ *time using Exact-IEBP when the genomes are circular,*
2. $O(m^2 n + n^7)$ *time using Exact-IEBP when the genomes are linear, and*
3. $O(m^2 n + \min\{n, m^2\} \log n)$ *time using IEBP (see [21]).*

4   Experiments

We now show the experimental study of different distance estimators. We compare the following five distance estimators on circular genomes: (1) BP, the breakpoint distance between two genomes, (2) INV [2], the minimum number of inversions needed to transform one genome into another, (3) IEBP [21], an approximation to the Exact-IEBP method with fast running time, (4) EDE [10], an estimation of the true evolutionary distance based on the INV distance, and (5) Exact-IEBP, our new method.

Software. We use PAUP* 4.0 [19] to compute the neighbor joining method and the false negative rates between two trees (which will be defined later). We have implemented a simulator [10, 21] for the GNT model. The input is a rooted leaf-labeled tree and the associated parameters (i.e. edge lengths, and the relative probabilities of inversions, transpositions, and inverted transpositions). On each edge, the simulator applies random rearrangement events to the circular genome at the ancestral node according to the model with given parameters α and β. We use tgen [6] to generate random trees. These trees have topologies drawn from the uniform distribution, and edge lengths drawn from the discrete uniform distribution on intervals $[a, b]$, where we specify a and b.

4.1   Accuracy of the Estimators

In this section we study the behavior of the Exact-IEBP distance by comparing it to the actual number of rearrangement events. We simulate the GNT model on a circular

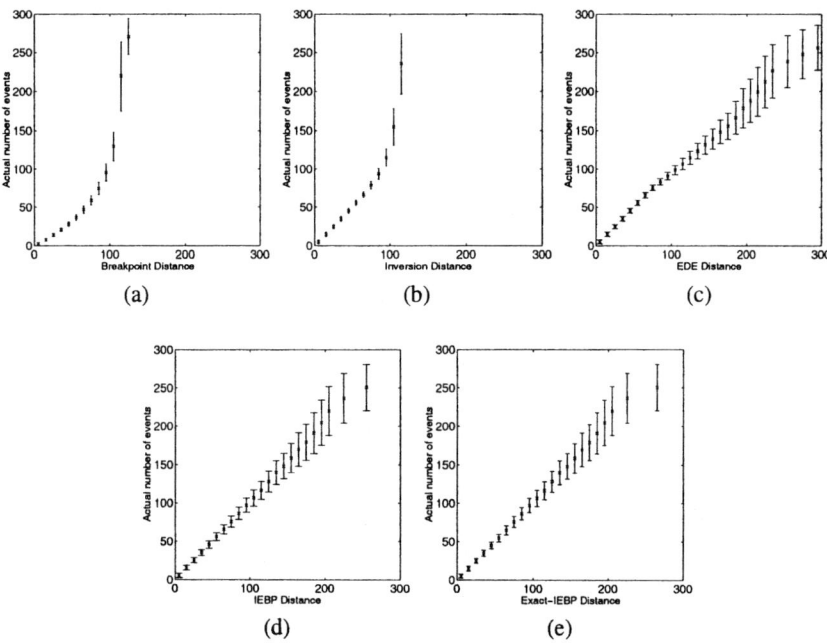

Fig. 2. Accuracy of the Estimators (See Section 4.1). The number of genes is 120. Each plot is a comparison between some distance measures and the actual number of rearrangements. We show the result for the inversion-only evolutionary model only. The x-axis is divided into 30 bins; the length of the vertical bars indicate the standard deviation. The distance estimators are (a) BP, (b) INV, (c) EDE, (d) IEBP, and (e) Exact-IEBP. The figures (a), (b), (d) are from [21], and the figure (d) is from [10].

genome with 120 genes, the typical number of genes in the plant chloroplast genomes [8]. Starting with the unrearranged genome G_0, we apply k events to it to obtain the genome G_k, where $k = 1, \ldots, 300$. For each value of k we simulate 500 runs. We then compute the five distances.

The simulation results under the inversion-only model are shown in Figure 2. Under the other two model settings, the simulation results show similar behavior (e.g. shape of curves and standard deviations). Note that both BP and INV distances underestimate the actual number of events, and EDE slightly overestimates the actual number of events when the number of events is high. The IEBP and Exact-IEBP distances are both unbiased — the means of the computed distances are equal to the actual number of rearrangement events — and have similar standard deviations. We then compare different distance estimators by the absolute difference in the measured distances and the actual number of events. Using the same data in the previous experiment, we generate the plots as follows. The x-axis is the actual number of events. For each distance estimator D we

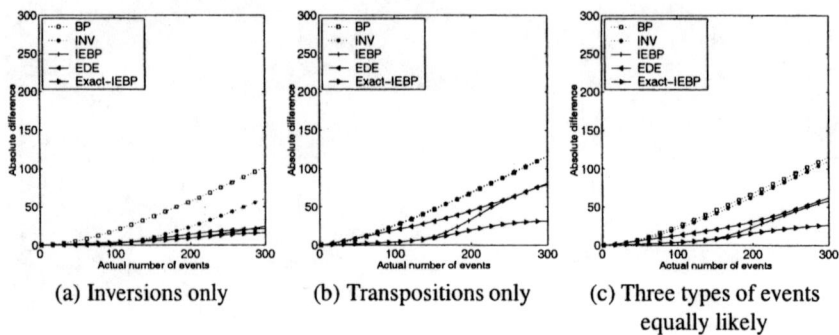

(a) Inversions only   (b) Transpositions only   (c) Three types of events
                                                             equally likely

Fig. 3. Accuracy of the estimators by absolute difference (See Section 4.1 for the details.). We simulate the evolution on 120 genes. The curves of BP, INV, IEBP, and EDE are published previously in [10]; they are included for comparative purposes.

plot the curve f_D, where $f_D(x)$ is the mean of the set $\{|\frac{1}{c}D(G_0, G_k) - k| : 1 \leq k \leq x\}$ over all observations G_k.[1]

The result is in Figure 3. The relative performance is the same for most cases: BP is the worst, followed by INV, IEBP, and EDE. Exact-IEBP has the best performance except for inversion-only scenarios, where EDE is slightly better only when the number of events is small. In most cases, IEBP has similar behavior as Exact-IEBP when the amount of evolution is small; the IEBP and Exact-IEBP curves are almost indistinguishable in (a). Yet, in (b) and (c) the IEBP curve is inferior than the Exact-IEBP curve by a large margin when the number of events is above about 200.

4.2 Accuracy of Neighbor Joining Using Different Estimators

In this section we explore the accuracy of the neighbor joining tree under different ways of calculating genomic distances. See Table 1 for the settings for the experiment.

Given an inferred tree, we compare its "topological accuracy" by computing "false negatives" with respect to the "true tree" [9, 5]. We begin by defining the true tree. During the evolutionary process, some edges of the model tree may have no changes (i.e. evolutionary events) on them. Since reconstructing such edges is at best guesswork, we are not interested in these edges. Hence, we define the true tree to be the result of *contracting* those edges in the model tree on which there are no changes.

We now define how we score an inferred tree, by comparison to the true tree. For every tree there is a natural association between every edge and the bipartition on the leaf set induced by deleting the edge from the tree. Let T be the true tree and let T' be the inferred tree. An edge e in T is "missing" in T' if T' does not contain an edge

---

[1] The constant c is to reduce the bias effect in different distances. For the IEBP and the Exact-IEBP distances $c = 1$ since they estimate the actual number of events. For the BP distance we let $c = 2(1 - \alpha - \beta) + 3(\alpha + \beta) = 2 + \alpha + \beta$ since this is the expected number of breakpoints created by each event in the model when the number of events is very low. Similarly for the INV and EDE distances we let $c = (1 - \alpha - \beta) + 3\alpha + 2\beta = 1 + 2\alpha + \beta$ since each transposition can be replaced by 3 inversions, and each inverted transposition can be replaced by 2 inversions.

Table 1. Settings for the Neighbor Joining Performance Simulation Study.

Parameter	Value
1. Number of genes	120
2. Number of leaves	10, 20, 40, 80, and 160
3. Expected number of	Discrete Uniform within the following intervals:
rearrangements in each edge	[1,3], [1,5], [1,10], [3,5], [3,10], and [5,10]
4. Probability settings: (α, β)[†]	(0,0) (Inversion only)
	(1, 0) (Transposition only)
	$(\frac{1}{3}, \frac{1}{3})$ (The three rearrangement classes are equally likely)
5. Datasets for each setting	100

† The probabilities that a rearrangement is an inversion, a transposition, or an inverted transposition are $1 - \alpha - \beta$, α, and β, respectively.

defining the same bipartition; such an edge is called a *false negative*. Note that the external edges (i.e. edges incident to a leaf) are trivial in the sense that they are present in every tree with the same set of leaves. The *false negative rate* is the number of false negative edges in T' with respect to T divided by the number of internal edges in T.

For each setting of the parameters (number of leaves, probabilities of rearrangements, and edge lengths), we generate 100 datasets of genomes as follows. First, we generate a random leaf-labeled tree (from the uniform distribution on topologies). The leaf-labeled tree and the parameter settings thus define a model tree in the GNT model. We run the simulator on the model tree, and produce a set of genomes at the leaves.

For each set of genomes, we compute the five distances. We then compute NJ trees on each of the five distance matrices, and compare the resultant trees to the true tree. The results of this experiment are in Figure 4. The x-axis is the maximum normalized inversion distance (as computed by the linear time algorithm for minimum inversion distances given in [2]) between any two genomes in the input. Distance matrices with some normalized edit distances close to 1 are said to be "saturated", and the recovery of accurate trees from such datasets is considered to be very difficult [7]. The y-axis is the false negative rate (i.e. the proportion of missing edges). False negative rates of less than 5% are excellent, but false negative rates of up to 10% can be tolerated. We use NJ(D) to denote the tree returned by NJ using distance D. Note that except for NJ(EDE), the relative orders of the NJ tree using different distances are very consistent with the orders of the accuracy of the distances in the absolute difference plot. NJ(BP) has the worst accuracy in all settings. NJ(INV) outperforms NJ(Exact-IEBP) and NJ(IEBP) by a small margin only when the amount of evolution is low; in the transposition-only scenario the accuracy of NJ(INV) degrades considerably. NJ(Exact-IEBP) has slightly better accuracy than NJ(IEBP) until the amount of evolution is high; after that the accuracy of NJ(IEBP) degrades, and NJ(Exact-IEBP) outperforms NJ(IEBP) by a larger margin. Despite the inferior accuracy of EDE in the experiments in Section 4.1, NJ using EDE

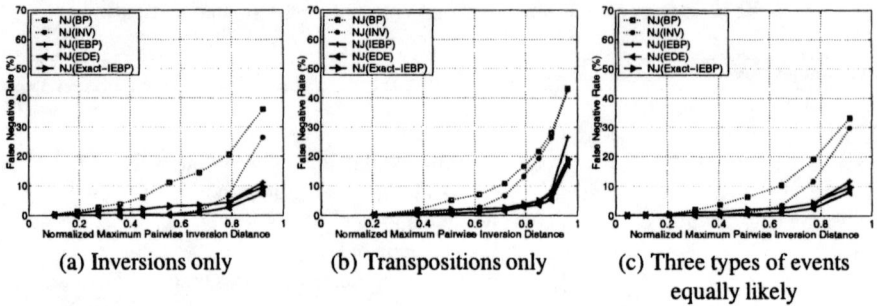

(a) Inversions only   (b) Transpositions only   (c) Three types of events
                                                       equally likely

Fig. 4. Neighbor Joining Performance under Several Distances (See Section 4.2). See Table 1 for the settings in the experiment. For comparative purposes, we include curves of NJ(BP), NJ(INV), NJ(IEBP) from [21], and the curve of NJ(EDE) from [10].

returns the most accurate tree on average[2] (especially in the inversion-only model), but the accuracy of the NJ tree using Exact-IEBP is comparable in most cases.

4.3   Robustness to Unknown Model Parameters

In this section we demonstrate the robustness of the Exact-IEBP estimator when the model parameters are unknown. The settings are the same in Table 1. The experiment is similar to the previous experiment, except here we use both the correct and the incorrect values of (α, β) for the Exact-IEBP distance. The results are in Figure 5. These results suggest that NJ(Exact-IEBP) is robust against errors in (α, β).

5   Conclusions

We have introduced Exact-IEBP, a new technique for estimating true evolutionary distances between whole genomes. This technique can be applied to signed circular and linear genomes with arbitrary relative probabilities between the three types of events in the GNT model. Our simulation study shows that the Exact-IEBP method improves upon the previous technique, IEBP [21], both in the distance estimation and the accuracy of the inferred tree when used in neighbor joining. The accuracy of the NJ trees using the new method is comparable with the best estimator so far, the EDE estimator [10]. These different methods are simple yet powerful and can be generalized easily to different models.

---

[2] We do not have a good explanation for the superior accuracy of NJ(EDE) due to the fact that the behavior of NJ is still not well understood.

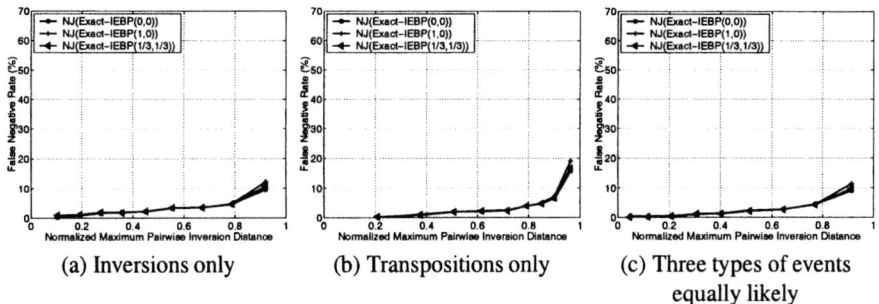

(a) Inversions only   (b) Transpositions only   (c) Three types of events
                                                              equally likely

Fig. 5. Robustness of the Exact-IEBP Method to Unknown Parameters (See Section 4.3). See Table 1 for the settings in the experiment. The two values in the legend are the α and β values used in the Exact-IEBP method. The probability a rearrangement event is an inversion, a transposition, or an inverted transposition are $1 - \alpha - \beta$, α, and β, respectively.

Acknowledgments

I would like to thank Tandy Warnow for her advice and guidance on this paper and Robert Jansen at the University of Texas at Austin for introducing the problem of genome rearrangement phylogeny to us. I am grateful to the three anonymous referees for their helpful comments.

References

1. K. Atteson. The performance of the neighbor-joining methods of phylogenetic reconstruction. *Algorithmica*, 25(2/3):251–278, 1999.
2. D.A. Bader, B.M.E. Moret, and M. Yan. A fast linear-time algorithm for inversion distance with an experimental comparison. In *Proc. 7th Workshop on Algs. and Data Structs. (WADS01)*, 2001. To appear.
3. M. Blanchette, G. Bourque, and D. Sankoff. Breakpoint phylogenies. In S. Miyano and T. Takagi, editors, *Genome Informatics*, pages 25–34. Univ. Acad. Press, 1997.
4. M. Blanchette, M. Kunisawa, and D. Sankoff. Gene order breakpoint evidence in animal mitochondrial phylogeny. *J. Mol. Evol.*, 49:193–203, 1999.
5. O. Gascuel. Personal communication, April 2001.
6. D. Huson. The tree library, 1999.
7. D. Huson, S. Nettles, K. Rice, T. Warnow, and S. Yooseph. The hybrid tree reconstruction method. *J. Experimental Algorithmics*, 4:178–189, 1999.
   http://www.jea.acm.org/.
8. R.K. Jansen. personal communication, October 3 2000.
9. S. Kumar. Minimum evolution trees. *Mol. Biol. Evol.*, 15:584–593, 1996.
10. B.M.E. Moret, L.-S. Wang, T. Warnow, and S. Wyman. New approaches for reconstructing phylogenies from gene order data. In *Proc. 9th Intl. Conf. on Intel. Sys. for Mol. Bio. ISMB 2001*. AAAI Press, 2001. To appear.
11. B.M.E Moret, S.K. Wyman, D.A. Bader, T. Warnow, and M. Yan. A new implementation and detailed study of breakpoint analysis. In *Proc. 6th Pacific Symp. Biocomputing (PSB 2001)*, pages 583–594, 2001.

12. J.H. Nadeau and B.A. Taylor. Lengths of chromosome segments conserved since divergence of man and mouse. *Proc. Nat'l Acad. Sci. USA*, 81:814–818, 1984.

13. R.G. Olmstead and J.D. Palmer. Chloroplast DNA systematics: a review of methods and data analysis. *Amer. J. Bot.*, 81:1205–1224, 1994.

14. J.D. Palmer. Chloroplast and mitochondrial genome evolution in land plants. In R. Herrmann, editor, *Cell Organelles*, pages 99–133. Wein, 1992.

15. L.A. Raubeson and R.K. Jansen. Chloroplast DNA evidence on the ancient evolutionary split in vascular land plants. *Science*, 255:1697–1699, 1992.

16. N. Saitou and M. Nei. The neighbor-joining method: A new method for reconstructing phylogenetic trees. *Mol. Biol. & Evol.*, 4:406–425, 1987.

17. D. Sankoff and M. Blanchette. Probability models for genome rearrangements and linear invariants for phylogenetic inference. *Proc. 3rd Int'l Conf. on Comput. Mol. Bio. (RECOMB99)*, pages 302–309, 1999.

18. D. Sankoff and J.H. Nadeau, editors. *Comparative Genomics : Empirical and Analytical Approaches to Gene Order Dynamics, Map Alignment and the Evolution of Gene Families*. Kluwer Academic Publishers, 2000.

19. D. Swofford. *PAUP* 4.0*. Sinauer Associates Inc, 2001.

20. D. Swofford, G. Olson, P. Waddell, and D. Hillis. Phylogenetic inference. In D. Hillis, C. Moritz, and B. Mable, editors, *Molecular Systematics*, chapter 11. Sinauer Associates Inc, 2 edition, 1996.

21. L.-S. Wang and T. Warnow. Estimating true evolutionary distances between genomes. In *Proc. 33th Annual ACM Symp. on Theory of Comp. (STOC 2001)*. ACM Press, 2001. To appear.

Finding an Optimal Inversion Median: Experimental Results

Adam C. Siepel[1] and Bernard M.E. Moret[2]

[1] Department of Computer Science, University of New Mexico
Albuquerque, NM 87131, USA
and National Center for Genome Resources
Santa Fe, NM 87505, USA
acs@ncgr.org
[2] Department of Computer Science, University of New Mexico
Albuquerque, NM 87131, USA
moret@cs.unm.edu

Abstract. We derive a branch-and-bound algorithm to find an optimal inversion median of three signed permutations. The algorithm prunes to manageable size an extremely large search tree using simple geometric properties of the problem and a newly available linear-time routine for inversion distance. Our experiments on simulated data sets indicate that the algorithm finds optimal medians in reasonable time for genomes of medium size when distances are not too large, as commonly occurs in phylogeny reconstruction. In addition, we have compared inversion and breakpoint medians, and found that inversion medians generally score significantly better and tend to be far more unique, which should make them valuable in median-based tree-building algorithms.

1 Introduction

Dobzhansky and Sturtevant [7] first proposed using the degree to which gene orders differ between species as an indicator of evolutionary distance that could be useful for phylogenetic inference, and Watterson *et al.* [23] first proposed the minimum number of chromosomal inversions necessary to transform one ordering into another as an appropriate distance metric. The 1992 study by Sankoff *et al.* [21] included a heuristic algorithm for finding rearrangement distance (which considered transpositions, insertions, and deletions, as well as inversions); it was the first large-scale application and experimental validation of rearrangement-based techniques for phylogenetic purposes and initiated what is now nearly a decade of intense interest in computational problems relating to genome rearrangement (see summaries in [16, 19, 22]).

While much of the attention given to rearrangement problems may be due to their intriguing combinatorial properties, rearrangement-based approaches to phylogenetic inference are of genuine biological interest in cases in which sequence-based approaches perform poorly, such as when species diverged early or are rapidly evolving [16]. In addition, rearrangement-based phylogenetic methods can suggest probable gene orderings of ancestral species [17, 18], while other methods cannot. Furthermore, mathematical models of genome rearrangement have applications beyond phylogeny (see [8, 20]).

O. Gascuel and B.M.E. Moret (Eds.): WABI 2001, LNCS 2149, pp. 189–203, 2001.

Recent work on rearrangement distance and sorting by inversions (or *reversals*, as they are often called in computer science) has produced a duality theorem and polynomial-time algorithm for inversion distance between two signed permutations [10], a duality theorem and polynomial-time algorithm for distance in terms of equally weighted translocations and inversions for signed permutations [11], polynomial-time algorithms for sorting by reversals [2, 12], and a linear-time algorithm for computing inversion distances [1]. Note that "signed permutations" correspond to genomes for which the direction of transcription of each gene is known as well as the ordering of the genes.

Much recent work on rearrangement-based phylogeny [5, 6, 14, 15, 18] stems from an algorithm by Sankoff and Blanchette [17] that iterates over a prospective tree, repeatedly finds medians of the three permutations adjacent to each internal vertex, and uses them to improve the tree until convergence occurs. This method finds locally optimal trees and simultaneously allows an estimation of the configurations of ancestral genomes. These studies have generally used *breakpoint distance* as the basis for finding medians, because it is more easily computable than inversion distance, it assumes no particular mechanism of rearrangement, and the problem of finding a breakpoint median has a straightforward reduction to the well known Travelling Salesman Problem (TSP) [17]. The number of breakpoints between two genomes is the number of genes that are adjacent in one but not the other genome; the breakpoint median of a set of genomes is the ordering of genes that minimizes the sum of the number of breakpoints with respect to each genome in the set.

Breakpoint distance is related to inversion distance (an inversion can remove at most two breakpoints) but the relationship is a loose one. Because it is believed that inversions are the primary mechanism of genome rearrangement for many taxa [13, 3], we seek a solution to the median problem based directly on inversion distance. Finding an *inversion median* is known to be NP-hard [4], and to date, no one has reported a reasonably efficient algorithm (approximate or exact) for this problem. (Although in one study [9], inversion medians were obtained for a particular data set using a bounded exhaustive search.)

In this paper, we present a simple yet effective branch-and-bound algorithm to solve the median of three problem exactly. Our approach does not depend on properties specific to inversions, but can be used with any rapidly computable metric. We have evaluated its effectiveness for the case of inversion medians, and found that it obtains optimal medians with reasonable computational effort for a range of parameters that include most realistic instances encountered in phylogenetic analysis. In addition, we have performed a comparison of inversion and breakpoint medians, and found that inversion medians score significantly better in terms of total induced edge length, and tend to be far more unique. These findings suggest that inversion medians, when used within the algorithm of Sankoff and Blanchette, will allow better trees to be computed in fewer iterations.

2   Notation and Definitions

We consider the case where all genomes have identical sets of n genes and inversion is the single mechanism of rearrangement. We represent each genome G_i as a permutation π_i of size n, and we let all pairs of genomes $G_i = (g_{i,1} \ldots g_{i,n})$ and $G_j = (g_{j,1} \ldots g_{j,n})$, in a set of genomes G, be represented by $\pi_i = (\pi_{i,1} \ldots \pi_{i,n})$ and $\pi_j = (\pi_{j,1} \ldots \pi_{j,n})$ such that $\pi_{i,k} = \pi_{j,l}$ iff $G_{i,k} = G_{j,l}$, and $\pi_{i,k} = -1 \cdot \pi_{j,l}$ iff $G_{i,k}$ is the reverse complement of $G_{j,l}$.

We define an *inversion* acting on permutation π from i to j, for $i \leq j$, as that operation which transforms π into $\phi = (\pi_1, \pi_2, \ldots, \pi_{i-1}, -\pi_j, -\pi_{j-1}, \ldots, -\pi_i, \pi_{j+1}, \ldots, \pi_n)$. The minimal number of inversions required to change one permutation π_i into another permutation π_j is the *inversion distance*, which we denote by $d(\pi_i, \pi_j)$ (sometimes abbreviated as $d_{i,j}$).

Let the *inversion median* M of a set of N permutations $\Pi = \{\pi_1, \pi_2, \ldots, \pi_N\}$ be the signed permutation that minimizes the sum $S(M, \Pi) = \sum_{i=1}^{N} d(M, \pi_i)$. Let this sum $S(M, \Pi) = S(\Pi)$ be called the *median score* of M with respect to Π.

For a given number of genes n, we can construct an undirected graph $G_n = (V, E)$ such that each vertex in V corresponds to a signed permutation of size n and two vertices are connected by an edge if and only if one of the corresponding permutations can be obtained from the other through a single inversion; formally, $E = \{\{v_i, v_j\} \mid v_i, v_j \in V \text{ and } d(\pi_i, \pi_j) = 1\}$. We will call G_n the *inversion graph* of size n. In this graph, the distance between any two vertices, v_i and v_j, is the same as the inversion distance between the corresponding permutations, π_i and π_j. Furthermore, finding the median of a set of permutations Π is equivalent to finding the minimum unweighted Steiner tree of the corresponding vertices in G_n. Note that G_n is very large ($|V| = n! \cdot 2^n$), so this representation does not immediately suggest a feasible graph-search algorithm, even for small n.

Definition 1. *A shortest path between two permutations of size n, π_1 and π_2, is a connected subgraph of the inversion graph G_n containing only the vertices v_1 and v_2 corresponding to π_1 and π_2, and the vertices and edges on a single shortest path between v_1 and v_2.*

Definition 2. *A median path of a set of permutations Π each of size n is a connected subgraph in the inversion graph of G_n containing only the vertices corresponding to permutations in Π, the vertex corresponding to a median M of Π, and a shortest path between M and each $\pi \in \Pi$.*

Definition 3. *A trivial median of a set of permutations Π is a median M that is a member of that set, $M \in \Pi$.*

Definition 4. *A trivial median path of a set of permutations Π is a median path that includes only the elements of Π and shortest paths between elements of Π.*

3   General Median Bounds

Because phylogenetic reconstruction algorithms generally work with binary trees in which each internal node has three neighbors, the special case of the median of three genomes is of particular interest. In this section we develop a general bound for the median-of-three problem, one that relies only on the metric property of the distance measure used.

Lemma 1. *The median score $S(\Pi)$ of a set of equally sized permutations $\Pi = \{\pi_1, \pi_2, \pi_3\}$, separated by pairwise distances $d_{1,2}, d_{1,3},$ and $d_{2,3}$, obeys these bounds:*

$$\left\lceil \frac{d_{1,2} + d_{1,3} + d_{2,3}}{2} \right\rceil \leq S(\Pi) \leq \min\left\{ (d_{1,2} + d_{2,3}), (d_{1,2} + d_{1,3}), (d_{2,3} + d_{1,3}) \right\}$$

Proof. The upper bound follows directly from the possibility of a trivial median, and the lower bound from properties of metric spaces (a median of lower score would necessarily violate the triangle inequality with respect to two of π_1, π_2, and π_3; see Figure 1).

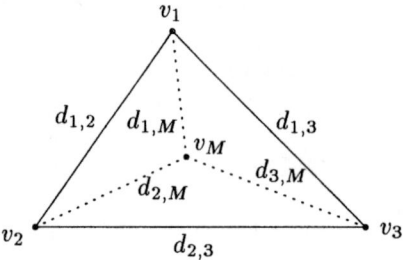

Fig. 1. Let vertices v_1, v_2, and v_3 correspond to permutations π_1, π_2, and π_3, and let vertex v_M correspond to a median M.

Lemma 2. *If three permutations π_1, π_2, and π_3 have a median M that is part of a trivial median path, then M must be a trivial median.*

Proof. Assume to the contrary that π_1, π_2, and π_3 have a trivial median path and have a median M that is not trivial. By Definition 4, M must be on a shortest path between two of π_1, π_2, and π_3. Without loss of generality, assume that the median path runs from π_1 to M to π_2 to π_3. Let $d_{1,2}$, $d_{1,3}$, and $d_{2,3}$ be the pairwise distances between $\{\pi_1, \pi_2\}$, $\{\pi_1, \pi_3\}$, and $\{\pi_2, \pi_3\}$, respectively, and let $d_{M,2} > 0$ be the distance of M from π_2. Then the median score of M is $(d_{1,2} - d_{M,2}) + d_{M,2} + (d_{M,2} + d_{2,3}) = d_{1,2} + d_{M,2} + d_{2,3}$. But this score is greater by $d_{M,2}$ than the score of a trivial median at π_2, so M cannot be a median.

Theorem 1. *Let π_1, π_2, and π_3 be permutations such that π_2 and π_3 are separated by distance $d_{2,3}$, and let ϕ be another permutation separated from π_1, π_2, and π_3 by distances $d_{1,\phi}, d_{2,\phi},$ and $d_{3,\phi}$, respectively. Suppose that ϕ is on a median path P_M of π_1, π_2, and π_3 such that ϕ is on a shortest path between π_1 and a median M. Then the score $S(M)$ of M obeys these bounds:*

$$d_{1,\phi} + \left\lceil \frac{d_{2,\phi} + d_{3,\phi} + d_{2,3}}{2} \right\rceil \leq S(M)$$

$$\leq d_{1,\phi} + \min\{ (d_{2,\phi} + d_{2,3}), (d_{2,\phi} + d_{3,\phi}), (d_{2,3} + d_{3,\phi}) \}$$

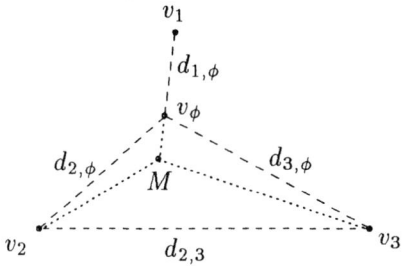

Fig. 2. A median path including v_ϕ can be constructed using a shortest path from v_1 to v_ϕ and any median path of v_ϕ, v_2, and v_3.

Proof. Let v_1, v_2, and v_3 be vertices corresponding to π_1, π_2, and π_3, in the inversion graph of the appropriate size. In addition, let there be a vertex v_ϕ corresponding to ϕ, as illustrated in Figure 2. We claim that a median path P_M including v_ϕ and M, such that v_ϕ is on a shortest path from v_1 to M, can be constructed by combining a shortest path between v_1 and v_ϕ and a median path of v_ϕ, v_2, and v_3. Assume to the contrary that there exists a shorter median path P_{short}, which also includes v_ϕ and M, but does not include the shortest path between v_1 and v_ϕ or does not include a median path of v_ϕ, v_2, and v_3. P_{short} has to include v_1 via a vertex other than v_ϕ and consequently other than M (because v_ϕ is on a shortest path between v_1 and M). By Definition 2, P_{short} must consist only of v_1, v_2, v_3, M, and vertices between them (including v_ϕ), so v_1 must be connected to P_{short} via v_2 or v_3. Consequently, M must be on a shortest path between v_2 and v_3; otherwise including M in P_{short} would result in a score greater than that of a trivial median. Therefore, M is part of a trivial median path, which means that by Lemma 2, M is a trivial median. In particular, M must be the vertex $v_i \in \{v_2, v_3\}$ to which v_1 is connected. Furthermore, our assumptions about ϕ require that v_ϕ be on the shortest path between v_1 and v_i. Then P_{short} includes both the shortest path between v_1 and v_ϕ and the median path of v_ϕ, v_2, and v_3, and we obtain the desired contradiction.

Because P_M can be constructed by combining a shortest path between v_1 and v_ϕ, and a median path of v_ϕ, v_2, and v_3, its score is equivalent to the sum of the distance between v_1 and v_ϕ ($d_{1,\phi}$), and the score of the median of v_ϕ, v_2, and v_3. By applying Lemma 1 to the latter, we obtain the desired bound.

4   The Algorithm

Algorithm find_inversion_median is presented below. It is essentially a branch-and-bound search for an optimal inversion median that uses Theorem 1 to prune a search tree based on the inversion graph and to prioritize among search branches.

Prioritization is managed using a *priority stack*—which always returns an item of highest priority, but returns items of equal priority in *last-in-first-out* order. Because the range of possible priorities is small, we use a fixed array of priority values, each pointing to a stack and so can execute push and pop operations in fast constant time. Using stacks rather than the more conventional queues in this application is not required for correctness, but, by inducing depth-first searching among alternatives of equal cost, rapidly produces a good upper bound for the search.

Algorithm 1: find_inversion_median

Input: Three signed permutations of size n: π_1, π_2, and π_3. Assume a function `distance`(π_i, π_j) that returns the inversion distance between π_i and π_j in linear time.

Output: An optimal inversion median M.

begin

$\quad$ $d_{1,2} \leftarrow$ `distance`(π_1, π_2);

$\quad$ $d_{1,3} \leftarrow$ `distance`(π_1, π_3);

$\quad$ $d_{2,3} \leftarrow$ `distance`(π_2, π_3);

1 $\quad$ $M_{min} \leftarrow \lceil \frac{d_{1,2}+d_{1,3}+d_{2,3}}{2} \rceil$;

2 $\quad$ $M_{max} \leftarrow \min\{(d_{1,2} + d_{2,3}), (d_{1,2} + d_{1,3}), (d_{2,3} + d_{1,3})\}$;

$\quad$ Initialize *priority stack* s for range M_{min} to M_{max};

$\quad$ $(\psi_{orig}, \psi_1, \psi_2) \leftarrow (\pi_i, \pi_j, \pi_k)$ such that $\{\pi_i, \pi_j, \pi_k\} = \{\pi_1, \pi_2, \pi_3\}$ and $d_{i,j} + d_{i,k} = M_{min}$;

3 $\quad$ create vertex v with $v_{label} = \psi_{orig}$, $v_{dist} = 0$, $v_{best} = M_{min}$, $v_{worst} = M_{max}$;

4 $\quad$ push(s, v);

5 $\quad$ $M \leftarrow \psi_{orig}$;

$\quad$ $d_{sep} \leftarrow d_{\psi_1, \psi_2}$;

$\quad$ $stop \leftarrow$ **false** ;

6 $\quad$ **while** s *is not empty and stop = false* **do**

$\quad\quad$ pop(s, v);

7 $\quad\quad$ **if** $v_{best} \geq M_{max}$ **then** $stop \leftarrow$ **true** ;

$\quad\quad$ **else**

8 $\quad\quad\quad$ **foreach** $\{w \mid w$ *is an unmarked neighbor of* $v\}$ **do**

$\quad\quad\quad\quad$ $w_{dist} \leftarrow$ `distance`(w_{label}, ψ_{orig});

9 $\quad\quad\quad\quad$ **if** $w_{dist} \leq v_{dist}$ **then** continue;

$\quad\quad\quad\quad$ mark w;

$\quad\quad\quad\quad$ $d_{\psi_1} \leftarrow$ `distance`(w_{label}, ψ_1);

$\quad\quad\quad\quad$ $d_{\psi_2} \leftarrow$ `distance`(w_{label}, ψ_2);

10 $\quad\quad\quad\quad$ $w_{best} \leftarrow w_{dist} + \lceil \frac{d_{\psi_1}+d_{\psi_2}+d_{sep}}{2} \rceil$;

11 $\quad\quad\quad\quad$ $w_{worst} \leftarrow w_{dist} + \min\{(d_{\psi_1} + d_{sep}), (d_{\psi_1} + d_{\psi_2}), (d_{sep} + d_{\psi_2})\}$;

12 $\quad\quad\quad\quad$ **if** $w_{worst} = M_{min}$ **then** $M \leftarrow w_{label}$; $stop \leftarrow$ **true** ;

$\quad\quad\quad\quad$ **else**

$\quad\quad\quad\quad\quad$ **if** $w_{best} < M_{max}$ **then** push(s, w, w_{best});

13 $\quad\quad\quad\quad\quad$ **if** $w_{worst} < M_{max}$ **then**

$\quad\quad\quad\quad\quad\quad$ $M \leftarrow w_{label}$; $M_{max} \leftarrow w_{worst}$;

end

The algorithm begins by establishing upper and lower bounds for the solution using Lemma 1 (steps 1 and 2) and priming the priority stack with a best-scoring vertex (steps

3 and 4). Then it enters a main loop (step 6) in which it repeatedly pops the "most promising" vertex from the priority stack, finds all of its as-yet-unvisited neighbors (step 8), and evaluates each one for feasibility. Neighbors are obtained by generating all $\binom{n}{2}$ possible permutations that can be produced from a vertex by a single inversion. Neighbors of a vertex v can be ignored if they are not farther from the origin than is v (step 9); such vertices will be examined as neighbors of another vertex if they can feasibly belong to a median path. The best possible score (i.e., lower bound) for a vertex w is is used as the basis for prioritization. Best and worst possible scores are calculated using the bounds of Theorem 1 (steps 10 and 11) and maintained for all vertices present in the priority stack. Vertices can be pruned when their best possible scores exceed the current global upper bound. The global upper bound can be lowered when a vertex is found that has a lesser upper bound (step 13). The search ends when no vertex in the queue has a best-possible score lower than the upper bound (step 7) or a score equal to the global lower bound is found (step 12).

The algorithm will return a permutation M only if M is a true median of the inputs $\pi_1, \pi_2,$ and π_3. Assume to the contrary that a permutation M' returned by the algorithm is not a true median. Because the algorithm returns the permutation having the lowest median score of all of the permutations (vertices) it visits (steps 5 and 13), it must not have visited some median. If the algorithm did not visit some median, then either it pruned all paths to medians or it exited before reaching any median.

Suppose the algorithm pruned all paths to medians. It only prunes vertices when their best possible scores are lower than the current global upper bound, M_{max}. Note that the global upper bound always corresponds to the actual median score of a vertex that has been visited (steps 2 and 13), so it cannot be wrong. Consider a median M with at least one median path P_M. By Definition 2, P_M must include at least one path between M and each of the vertices $v_1, v_2,$ and v_3 corresponding to $\pi_1, \pi_2,$ and π_3. The algorithm proceeds by examining neighbors of an origin $\psi_{orig} \in \{\pi_1, \pi_2, \pi_3\}$. Therefore, if the algorithm pruned all paths to M, then it must have pruned a vertex on the path between ψ_{orig} and M. But the best scores of such vertices are calculated using the lower bound of Theorem 1 (step 10), which we have shown to be correct. Therefore, the algorithm cannot have pruned the shortest paths to medians.

Suppose instead that the algorithm exited before reaching a median. The algorithm can exit for one of three reasons:

1. The priority stack s becomes empty (step 6);
2. The next item returned from s has a best possible score greater than or equal to the current global upper bound (step 7);
3. A vertex w is found with a worst possible score equal to the global lower bound (step 12);

Case 1 can occur only if all vertices have been visited, or if all remaining neighbors have been pruned (because except when the algorithm stops for another reason, each new neighbor is either pruned or pushed onto s). If all vertices have been visited, then a median must have been visited. We have shown above that all neighbors on paths to a median cannot have been pruned. Because s always returns a vertex v such that no other vertex in s has a lower best-possible score than v, and because all neighbors that are not pruned are added to s, case 2 can only occur if a median has been visited or if all paths to medians have been pruned. We have shown that all paths to medians cannot have been pruned. Therefore, if case 2 occurs, a median must have been visited. In case 3, w must be a median, since the global lower bound is set directly according to Lemma 1 (step 1), which we have shown to be correct.

Thus, none of these three cases can arise before a median has been found, and the algorithm must return a median. The worst-case running time of the algorithm is $O(n^{3d})$, with $d = \min\{d_{1,2}, d_{2,3}, d_{1,3}\}$, but as would be expected with a branch-and-bound algorithm, the average running time appears to be much better.

5   Experimental Method

We implemented find_inversion_median in C, reusing the linear-time distance routine (as well as some auxiliary code) from GRAPPA [1], and we evaluated its performance on simulated data. All test data was generated by a simple program that creates multiple sets of three permutations by applying random inversions to the identity permutation, such that each set of three permutations represents three taxa derived from a common ancestor under an inversions-only model of evolution. In addition to the number of genes n to model and the number of sets s to create, this program accepts a parameter i that determines how many random inversions to apply in obtaining the permutation for each taxon. Thus, if $n = 100$, $i = 10$, and $s = 10$, the program generates 10 sets of 3 signed permutations, each of size 100, and obtains each permutation by applying 10 random inversions to the permutation $+1, +2, \ldots, +100$. A random inversion is defined as an inversion between two random positions i and j such that $1 \leq i, j \leq n$ (if $i = j$, a single gene simply changes its sign). When i is small compared to n, each permutation in a set tends to be a distance of $2i$ from each other.

We used several algorithmic engineering techniques to improve the efficiency of find_inversion_median. For example, we avoided dynamic memory allocation and reused records representing graph vertices. We were able to gain a significant speedup by optimizing the hash table used for marking vertices: a custom hash table offered a fourfold increase in the overall speed of the program, as compared with UNIX's *db* implementation. With circular genomes, we achieved a further improvement in performance by hashing on the *circular identity* of each permutation rather than on the permutation itself. We define the circular identity of a permutation as that equivalent permutation that begins with the gene labeled $+1$. By hashing on circular identities, we reduced the number of vertices to visit and the number of permutations to mark by approximately a factor of $2n$.

To improve performance further, we adapted our sequential implementation to run in parallel on shared-memory architectures. Two steps in the algorithm are readily parallelizable: the major loop (step 6), during each iteration of which a new vertex is popped from the priority stack, and the minor loop (step 8), in which the neighbors of a vertex v are generated, examined for marks, and evaluated for feasibility as medians. We enabled parallel processing at both levels, using pthreads for maximum portability across shared-memory architectures. With careful use of semaphores and pthreads mutex functions, we were able to reduce the cost of synchronization among threads to an acceptable level.

6   Experimental Results

6.1   Performance of Bounds

Being especially concerned with the effectiveness of the pruning strategy, we have chosen as a measure of performance the number of vertices V of the inversion graph that
the algorithm visited. In particular, we have taken V to be the number of times the
program executed the loop at step 8 of the algorithm. Note that the number of calls to
distance is approximately $3V$. We recorded the distribution of V over many experiments, in which we used various values for the number of genes n and the number of
inversions per tree edge i. Figure 3 is typical of our results. It summarizes 500 experi-

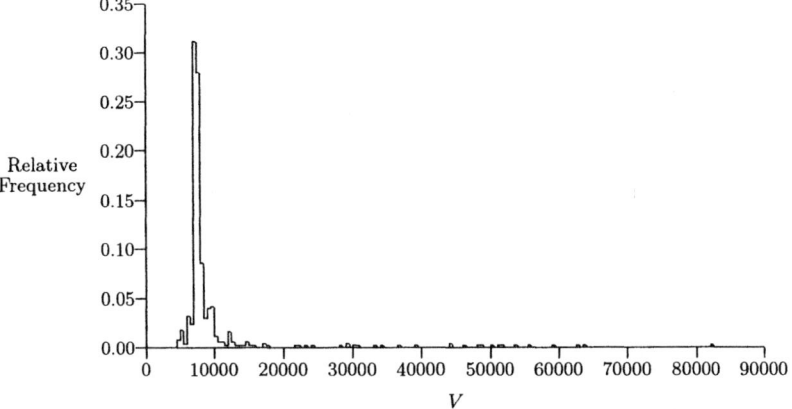

Fig. 3. Distribution of the number of vertices visited in the course of 500 experiments with $n = 50$
and $i = 7$.

ments with $n = 50$ and $i = 7$ and shows a roughly exponential distribution, with high
relative frequencies in a few intervals having small V: in 87% of the experiments, fewer
than 10,000 vertices were visited, and in 95%, fewer than 20,000 were visited. This figure demonstrates that the algorithm generally finds a median rapidly, but occasionally
becomes mired in an unprofitable region of the search space. We have observed that
the tail of the exponential distribution becomes more substantial as i grows larger with
respect to n.

In order to characterize typical performance, we recorded the statistical medians
of V as n and i varied independently. The results are shown in Figures 4 and 5. For
comparison, we have also plotted the mean values of V. Note that, at least for $i = 5$,
the median and mean of V appear to grow quadratically over a significant range of
values for n; a simple fit yields $f(n) = 2.1n^2$ for the median values. Note also that,
for $n = 50$, the median of V grows approximately linearly with i, at least as long as i
remains small (mean V grows somewhat faster than median V). To put the observed rate
of growth into perspective, note that in the theoretical worst-case of $O(n^{3d})$, because

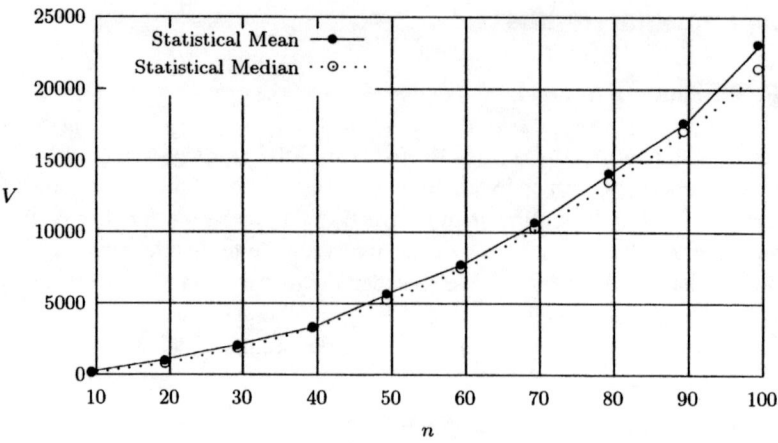

Fig. 4. Statistical median of the number of vertices visited V for $i = 5$ and $10 \leq n \leq 100$, over 50 experiments for each value of n.

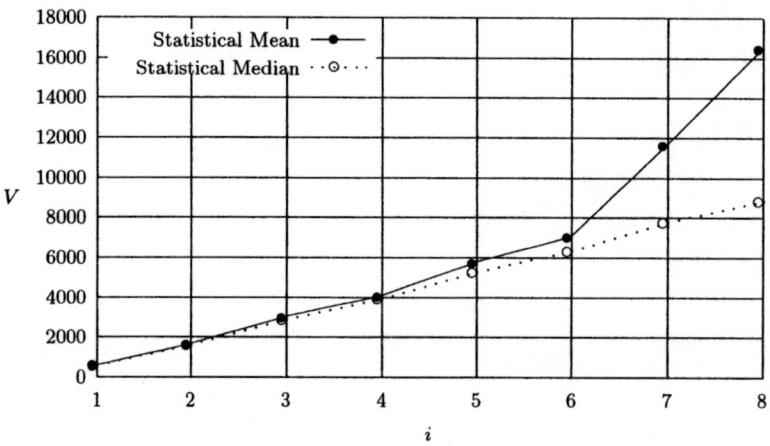

Fig. 5. Statistical median of the number of vertices visited V for $n = 50$ and $1 \leq i \leq 8$, plotted with mean of V. The number of experiments for each value of i is 50.

$d \approx 2i$ and $V = O(\frac{n^{3d}}{n}) = O(n^{(6i-1)})$, one would see (given $i = 5$ and $n = 50$) growth of V with n^{29} and 50^{6i-1}.

6.2   Running Time and Parallel Speedup

We have tested program find_inversion_median sequentially on a 700 MHz Intel Pentium III with 128MB of memory, and using various levels of parallelism on a Sun E10000 with 64 333 MHz UltraSPARC processors and 64GB of memory. Figure 6 shows average running times for $i = 5$ and n between 50 and 125. Sequential running

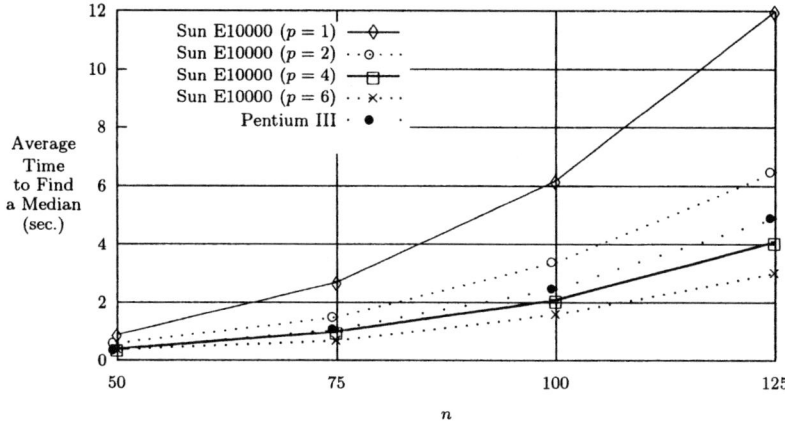

Fig. 6. Sequential and parallel running times for $i = 5$ and $n \in \{50, 75, 100, 125\}$. Each data point represents an average taken over 10 experiments. Parallel configurations used parallelism only in the minor loop of the algorithm.

times are shown for the Sun and Intel processors and parallel running times for the Sun with the number of processors $p \in \{1, 2, 4, 6\}$. In all cases, the average time to find a median is about 12 seconds or less. Observe that for $n = 100$ (a realistic size for chloroplast or mitochondrial genomes) medians can generally be found in an average of about 2 seconds using a reasonably fast computer. We should note that the memory requirements for the program are considerable, and that the level of performance shown here is partly a consequence of the large amount of RAM available on the Sun.

It is evident from Figure 6 that we achieve a good parallel speedup for small p, but that the benefits of parallelization begin to erode between $p = 4$ and $p = 6$ (this tendency becomes more pronounced at $p = 8$, which we have not plotted here for clarity of presentation). Anecdotal evidence suggests that the cause of this trend is a combination of the overhead of synchronization and uneven load balancing among the computing threads. We also observed that parallelism in the minor loop of the algorithm was far more effective than parallelism in the major loop, presumably because the heuristic for prioritization is sufficiently effective that the latter strategy results in a large amount of unnecessary work.

6.3   Inversion Medians vs. Breakpoint Medians

Using program find_inversion_median, we evaluated the significance of inversion medians, by comparing them with breakpoint medians, trivial medians, and "actual" medians (i.e., the ancestral permutations from which observed taxa actually arose - in this case, always equal to the identity permutation). Figure 7, which shows results over $1 \leq i \leq 5$ for $n = 25$, is typical of what we observed. It demonstrates that

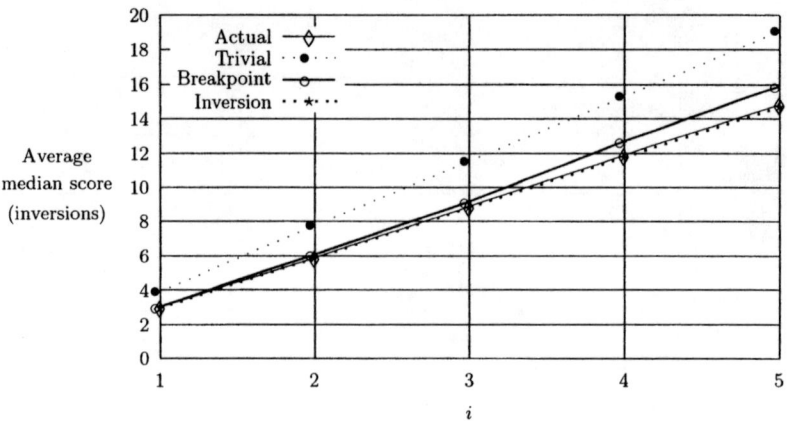

Fig. 7. Comparison of inversion medians with breakpoint medians, trivial medians, and actual medians, for $n = 25$. Averages were taken over 50 experiments.

inversion medians achieve comparable scores to actual medians[1] and that breakpoint medians, when scored in terms of inversion distance, perform significantly worse. A comparison in terms of inversion median scores is clearly biased in favor of inversion medians; however, if it is true that inversion distances are (in at least some cases) more meaningful than breakpoint distances, then these results suggest that inversion medians are worth obtaining.

We used a slight modeification of program find_inversion_median to find *all* optimal medians and thus to characterize the extent to which inversion medians are unique. An example of our results is shown in Figure 8, which describes the number of optimal inversion medians for $n = 15$ and $1 \leq i \leq 5$, over 50 experiments for each value of i. Observe that, when i is small compared to n (roughly $i \leq 0.15n$), the inversion median is virtually always unique; and even when i is moderately large with respect to n (roughly $0.15n < i \leq 0.3n$)[2], the inversion median is unique or nearly unique most of the time. This finding stands in stark contrast with breakpoint medians, which are only very rarely unique.

In addition, we observed a strong relationship between unique inversion medians and actual medians. For example, with $n = 15$ and $i = 1$, for which all inversion medians were unique, 49 out of 50 inversion medians were identical to actual medians; similarly, for $n = 15$ and $i = 2$, 48 out of 50 were identical to actual medians (in both cases the exceptional inversion medians differed from actual medians by a single inversion). As i becomes greater compared to n, this relationship weakens but remains significant. For example, with $n = 15$ and $i = 4$, 38 out of 50 inversion medians

---

[1] Inversion medians are slightly better than actual medians when i becomes large with respect to n, because saturation begins to cause convergence between taxa.

[2] Recall that the distance between permutations is approximately $2i$ and that random permutations tend to be separated by a distance of approximately n. The effects of saturation are evident at $i = 0.2n$ and are pronounced at $i = 0.3n$.

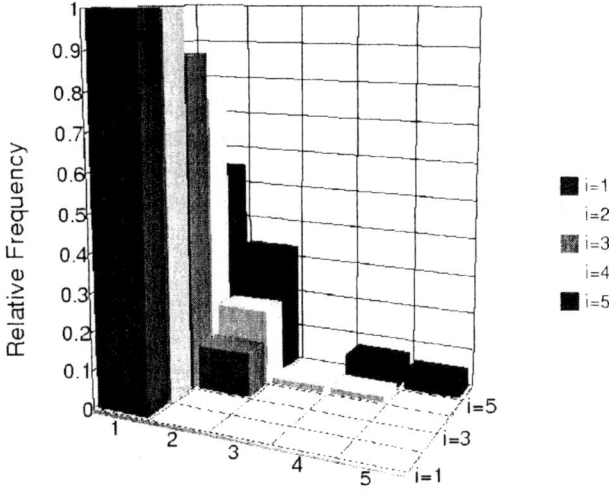

Fig. 8. Distribution of number of optimal medians in the course of 50 experiments for $n = 15$ and $1 \leq i \leq 5$.

were unique, and 22 of those 38 were identical to actual medians (an additional 10 non-unique inversion medians equaled actual medians).

7 Future Work

The strength and weakness of the current algorithm both lie in its generality. On the one hand, our approach depends only on elementary properties of metric spaces and thus extends easily to the case of equally weighted inversions, translocations, fissions, and fusions; furthermore, it could also be used with weighted rearrangement distances. (One should note, however, that the running time is a direct function of the cost of evaluating distances; we can compute exact breakpoint and inversion distances, but no efficient algorithm is yet known for more complex distance computations.) On the other hand, our approach does not exploit the unique structure of the inversion problem; as shown elsewhere in this volume by A. Caprara, restricting the algorithm to inversion distances only and using aspects of the Hannenhalli-Pevzner theory enables the derivation of tighter bounds and thus also the solution of larger instances of the inversion median problem.

Many simple changes to our current implementation will considerably reduce the running time. For example, the current implementation does not "condense" genomes before processing them—i.e., it does not convert subsequences of genes shared among all three genomes to single "supergenes". Preliminary experiments indicate that condensing genomes yields very significant improvements in performance when i is small relative to n. Distance computations themselves, while already fast, can be further improved by reusing previous computations, since a move by the algorithm makes only minimal changes to the candidate permutation. Finally, we can use the Kaplan-Shamir-Tarjan algorithm, in combination with metric properties, to prepare better initial solutions (by walking halfway through shortest paths between chosen permutations), thus considerably decreasing the search space to be explored.

References

1. D.A. Bader, B.M.E. Moret, and M. Yan. A linear-time algorithm for computing inversion distance between signed permutations with an experimental study. In *Proceedings 7th Workshop on Algorithms and Data Structures WADS91*. Springer Verlag, 2001. to appear in LNCS.

2. P. Berman and S. Hannenhalli. Fast sorting by reversal. In D. Hirschberg and E. Myers, editors, *Proceedings of the 7th Annual Symposium on Combinatorial Pattern Matching*, pages 168–185, 1996.

3. M. Blanchette, T. Kunisawa, and D. Sankoff. Parametric genome rearrangement. *Gene*, 172:GC11–GC17, 1996.

4. A. Caprara. Formulations and complexity of multiple sorting by reversals. In S. Istrail, P.A. Pevzner, and M.S. Waterman, editors, *Proceedings of the Third Annual International Conference on Computational Molecular Biology (RECOMB-99)*, pages 84–93, Lyon, France, April 1999.

5. M.E. Cosner, R.K. Jansen, B.M.E. Moret, L.A. Raubeson, L.-S. Wang, T. Warnow, and S. Wyman. An empirical comparison of phylogenetic methods on chloroplast gene order data in *Campanulaceae*. In D. Sankoff and J.H. Nadeau, editors, *Comparative Genomics*, pages 99–122. Kluwer Academic Press, 2000.

6. M.E. Cosner, R.K. Jansen, B.M.E. Moret, L.A. Raubeson, L.-S. Wang, T. Warnow, and S. Wyman. A new fast heuristic for computing the breakpoint phylogeny and experimental phylogenetic analyses of real and synthetic data. In *Proceedings of the 8th International Conference on Intelligent Systems for Molecular Biology ISMB-2000*, pages 104–115, 2000.

7. T. Dobzhansky and A.H. Sturtevant. Inversions in the chromosomes of *Drosophila pseudoobscura*. *Genetics*, 23:28–64, 1938.

8. D.S. Goldberg, S. McCouch, and J. Kleinberg. Algorithms for constructing comparative maps. In D. Sankoff and J.H. Nadeau, editors, *Comparative Genomics*. Kluwer Academic Press, 2000.

9. S. Hannenhalli, C. Chappey, E.V. Koonin, and P.A. Pevzner. Genome sequence comparison and scenarios for gene rearrangements: A test case. *Genomics*, 30:299–311, 1995.

10. S. Hannenhalli and P.A. Pevzner. Transforming cabbage into turnip (polynomial algorithm for sorting signed permutations by reversals). In *Proceedings of the 27th Annual ACM Symposium on the Theory of Computing*, pages 178–189, 1995.

11. S. Hannenhalli and P.A. Pevzner. Transforming men into mice (polynomial algorithm for genomic distance problem). In *Proceedings of the 36th Annual IEEE Symposium on Foundations of Computer Science*, pages 581–592, 1995.

12. H. Kaplan, R. Shamir, and R.E. Tarjan. A faster and simpler algorithm for sorting signed permutations by reversals. *SIAM Journal of Computing*, 29(3):880–892, 1999.

13. A. McLysaght, C. Seoighe, and K.H. Wolfe. High frequency of inversions during eukaryote gene order evolution. In D. Sankoff and J.H. Nadeau, editors, *Comparative Genomics*, pages 47–58. Kluwer Academic Press, 2000.

14. B.M.E. Moret, D.A. Bader, S. Wyman, T. Warnow, and M. Yan. A new implementation and detailed study of breakpoint analysis. In R.B. Altman, A.K. Dunker, L. Hunter, K. Lauderdale, and T.E. Klein, editors, *Pacific Symposium on Biocomputing 2001*, pages 583–594. World Scientific Publ. Co., 2001.

15. B.M.E. Moret, L.-S. Wang, T. Warnow, and S.K. Wyman. New approaches for reconstructing phylogenies from gene-order data. In *Proceedings 9th International Conference on Intelligent Systems for Molecular Biology ISMB-2001*, 2001. to appear in *Bioinformatics*.

16. Pavel A. Pevzner. *Computational Molecular Biology: An Algorithmic Approach*. MIT Press, 2000.

17. D. Sankoff and M. Blanchette. Multiple genome rearrangement and breakpoint phylogeny. *Journal of Computational Biology*, 5(3):555–570, 1998.

18. D. Sankoff, D. Bryant, M. Deneault, B.F. Lang, and G. Burger. Early eukaryote evolution based on mitochondrial gene order breakpoints. *Journal of Computational Biology*, 7(3/4):521–535, 2000.

19. D. Sankoff and N. El-Mabrouk. Genome rearrangement. In T. Jiang, Y. Xu, and M. Zhang, editors, *Topics in Computational Biology*. MIT Press, 2001.

20. D. Sankoff, V. Ferretti, and J.H. Nadeau. Conserved segment identification. *Journal of Computational Biology*, 4(4):559–565, 1997.

21. D. Sankoff, G. Leduc, N. Antoine, B. Paquin, and B.F. Lang. Gene order comparisons for phylogenetic inference: Evolution of the mitochondrial genome. *Proceedings of the National Academy of Sciences*, 89:6575–6579, 1992.

22. J. Setubal and J. Meidanis. *Introduction to Computational Molecular Biology*. PWS Publishing, 1997.

23. G.A. Watterson, W.J. Ewens, and T.E. Hall. The chromosome inversion problem. *Journal of Theoretical Biology*, 99:1–7, 1982.

Analytic Solutions for Three-Taxon ML_{MC} Trees with Variable Rates Across Sites

Benny Chor[1], Michael Hendy[2], and David Penny[3]

[1] Computer Science Department, Technion
Haifa 32000, Israel
benny@cs.technion.ac.il
[2] Institute of Fundamental Sciences, Massey University
Palmerston North, New Zealand
M.Hendy@massey.ac.nz
[3] Institute of Molecular Biosciences, Massey University
Palmerston North, New Zealand
D.Penny@massey.ac.nz

Abstract. We consider the problem of finding the maximum likelihood rooted tree under a molecular clock (ML_{MC}), with three species and 2-state characters under a symmetric model of substitution. For identically distributed rates per site this is probably the simplest phylogenetic estimation problem, and it is readily solved numerically. Analytic solutions, on the other hand, were obtained only recently (Yang, 2000).

In this work we provide analytic solutions for any distribution of rates across sites (provided the moment generating function of the distribution is strictly increasing over the negative real numbers). This class of distributions includes, among others, identical rates across sites, as well as the Gamma, the uniform, and the inverse Gaussian distributions. Therefore, our work generalizes Yang's solution. In addition, our derivation of the analytic solution is substantially simpler. We employ the Hadamard conjugation (Hendy and Penny, 1993) and convexity of an entropy–like function.

1 Introduction

Maximum likelihood (Felsenstein, 1981) is increasingly used as an optimality criterion for selecting evolutionary trees, but finding the global optimum is difficult computationally, even on a single tree. Because no general analytical solution is available, it is necessary to use numeric techniques, such as hill climbing or expectation maximization (EM), in order to find optimal values. Two recent developments are relevant when considering analytical solutions for simple substitution models with a small number of taxa. Yang (2000) has reported an analytical solution for three taxa with two state characters under a molecular clock. Thus in this special case the tree and the edge lengths that yield maximum likelihood values can now be expressed analytically, allowing the most likely tree to be positively identified. Yang calls this case the "simplest phylogeny estimation problem".

A second development is in Chor *et al.* (2000), who used the Hadamard conjugation for unrooted trees on four taxa, again with two state characters. As part of that study

O. Gascuel and B.M.E. Moret (Eds.): WABI 2001, LNCS 2149, pp. 204–213, 2001.

analytic solutions were found for some families of observed data. It was reported that multiple optima on a single tree occurred more frequently with maximum likelihood than has been expected. In one case, the best tree had a local (non global) optimum that was less likely than the optimum value on a different, inferior tree. In such a case, a hill climbing heuristic could misidentify the "optimal" tree. Such examples reinforce the desirability of analytical solutions that guarantee to find the global optima for any tree.

Even though three taxon, two state characters models under a molecular clock is the "simplest phylogeny estimation problem", it is still potentially an important case to solve analytically. It can allow a "rooted triplet" method for inferring larger rooted trees by building them up from the triplets. This would be analogous to the use of un-rooted quartets for building up unrooted trees. Trees from quartets methods are already used extensively in various studies (Bandelt and Dress 1986, Strimmer and von Hae-seler 1996, Wilson 1998, Ben-Dor *et al.* 1998, Erdös *et al.* 1999). The fact that general analytical solutions are not yet available for unrooted quartets only emphasizes the im-portance of analytical solutions to the rooted triplets case.

In this work we provide analytic solutions for three taxon ML$_{MC}$ trees under *any* distribution of variable rates across sites (provided the moment generating function of the distribution is strictly increasing over the negative real numbers). This class of distributions includes, as a special case, identical rates across sites. It also includes the Gamma, the uniform, and the inverse Gaussian distributions. Therefore, our work generalizes Yang's solution of identical rates across sites. In addition, our derivation of the analytic solution is substantially simpler. We employ the Hadamard conjugation (Hendy and Penny 1993, Hendy, Penny, and Steel 1994) and convexity of an entropy-like function.

The remainder of this paper is organized as follows: In subsection 2 we explain the Hadamard conjugation and its relation to maximum likelihood. In Section 3 we state and prove our main technical theorem. Section 4 applies the theorem to solve ML$_{MC}$ analytically on three species trees. Finally, Section 5 presents some implications of this work and directions for further research.

2   Hadamard Conjugation and ML

The Hadamard conjugation (Hendy and Penny 1993, Hendy, Penny, and Steel 1994) is an invertible transformation linking the probabilities of site substitutions on edges of an evolutionary tree T to the probabilities of obtaining each possible combination of characters. It is applicable to a number of simple models of site substitution: Neyman 2 state model (Neyman 1971), Jukes–Cantor model (Jukes and Cantor 1969), and Kimura 2ST and 3ST models (Kimura 1983). For these models, the transformation yields a powerful tool which greatly simplifies and unifies the analysis of phylogenetic data. In this section we explain the Hadamard conjugate and its relationships to ML.

We now introduce a notation that we will use for labeling the edges of unrooted binary trees. (For simplicity we use four taxa, but the definitions extend to any n.) Suppose the four species, $1, 2, 3$ and 4, are represented by the leaves of the tree T'. A *split* of the species is any partition of $\{1, 2, 3, 4\}$ into two disjoint subsets. We will identify each split by the subset which does not contain 4 (in general n), so that for example the split $\{\{1,2\}, \{3,4\}\}$ is identified by the subset $\{1,2\}$. Each edge e of T induces a split of the taxa, namely the two sets of leaves on the two components of T resulting from the deletion of e. Hence the central edge of the tree $T' = (12)(34)$ in the brackets notation induces the split identified by the subset $\{1,2\}$. For brevity we will label this edge by e_{12} as a shorthand for $e_{\{1,2\}}$. Thus $E(T') = \{e_1, e_2, e_{12}, e_3, e_{123}\}$ (see Figure 1).

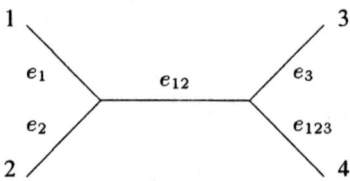

Fig. 1. The Tree $T' = (12)(34)$ and Its Edges.

We use a similar indexing scheme for splits at a site in the sequences: For a subset $\alpha \subseteq \{1, ..., n-1\}$, we say that a given site i is an α-split pattern if α is the set of sequences whose character state at position i differs from the i-th position in the n-th sequence. Given a tree T with n leaves and edge lengths $\mathbf{q} = [q_e]_{e \in E(T)}$ $(0 \leq q_e < \infty)$ (where q_e is the expected number of substitutions per site, across the edge e), the expected probability (averaged over all sites) of generating an α-split pattern $(\alpha \subseteq \{1, ..., n-1\})$ is well defined (this probability may vary across sites, depending on the distribution of rates). Denote this expected probability by $s_\alpha = Pr(\alpha\text{-split}|T, \mathbf{q})$. We define the *expected sequence spectrum* $\mathbf{s} = [s_\alpha]_{\alpha \subseteq \{1,...,n-1\}}$. Having this spectrum at hand greatly facilitates the calculation and analysis of the likelihood, since the likelihood of observing a sequence with splits described by the vector $\hat{\mathbf{s}}$ given the sequence spectrum $\mathbf{s}$ equals

$$L(\hat{\mathbf{s}}|\mathbf{s}) = \prod_{\alpha \subseteq \{1,...,n-1\}} Pr(\alpha\text{-split}\,|\,\mathbf{s})^{\hat{s}_\alpha} = \prod_{\hat{s}_\alpha > 0} s_\alpha^{\hat{s}_\alpha}.$$

Definition 1. *A Hadamard matrix of order ℓ is an $\ell \times \ell$ matrix A with ± 1 entries such that $A^t A = \ell I_\ell$.*

We will use a special family of Hadamard matrices, called Sylvester matrices in Mac/-Williams and Sloan (1977, p. 45), defined inductively for $n \geq 0$ by $H_0 = [1]$ and $H_{n+1} = \begin{bmatrix} H_n & H_n \\ H_n & -H_n \end{bmatrix}$. For example,

$$H_1 = \begin{bmatrix} 1 & 1 \\ 1 & -1 \end{bmatrix} \text{ and } H_2 = \begin{bmatrix} 1 & 1 & 1 & 1 \\ 1 & -1 & 1 & -1 \\ 1 & 1 & -1 & -1 \\ 1 & -1 & -1 & 1 \end{bmatrix}.$$

It is convenient to index the rows and columns of H_n by lexicographically ordered subsets of $\{1, \ldots, n\}$. Denote by $h_{\alpha, \gamma}$ the (α, γ) entry of H_n, then $h_{\alpha, \gamma} = (-1)^{|\alpha \cap \gamma|}$. This implies that H_n is symmetric, namely $H_n^t = H_n$, and thus by the definition of Hadamard matrices $H_n^{-1} = \frac{1}{2^n} H_n$.

The length of an edge q_e, $e \in E(T)$ in the tree T was defined as the expected number of substitutions (changes) per site along that edge. The *edge length* spectrum of a tree T be with n leaves is the 2^{n-1} dimensional vector $\mathbf{q} = [q_\alpha]_{\alpha \subseteq \{1, \ldots, n-1\}}$, defined for any subset $\alpha \subseteq \{1, \ldots, n-1\}$ by

$$
q_\alpha = \begin{cases} q_e & \text{if } e \in E(T) \text{ induces the split } \alpha, \\ -\sum_{e \in E(T)} q_e & \text{if } \alpha = \emptyset, \\ 0 & \text{otherwise.} \end{cases}
$$

The Hadamard conjugation specifies a relation between the expected sequence spectrum s and the edge lengths spectrum q of the tree.

Proposition 1. *(Hendy and Penny 1993) Let T be a phylogenetic tree on n leaves with finite edge lengths ($q_e < \infty$ for all $e \in E(T)$). Assume that sites mutate according to a symmetric substitution model, with equal rates across sites. Let s be the expected sequence spectrum. Then for $H = H_{n-1}$ we have:*

$$
\mathbf{s} = \mathbf{s}(\mathbf{q}) = H^{-1} \exp(H\mathbf{q}),
$$

where the exponentiation function exp is applied element wise to the vector $\rho = H\mathbf{q}$. That is, for $\alpha \subseteq \{1, \ldots, n-1\}$, $s_\alpha = 2^{-(n-1)} \sum_\gamma h_{\alpha, \gamma} \left(\exp \left(\sum_\delta h_{\gamma\delta} q_\delta \right) \right)$.

This transformation is called the *Hadamard conjugation*.

For the case of unequal rates across sites, the following generalization applies:

Proposition 2. *(Waddell, Penny, and Moore 1997) Let T be a phylogenetic tree on n leaves with finite edge lengths ($q_e < \infty$ for all $e \in E(T)$). Assume that sites mutate according to a symmetric substitution model, with unequal rates across sites, so that $M : R \to R$ be the moment generating function of the rate distribution. Let s be the expected sequence spectrum. Then with $H = H_{n-1}$,*

$$
\mathbf{s} = \mathbf{s}(\mathbf{q}) = H^{-1}(M(H\mathbf{q})),
$$

where the function M is applied element wise to the vector $\rho = H\mathbf{q}$.

This transformation is called the *Hadamard conjugation of M*. Specific examples of the moment generating function include

- For equal rates across sites, $M(\rho) = e^\rho$.
- For the uniform distribution in the interval $[1 - b, 1 + b]$ with parameter b ($1 \geq b > 0$),
  $M(\rho) = \frac{1}{2b\rho} \left(e^{(1+b)\rho} - e^{(1-b)\rho} \right)$.
- For the Γ distribution with parameter k ($k > 0$), $M(\rho) = (1 - \rho/k)^{-k}$.
- For the inverse Gaussian distribution with parameter d ($d > 0$), $M(\rho) = e^{d(1-\sqrt{1-2\rho/d})}$.

Notice that for $k \to \infty$, the Γ distribution converges to the equal rates distribution.

3   Technical Results

Under a molecular clock, a tree on n taxa has at least two sister taxa i and j whose pendant edges q_i and q_j are of equal length ($q_i = q_j$). Our first result states that if $q_i = q_j$, then the corresponding split probabilities are equal as well ($s_i = s_j$). Knowing that a pair of these variables attains the same value simplifies the analysis of the maximum likelihood tree in general, and in particular makes it possible for the case of $n = 3$ taxa. Furthermore, if $q_i > q_j$ and the moment generating function M is strictly increasing in the range $(-\infty, 0]$, then the corresponding split probabilities satisfy $s_i > s_j$ as well.

3.1   Main Technical Theorem

Theorem 1. *Let i and j be sister taxa on a phylogenetic tree T on n leaves, with edge weights q. Let s be the expected sequence spectrum, let $H = H_{n-1}$, and let M be a real valued function such that*

$$\mathbf{s} = H^{-1}M(H\mathbf{q}),$$

then:

$$q_i = q_j \Longrightarrow s_i = s_j;$$

and if the function M is strictly monotonic ascending in the range $\rho \in (-\infty, 0]$ then:

$$q_i > q_j \Longrightarrow s_i > s_j.$$

Proof. Let $X = \{1, 2, \ldots, n\}$ be the taxa set with reference element n, and let $X' = X - \{n\}$. Without loss of generality $i, j \neq n$. For $\alpha \subseteq X'$, let $\alpha' = \alpha \Delta \{i, j\}$ (where $\alpha \Delta \beta = (\alpha \cup \beta) - (\alpha \cap \beta)$ is the symmetric difference of α and β). The mapping $\alpha \to \alpha'$ is a bijection between

$$X_i = \{\alpha \subseteq X' | i \notin \alpha, j \in \alpha\}$$

and

$$X_j = \{\alpha \subseteq X' | i \in \alpha, j \notin \alpha\}.$$

Note that the two sets X_i and X_j are disjoint. Writing $h_{\alpha,i}$ for $h_{\alpha,\{i\}}$ we have

$$\alpha \in X_i \Longrightarrow h_{\alpha,i} = 1, h_{\alpha,j} = -1, h_{\alpha',i} = -1, h_{\alpha',j} = 1.$$

On the other hand, if $\alpha \notin X_i \cup X_j$ then $h_{\alpha,i} = h_{\alpha,j}$. Hence

$$
\begin{aligned}
s_i - s_j &= 2^{-(n-1)} \sum_{\alpha \subseteq X'} (h_{\alpha,i} - h_{\alpha,j}) M(\rho_\alpha) \\
&= 2^{-(n-1)} \Big(\sum_{\alpha \notin X_i \cup X_j} (h_{\alpha,i} - h_{\alpha,j}) M(\rho_\alpha) + \sum_{\alpha \in X_i} (h_{\alpha,i} - h_{\alpha,j}) M(\rho_\alpha) \\
&\qquad\qquad + \sum_{\alpha \in X_j} (h_{\alpha,i} - h_{\alpha,j}) M(\rho_\alpha) \Big) \\
&= 2^{-(n-1)} \Big(\sum_{\alpha \in X_i} (h_{\alpha,i} - h_{\alpha,j}) M(\rho_\alpha) + \sum_{\alpha \in X_j} (h_{\alpha,i} - h_{\alpha,j}) M(\rho_\alpha) \Big) \\
&= 2^{-(n-1)} \Big(\sum_{\alpha \in X_i} (h_{\alpha,i} - h_{\alpha,j}) M(\rho_\alpha) + \sum_{\alpha \in X_i} (h_{\alpha',i} - h_{\alpha',j}) M(\rho_{\alpha'}) \Big) \\
&= 2^{-(n-1)} \Big(\sum_{\alpha \in X_i} 2M(\rho_\alpha) - \sum_{\alpha \in X_i} 2M(\rho_{\alpha'}) \Big) \\
&= 2^{-(n-2)} \sum_{\alpha \in X_i} (M(\rho_\alpha) - M(\rho_{\alpha'})).
\end{aligned}
$$

By the definition of the Hadamard conjugate,

$$\rho_\alpha = \sum_{\beta \subseteq X'} h_{\alpha,\beta} q_\beta \quad , \text{so} \quad \rho_\alpha - \rho_{\alpha'} = \sum_{\beta \subseteq X'} (h_{\alpha,\beta} - h_{\alpha',\beta}) q_\beta \,.$$

Now for $\beta = \emptyset$ we have $h_{\alpha,\beta} = h_{\alpha',\beta} = 1$ so the contribution of $\beta = \emptyset$ to $\rho_\alpha - \rho_{\alpha'}$ is zero. Likewise, for any split $\beta \subseteq X'$ ($\beta \neq \emptyset$), which does not correspond to an edge $e \in E(T)$, $q_\beta = 0$. So the only contributions to $\rho_\alpha - \rho_{\alpha'}$ may come from splits β corresponding to edges in T. Now since i and j are sister taxa in T, every edge $e \in E(T)$ that is not pendant upon i or j does not separate i from j. Thus the split β corresponding to such edge e satisfies $\beta \notin X_i \cup X_j$. Therefore the parities of $|\alpha \cap \beta|$ and $|\alpha' \cap \beta|$ are the same, so

$$h_{\alpha,\beta} = (-1)^{|\alpha \cap \beta|} = (-1)^{|\alpha' \cap \beta|} = h_{\alpha',\beta} \,.$$

Thus the only contributions to $\rho_\alpha - \rho_{\alpha'}$ may come from the two edges pendant upon i and j, namely

$$\rho_\alpha - \rho_{\alpha'} = (h_{\alpha,i} - h_{\alpha',i}) q_i + (h_{\alpha,j} - h_{\alpha',j}) q_j \,,$$

and for $\alpha \in X_i$ we get $\rho_\alpha - \rho_{\alpha'} = 2(q_i - q_j)$.

Thus if $q_i = q_j$ then for every $\alpha \in X_i$ we have $\rho_\alpha = \rho_{\alpha'}$, so $M(\rho_\alpha) = M(\rho_{\alpha'})$ and thus $s_i - s_j = 2^{-(n-2)} \sum_{\alpha \in X_i} (M(\rho_\alpha) - M(\rho_{\alpha'})) = 0$, namely $s_i = s_j$. Further if $q_i > q_j$ then for every $\alpha \in X_i$ we have $\rho_\alpha > \rho_{\alpha'}$. Now $q_\emptyset = -\sum_{e \in E(T)} q_e$, and for every $e \in E(T)$, $q_e \geq 0$. Since $\rho_\alpha = \sum_{\beta \subseteq X'} h_{\alpha,\beta} q_\beta$ and $h_{\alpha,\emptyset} = 1$ we conclude that $\rho_\alpha \leq 0$ for all $\alpha \subseteq X'$. Therefore, if M is strictly monotone ascending in $(-\infty, 0]$ then $M(\rho_\alpha) > M(\rho_{\alpha'}), \forall \alpha \in X_i$. Since $s_i - s_j = 2^{-(n-2)} \sum_{\alpha \in X_i} (M(\rho_\alpha) - M(\rho_{\alpha'}))$, we have $s_i > s_j$. □

We remark that the moment generating function M in the four examples of Section 2 (equal rates across sites, uniform distribution with parameter b, $0 < b \leq 1$, Gamma distribution with parameter k, $0 < k$, and inverse Gaussian distribution with parameter d, $0 < d$) are strictly increasing in the range $\rho \in (-\infty, 0]$.

4 Three Taxa ML$_{MC}$ Trees

We first note that for three taxa, the problem of finding analytically the ML trees without the constraint of a molecular clock is trivial. This is a special case of unconstrained likelihood for the multinomial distribution. On the other hand, adding a molecular clock makes the problem interesting even for $n = 3$ taxa, which is the case we treat in this section.

For $n = 3$, let s_0 be the probability of observing the constant site pattern (xxx or yyy). Let s_1 be the probability of observing the site pattern which splits 1 from 2 and 3 (xyy or yxx). Similarly, let s_2 be the probability of observing the site pattern which splits 2 from 1 and 3 (yxy or xyx), and let s_3 be the probability of observing the site pattern which splits 3 from 1 and 2 (xxy or yyx).

Consider unrooted trees on the taxa set $X = \{1, 2, 3\}$ that have two edges of the same length. Let $\mathcal{T}_1$ denote the family of such trees with edges 2 and 3 of the same length ($q_2 = q_3$), $\mathcal{T}_2$ denote the family of such trees with edges 1 and 3 of the same length ($q_1 = q_3$), and $\mathcal{T}_3$ denote the family of such trees with edges 2 and 1 of the same length ($q_2 = q_1$). Finally, let $\mathcal{T}_0$ denotes the family of trees with $q_1 = q_2 = q_3$. We first see how to determine the ML tree for each family.

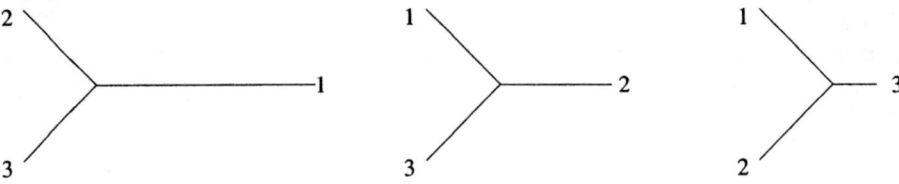

Fig. 2. Three Trees in the Families $\mathcal{T}_1, \mathcal{T}_2, \mathcal{T}_3$, respectively.

Given an observed sequence of m sites, let m_0 be the number of sites where all three nucleotides are equal, and let m_i $(i = 1, 2, 3)$ be the number of sites where the character in sequence i differs from the state of the other sequences. Then $m = m_0 + m_1 + m_2 + m_3$, and $f_i = m_i/m$ is the frequency of sites with the corresponding character state pattern.

Theorem 2. *Let* (m_0, m_1, m_2, m_3) *be the observed data. The ML tree in each family is obtained at the following point:*

- *For the family* $\mathcal{T}_0$, *the likelihood is maximized at* T_0 *with* $s_0 = f_0$, $s_1 = s_2 = s_3 = (1 - f_0)/3$.
- *For the family* $\mathcal{T}_1$, *the likelihood is maximized at* T_1 *with* $s_0 = f_0$, $s_1 = f_1$, $s_2 = s_3 = (f_2 + f_3)/2$.
- *For the family* $\mathcal{T}_2$, *the likelihood is maximized at* T_2 *with* $s_0 = f_0$, $s_2 = f_2$, $s_1 = s_3 = (f_1 + f_3)/2$.
- *For the family* $\mathcal{T}_3$, *the likelihood is maximized at* T_3 *with* $s_0 = f_0$, $s_3 = f_3$, $s_1 = s_2 = (f_1 + f_2)/2$.

Proof. The log likelihood function equals

$$l(m_0, m_1, m_2, m_3|\mathbf{s}) = \sum_{i=0}^{3} m_i \log s_i,$$

and for the normalized function $\ell = l/m$ we have

$$\ell(m_0, m_1, m_2, m_3|\mathbf{s}) = \sum_{i=0}^{3} f_i \log s_i .$$

Consider, without loss of generality, the case of the $\mathcal{T}_1$ family. We are interested in maximizing ℓ under the constraint $q_2 - q_3 = 0$. By Theorem 3.1, this implies $s_2 - s_3 = 0$. Therefore, using Lagrange multipliers, a maximum point of the likelihood must satisfy

$$\frac{\partial \ell}{\partial s_i} = \mu \frac{\partial (s_2 - s_3)}{\partial s_i} \quad (i = 1, 2, 3),$$

implying

$$\frac{f_1}{s_1} = \frac{f_0}{s_0},$$

$$\frac{f_2}{s_2} = \mu + \frac{f_0}{s_0},$$

$$\frac{f_3}{s_3} = -\mu + \frac{f_0}{s_0}.$$

Denote $d = f_0/s_0$, then by adding the last two equations and substituting $s_3 = s_2$ we have $f_2 + f_3 = 2ds_2$. Adding the right hand sides and left hand sides of this equality to these of $f_1 = ds_1$ and $f_0 = ds_0$, we get

$$f_0 + f_1 + f_2 + f_3 = d(s_0 + s_1 + 2s_2).$$

Since both $f_0 + f_1 + f_2 + f_3 = 1$ and $s_0 + s_1 + 2s_2 = 1$, we get $d = 1$. So the ML point for the family $\mathcal{T}_1$ is attained at the tree T_1 with parameters

$$s_0 = f_0, s_1 = f_1, s_2 = s_3 = (f_2 + f_3)/2.$$

We denote by T_2, T_3, T_0 the three corresponding trees that maximize the function ℓ for the families $\mathcal{T}_2, \mathcal{T}_3, \mathcal{T}_0$. The weights of these three trees can be obtained in a similar fashion to T_1. □

Theorem 3. *Assume* $m_3 \leq m_2 \leq m_1$. *Then the ML_{MC} tree equals T_1.*

Proof. By Theorem 2, the maximum likelihood tree under the condition that two edges have the same length is one of the trees $T_1, T_2,$ or T_3. Let

$$G(p) = f_0 \log f_0 + p \log p + (1 - f_0 - p) \log \frac{(1 - f_0 - p)}{2}.$$

Substituting the values s_0, s_1, s_2, s_3 for each tree in the expression

$$\ell(m_0, m_1, m_2, m_3 | \mathbf{s}) = \sum_{i=0}^{3} f_i \log s_i,$$

and somewhat abusing the notation, we get the following values for the function ℓ on the three trees

$$\ell(T_1) = G(f_1),$$
$$\ell(T_2) = G(f_2),$$
$$\ell(T_3) = G(f_3).$$

The function $G(p)$ behaves similarly to minus the binary entropy function (Gallager, 1968)

$$-H(p) = p \log p + (1 - p) \log(1 - p) .$$

The range where $G(p)$ is defined is $0 \leq p \leq 1 - f_0$. In this interval, $G(p)$ is negative and ∪-convex, just like $-H(p)$. So G has a single minimum at the point p_0 where its derivative is zero, $dG(p)/dp = 0$. Solving for p we get $p_0 = (1 - f_0)/3$.

Now $f_3 \leq f_2 \leq f_1$ and $G(p)$ is ∪-convex. Therefore, out of the three values $G(f_1), G(f_2), G(f_3)$, the maximum is attained at either $G(f_3)$ or at $G(f_1)$, but not at $G(f_2)$ (unless $f_2 = f_1$ or $f_2 = f_3$).

Since $f_3 + f_2 + f_1 = 1 - f_0$ and $f_3 \leq f_2 \leq f_1$, we have $f_3 \leq (1 - f_0)/3 \leq f_1$, namely the two "candidates" for ML points are on different sides of the minimum point. The point f_3 is strictly to the left and the point f_1 is strictly to the right (except the case where $f_3 = f_1$ and the two points coincide). If $G(f_1) \geq G(f_3)$, then the tree T_1 is the obvious candidate for ML_{MC} tree. Indeed, T_1 satisfies $s_3 = s_2 < s_1$, so by Theorem 2, $q_3 = q_2 < q_1$. Thus, a root can be placed on the edge q_1 so that the molecular clock assumption is satisfied.

We certainly could have a case where $G(f_3) > G(f_1)$. However, the tree T_3 has $s_3 < s_1 = s_2$, implying (by Theorem 2) $q_3 < q_1 = q_2$. Therefore there is no way to place a root on an edge of T_3 so as to satisfy a molecular clock. In fact, any tree with edge lengths $q_3 < q_1 = q_2$ does not satisfy a molecular clock. So the remaining possibilities could be either the tree T_0 (where $s_1 = s_2 = s_3 = (1 - f_0)/3$) or the tree T_1. As T_0 attains the minimum over the function G, we are always better off taking the tree T_1 (except in the redundant case $f_1 = f_3$, where all these trees collapse to T_0). This completes the proof of Theorem 3.   □

The case $m_2 < m_3 < m_1$ and its other permutations can clearly be handled similarly.

5 Discussion and Open Problems

In the case where $G(f_3) > G(f_1)$, T_1 is still the ML_{MC} tree. However, if the difference between the two values is significant, it may give a strong support for rejecting a molecular clock assumption for the given data m_0, m_1, m_2, m_3. This would be the case, for example, when $0 \approx m_3 \ll m_1 \approx m_2$.

Two natural directions for extending this work are to consider four state characters and to extend the number of taxa to $n = 4$ and beyond. The question of constructing rooted trees from rooted triplets is an interesting algorithmic problem, analogous to that of constructing unrooted trees from unrooted quartets. The biological relevance of triplets based reconstruction methods is also of interest.

Acknowledgments

Thanks to Sagi Snir for helpful comments on earlier versions of this manuscript. Benny Chor was supported by ISF grant 418/00 and did part of this work while visiting Massey University.

References

Bandelt, H.-J., and A. Dress, 1986. Reconstructing the shape of a tree from observed dissimilarity data. *Advances in Applied Mathematics*, 7:309–343.

Ben-Dor, A., B. Chor, D. Graur, R. Ophir, and D. Pelleg, 1998. Constructing phylogenies from quartets: Elucidation of eutherian superordinal relationships. *Jour. of Comput. Biology*, 5(3):377–390.

Chor, B., M. D. Hendy, B. R. Holland , and D. Penny, 2000. Multiple Maxima of Likelihood in Phylogenetic Trees: An Analytic Approach. *Mol. Biol. Evol.*, Vol. 17, No.10, September 2000, pp. 1529–1541.

Erdos, P., M. Steel, L. Szekely, and T. Warnow, 1999. A few logs suffice to build (almost) all trees (i). *Random Structures and Algorithms*, 14:153–184.

Felsenstein, J., 1981. Evolutionary trees from DNA sequences: A maximum likelihood approach. *J. Mol. Evol.*, 17:368–376.

Gallager, R.G. Information Theory and Reliable Communication, Wiley, New York (1968).

Hendy, M. D., and D. Penny, 1993. Spectral analysis of phylogenetic data. *J. Classif.*, 10:5–24.

Hendy, M. D., D. Penny, and M.A. Steel, 1994. Discrete fourier analysis for evolutionary trees. *Proc. Natl. Acad. Sci. USA.*, 91:3339–3343.

Neyman, J., 1971. Molecular studies of evolution: A source of novel statistical problems. In S. Gupta and Y. Jackel, editors, *Statistical Decision Theory and Related Topics*, pages 1–27. Academic Press, New York.

Strimmer, K., and A. von Haeseler, 1996. Quartet puzzling: A quartet maximum-likelihood method for reconstructing tree topologies. *Molecular Biology and Evolution*, 13(7):964–969.

Waddell, P., D. Penny, and T. Moore, 1997. Hadamard conjugations and Modeling Sequence Evolution with Unequal Rates across Sites. *Molecular Phylogenetics and Evolution*, 8(1):33–50.

Wilson, S.J., 1998. Measuring inconsistency in phylogenetic trees. *Journal of Theoretical Biology*, 190:15–36.

Yang, Z., 2000. Complexity of the simplest phylogenetic estimation problem. *Proc. R. Soc. Lond. B*, 267:109–119.

The Performance of Phylogenetic Methods on Trees of Bounded Diameter

Luay Nakhleh[1], Usman Roshan[1], Katherine St. John[1,2],
Jerry Sun[1], and Tandy Warnow[1]

[1] Department of Computer Sciences
University of Texas, Austin, TX 78712
{nakhleh,usman,jsun,stjohn,tandy}@cs.utexas.edu
[2] Lehman College & the Graduate Center, City U. of New York

Abstract. We study the convergence rates of neighbor-joining and several new phylogenetic reconstruction methods on families of trees of bounded diameter. Our study presents theoretically obtained convergence rates, as well as an empirical study based upon simulation of evolution on random birth-death trees. We find that the new phylogenetic methods offer an advantage over the neighbor-joining method, except at low rates of evolution where they have comparable performance. The improvement in performance of the new methods over neighbor-joining increases with the number of taxa and the rate of evolution.

1 Introduction

Phylogenetic trees (that is, evolutionary trees) form an important part of biological research. As such, there are many algorithms for inferring phylogenetic trees. The majority of these methods are designed to be used on biomolecular (i.e., DNA, RNA, or amino-acid) sequences. Methods for inferring phylogenies from biomolecular sequence data are studied (both theoretically and empirically) with respect to the topological accuracy of the inferred trees. Such studies evaluate the effects of various model conditions (such as the sequence length, the rates of evolution on the tree, and the tree "shape") on the performance of various methods.

The *sequence length requirement* of a method is the sequence length needed by the method in order to obtain (with high probability) the true tree topology. Earlier studies established analytical upper bounds on the sequence length requirements of various methods (including the popular neighbor-joining [18] method). These studies showed that standard methods, such as neighbor-joining, recover the true tree (with high probability) from sequences of lengths that are exponential in the evolutionary diameter of the true tree. Based upon these studies, in [5,6] we defined a parameterization of model trees in which the longest and shortest edge lengths are fixed, so that the sequence length requirement of a method can be expressed as a function of the number of taxa, n. This parameterization leads to the definition of "fast-converging" methods, which are methods that recover the true tree from sequences of lengths bounded by a polynomial in n once f, the minimum edge length, and g, the maximum edge length, are bounded. Several fast-converging methods were developed [3,4,8,21]. We and others analyzed

O. Gascuel and B.M.E. Moret (Eds.): WABI 2001, LNCS 2149, pp. 214–226, 2001.
© Springer-Verlag Berlin Heidelberg 2001

the sequence length requirement of standard methods, such as neighbor-joining (NJ), under the assumptions that f and g are fixed. These studies [1, 6] showed that neighbor-joining and many other methods can be proven to be "exponentially-converging", that is, they recover the true tree with high probability from sequences of lengths bounded by a function that grows exponentially in n. So far, none of these standard methods are known to be "fast-converging."

In this paper, we consider a different parameterization of the model tree space, where we fix the evolutionary diameter of the tree, and let the number of taxa vary. This parameterization, suggested by John Huelsenbeck [personal communication], allows us to examine the differential performance of methods with respect to "taxon sampling" strategies [7]. In this case, the shortest edges can be arbitrarily short, forcing the method to require unboundedly long sequences in order to recover these shortest edges. Hence, the sequence length requirements of all methods cannot be bounded. However, for a natural class of model trees, it can be assumed that $f = \Theta(1/n)$ (for example, random birth-death trees fall into this class). In this case even very simple polynomial time methods converge to the true tree from sequences whose lengths are bounded by a polynomial in n. Furthermore, the degrees of the polynomials bounding the convergence rates of neighbor-joining and the "fast-converging" methods are identical – they differ only with respect to the leading constants. Therefore, with respect to this parameterization, there is no significant theoretical advantage between standard methods and the "fast-converging" methods. We then evaluate two methods, neighbor-joining and DCM-NJ+MP (a method introduced in [14]) with respect to their performance on simulated data, obtained on random birth-death trees with bounded deviation from ultrametricity. We find that DCM-NJ+MP obtains an advantage over neighbor-joining throughout most of the parameter space we examine, and is never worse. That advantage increases as the deviation from ultrametricity increases or as the number of taxa increases.

The rest of the paper is organized as follows. In Section 2, we present the basic definitions, models of evolution, methods, and terms, upon which the rest of the paper is based. In Section 3, we present the theory behind convergence rate bounds for both neighbor-joining and "fast-converging" methods. We derive bounds on the convergence rates of various methods for trees in which the evolutionary diameter (but not the shortest edge lengths) is fixed. We then derive bounds on the convergence rates of these methods for random trees drawn from the distribution on birth-death trees described above. In Section 5, we describe our experimental study comparing the performance of neighbor-joining and DCM-NJ+MP. In Section 6, we conclude with a discussion and open problems.

2   Basics

In this section, we present the basic definitions, models of evolution, methods, and terms, upon which the rest of the paper is based.

2.1   Model Trees

The first step of every simulation study for phylogenetic reconstruction methods is to generate *model trees*. Sequences are then evolved down these trees, and these sequences are used, by the methods in question, to estimate the model tree. The accuracy of the method is determined by how well the method reproduces the model tree. Model trees are often taken from some underlying distribution on all rooted binary trees with n leaves. Some possible distributions include the uniform (all binary trees on n leaves are equiprobable) and the Yule-Harding distribution (a distribution based upon a model of speciation).

In this paper, we use random birth-death trees with n leaves as our underlying distribution. To generate these trees, we view speciation and extinction events occurring over a continuous interval. During a short time interval, Δt, a species can split into two with probability $b(t)\Delta t$, and a species can become extinct with probability $d(t)\Delta t$. The values of $b(t)$ and $d(t)$ depend on how much time has passed in the model. To generate a tree with n taxa, we begin this process with a single node and continue until we have a tree with n taxa (with some non-zero probability some processes will not produce a tree of the desired size since all nodes could go "extinct" before n species are generated; if this happens, we repeat the process, until a tree of the desired size is generated). Under this distribution, trees have a natural length assigned to each edge– that is the time t between the speciation event that began that edge and the event (which could be either speciation or extinction) that ended that edge.

Birth-death trees are inherently ultrametric, that is, the branch lengths are proportional to time. In all of our experiments we modified each edge length to deviate from this assumption that sites evolve under the strong molecular clock. To do this, we multiplied each edge by a random number within a range $[1/c, c]$, where we set c to be some small constant. We call this constant the *deviation factor*.

2.2   Models of Evolution

Under the *Kimura 2-Parameter* (K2P) model [10], each site evolves down the tree under the Markov assumption, but there are two different types of nucleotide substitutions: transitions and transversions. A transition is a substitution of a purine (an adenine or guanine nucleotide) for a purine, or a pyrimidine (a cytosine or thymidine nucleotide) for a pyrimidine; a transversion is a substitution of a purine for a pyrimidine or vice versa. The probability of a given nucleotide substitution depends on the edge and upon the type of substitution. A K2P tree is defined by the triplet $(T, \{\lambda(e)\}, ts/tv)$, where $\lambda(e)$ is the expected number of times a random site will change its nucleotide on e, and ts/tv is the transition/transversion ratio. In our experiments, we fix this ratio to 2, one of the standard settings.

It is sometimes assumed that the sites evolve identically and independently down the tree. However, we can also assume that the sites have different rates of evolution, and that these rates are drawn from a known distribution. One popular assumption is that the rates are drawn from a gamma distribution with shape parameter α, which is the inverse of the coefficient of variation of the substitution rate. We use $\alpha = 1$ for our experiments under K2P+Gamma. With these assumptions, we can specify a K2P+Gamma tree just by the pair $(T, \{\lambda(e)\})$.

2.3   Statistical Performance Issues

A phylogenetic reconstruction method is *statistically consistent* under a model of evolution if for every tree in that model the probability that the method reconstructs the tree tends to 1 as the sequence length increases. Under the assumption of a K2P+Gamma evolutionary process, if the transition/transversion ratio and shape parameter are known, it is possible to define pairwise distances between taxa so that distance-based methods (such as neighbor-joining) are statistically consistent [11]. Real biomolecular sequences are of limited length. Therefore, the length k of the sequences affects the performance of the method M significantly. The *convergence rate* of a method M is the rate at which it converges to 100% accuracy as a function of the sequence length.

2.4   Phylogenetic Reconstruction Methods

We briefly discuss the two phylogenetic methods we use in our empirical studies: neighbor-joining and DCM-NJ+MP. Both methods have polynomial running time.

Neighbor-Joining: Neighbor-joining [18] is one of the most popular distance based methods. Neighbor-joining takes a distance matrix as input and outputs a tree. For every two taxa, it determines a score, based on the distance matrix. At each step, the algorithm joins the pair with the minimum score, making a subtree whose root replaces the two chosen taxa in the matrix. The distances are recalculated to this new node, and the "joining" is repeated until only three nodes remain. These are joined to form an unrooted binary tree.

DCM-NJ+MP: The DCM-NJ+MP method is a variant of a provably fast-converging method that has performed very well in previous studies [14]. In these simulation studies, DCM-NJ+MP outperforms, in terms of topological accuracy, the methods DCM^*-NJ (of which it is a variant) and neighbor-joining.

The method works as follows: let d_{ij} be the distance between taxa i and j.

- *Phase 1:* For each $q \in \{d_{ij}\}$, compute a binary tree T_q, by using the Disk-Covering Method from [6], followed by a heuristic for refining the resultant tree into a binary tree. Let $\mathcal{T} = \{T_q : q \in \{d_{ij}\}\}$. (Readers interested in more details of how Phase I is handled should see [6].)
- *Phase 2:* Select the tree from $\mathcal{T}$ which optimizes the parsimony criterion.

If we consider all $\binom{n}{2}$ thresholds in Phase 1, DCM-NJ+MP takes $O(n^6)$ time. However, if we consider only a fixed number p of thresholds, DCM-NJ+MP takes $O(pn^4)$.

2.5   Measures of Accuracy

There are many ways of measuring error between trees. We use the *Robinson-Foulds* (RF) distance [16] which is defined as follows. Every edge e in a leaf-labeled tree T defines a bipartition π_e on the leaves (induced by the deletion of e), and hence the tree T is uniquely encoded by the set $C(T) = \{\pi_e : e \in E(T)\}$, where $E(T)$ is the set of all internal edges of T. If T is a model tree and T' is the tree obtained by a phylogenetic reconstruction method, then the error in the topology can be calculated as follows:

- *False Positives:* $C(T') - C(T)$.
- *False Negatives:* $C(T) - C(T')$.

The RF distance is $\frac{|C(T)\triangle C(T')|}{2(n-3)}$, i.e., the average of the false positive and the false negative rates.

3   Theoretical Results on Convergence Rates

In [1], the sequence length requirement for the neighbor-joining method under the Cavender-Farris model was bounded from above, and extended to the General Markov model in [5]. We state the result here:

Theorem 1. *([1, 5]) Let* (T, M) *be a model tree in the General Markov model. Let*

$$\lambda(e) = -\log|det(M_e)|, \text{ and set } \lambda_{ij} = \sum_{e \in P_{ij}} \lambda(e).$$

Assume that f *is fixed with* $0 < f \leq \lambda(e)$ *for all edges* $e \in T$. *Let* $\varepsilon > 0$ *be given. Then, there are constants* C *and* C' *(that do not depend upon* f*) such that, for*

$$k = \frac{C}{f^2} \log n e^{C'(\max \lambda_{ij})}$$

then with probability at least $1 - \epsilon$, *neighbor-joining on* S *returns the true tree, where* S *is a set of sequences of length* k *generated on* T. *The same sequence length requirement applies to the* Q^* *method of [2].*

From Theorem 1 we can see that as the edge length gets smaller, the sequence length has to be larger in order for neighbor-joining to return the true tree with high probability. Note that the diameter of the tree and the sequence length are "exponentially" related.

3.1   Fixed-Parameter Analyses of the Convergence Rate

Analysis when both f *and* g *Are Fixed:* In [8,21], the convergence rate of neighbor-joining was analyzed when both f and g are fixed (recall that f is the smallest edge length, and g is the largest edge length). In this setting, by Theorem 1 and because $\max \lambda_{ij} = O(gn)$, we see that neighbor-joining recovers the true tree, with probability $1 - \epsilon$, from sequences that grow exponentially in n. An average case analysis of tree topologies under various distributions shows that $\max \lambda_{ij} = \Theta(g\sqrt{n})$ for the uniform distribution and $\Theta(g \log n)$ for the Yule-Harding distribution. Hence, neighbor-joining has an average case convergence rate which is polynomial in n under the Yule-Harding distribution, but not under the uniform distribution.

By definition, "fast-converging" methods are required to converge to the true tree from polynomial length sequences, when f and g are fixed. The convergence rates of fast-converging methods have a somewhat different form. We show the analysis for the DCM^*-NJ method (see [21]):

Theorem 2. *([21]) Let (T, M) be a model tree in the General Markov model. Let*

$$\lambda(e) = -\log|det(M_e)|, \text{ and set } \lambda_{ij} = \sum_{e \in P_{ij}} \lambda(e).$$

Assume that f is fixed with $0 < f \leq \lambda(e)$ for all edges $e \in T$. Let $\varepsilon > 0$ be given. Then, there are constants C and C' (that do not depend upon f) such that, for

$$k = \frac{C}{f^2} \log n e^{C'(width(T))}$$

then with probability at least $1 - \epsilon$, DCM^*-NJ on S returns the true tree, where S is a set of sequences of length k generated on T, and $width(T)$ is a topologically defined function which is bounded from above by $\max \lambda_{ij}$ and is also $O(g \log n)$.

Consequently, fast-converging methods recover the true tree from polynomial length sequences when both f and g are fixed.

Analysis when $\max \lambda_{ij}$ Is Fixed: Suppose now that we fix $\max \lambda_{ij}$ but not f. In this case, neither neighbor-joining nor the "fast-converging" methods will recover the true tree from sequences whose lengths grow polynomially in n, because as $f \to 0$, the sequence length requirement increases without bound. However, for "random" birth-death trees, the expected minimum edge length is $\Theta(1/n)$. Hence, suppose that in addition to fixing $\max \lambda_{ij}$ we also require that $f = \Theta(1/n)$. In this case, application of Theorem 1 and Theorem 2 shows that neighbor-joining and the "fast-converging" methods all recover the true tree with high probability from $O(n^2 \log n)$-length sequences. The theoretically obtained convergence rates differ only in the leading constant, which in neighbor-joining's case depends exponentially on $\max \lambda_{ij}$, while in the case of DCM^*-NJ's this rate depends exponentially on $width(T)$. Thus, the performance advantage of a fast-converging method– from a theoretical perspective– depends upon the difference between these two values. We know that $width(T) \leq \max \lambda_{ij}$ for all trees. Furthermore, the two values are essentially equal only when the strong molecular clock assumption holds. Note also that when the tree has a low evolutionary diameter (i.e., when $\max \lambda_{ij}$ is small), then the predicted performance of these methods suggests that they will be approximately identical. Only for large evolutionary diameters should we obtain a performance advantage by using the fast-converging methods instead of neighbor-joining.

In the next section we discuss the empirical performance of these methods.

4   Earlier Performance Studies Comparing DCM-NJ+MP to NJ on Random Trees

In an earlier study [14], we studied the performance of the neighbor-joining (NJ) method, and several new variants of the disk-covering method. The DCM-NJ+MP method was one of these new variants we tested. Our experiments (some of which we present here) showed that for random trees (from the uniform distribution on binary tree topologies) with random branch lengths (also drawn from the uniform distribution within some specified range), the DCM-NJ+MP method was a clear improvement upon the NJ method with respect to topological accuracy. The DCM-NJ+MP method was also more accurate in many of our experiments than the other variants we tested, leading us to conclude that the improved performance on random trees might extend to other distributions on model trees.

Later in this paper we will present new experiments, testing this conclusion on random birth-death trees with a moderate deviation from ultrametricity. Here we present a small sample of our earlier experiments, which shows the improved performance and indicates how DCM-NJ+MP obtains this improved performance.

Recall that the DCM-NJ+MP method has two phases. In the first phase, a collection of trees is obtained, one for each setting of the parameter q. This inference is based upon dividing the input set into overlapping subsets, each of diameter bounded from above by q. The NJ method is then used on each subset to get a subtree for the subset, and these subtrees are merged into a single supertree. These trees are constructed to be binary trees, and hence do not need to be further resolved. This first phase is the "DCM-NJ" portion of the method. In the second phase, we select a single tree from the collection of trees $\{T_q : q \in d_{ij}\}$, by selecting the tree which has the optimal parsimony score (i.e., the fewest changes on the tree).

The accuracy of this two-phase method depends upon two properties: first, the first phase must produce a set of trees so that at least some of these trees are better than the NJ tree, and second, the technique (in our case, maximum parsimony) used in the second phase must be capable of selecting a better tree than the NJ tree. Thus, the first property depends upon the DCM-NJ method providing an improvement, and the second property depends upon the performance of the maximum parsimony criterion as a technique for selecting from the set $\{T_q\}$. In the following figures we show that both properties hold for random trees under the uniform distribution on tree topologies and branch lengths.

In Figure 1, we show the results of an experiment in which we scored each of the different trees T_q for topological accuracy. This experiment is based upon random trees from the uniform distribution. Note that the best trees are significantly better than the NJ tree. Thus, the DCM-NJ method itself is providing an advantage over the NJ method.

In Figure 2 we show the result of a similar experiment in which we compared several different techniques for the second phase (i.e., for selecting a tree from the set $\{T_q\}$). This figure shows that the Maximum Parsimony (MP) technique obtains better trees than the Short Quartet Support Method, which is the technique used in the second phase of the DCM^*-NJ method. Furthermore, both DCM-NJ+MP and DCM^*-NJ improve upon NJ, and this improvement increases with the number of taxa.

Thus, for random trees from the uniform distribution on tree topologies and branch lengths, DCM-NJ+MP improves upon NJ, and this improvement is due to both the decomposition strategy used in Phase 1, and the selection criterion used in Phase 2.

Note however that DCM-NJ+MP is not statistically consistent, even under the simplest models, since the maximum parsimony criterion can select the wrong tree with probability going to 1 as the sequence length increases.

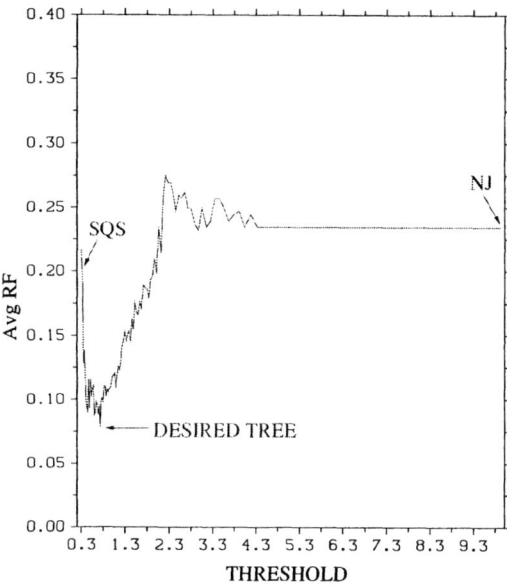

Fig. 1. The accuracy of the T_q's for different values of q on a randomly generated tree with 100 taxa, sequence length 1000, and an average branch length of 0.05.

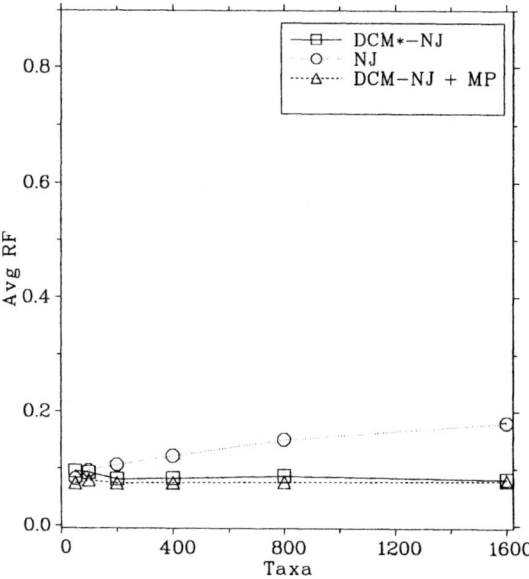

Fig. 2. DCM-NJ+MP vs. DCM^*-NJ vs. NJ on random trees (uniform distribution on tree topologies and branch lengths) with sequence evolution under the K2P+Gamma model. Sequence length is 1000. Average branch length is 0.05.

5   New Performance Studies under Birth-Death Trees

5.1   Introduction

In this paper we focused upon the question of whether the improvement in performance over NJ that we saw in DCM-NJ+MP was a function of the distribution on tree topologies and branch lengths (both uniform), or whether we would continue to see an improvement in performance, by comparison to NJ, when we restrict our attention to a more biologically based distribution on model trees. Hence we focus on random birth-death trees, with some deviation from ultrametricity added (so that the strong molecular clock does not hold). As we will show, the improvement in performance is still visible, and our earlier claims extend to this case.

5.2   Experimental Platform

Machines: The experiments were run on the SCOUT cluster at University of Texas, which contains approximately 130 different processors running the Debian Linux operating system. We also had nighttime use of approximately 150 Pentium III processors located in public undergraduate laboratories.

Software: We used Sanderson's r8s package for generating birth-death trees [17] and the program Seq-Gen [15] to randomly generate a DNA sequence for the root and evolve it through the tree under K2P+Gamma model of evolution. We calculated evolutionary distances appropriately for the model (see [11]). In the presence of saturation (that is, datasets in which some distances could not be calculated because the formula did not apply), we used the "fix-factor 1" technique, as defined in [9]. In this technique, the distances that cannot be set using the standard technique are all assigned the largest corrected distance in the matrix.

The software for DCM-NJ was written by Daniel Huson. To calculate the maximum parsimony scores of the trees we used PAUP* 4.0 [19]. For job management across the cluster and public laboratory machines, we used the Condor software package [20]. We generated the rest of this software (a combination of C++ programs and Perl scripts) explicitly for these experiments.

5.3   Bounded Diameter Trees

We performed experiments on bounded diameter trees, and observed how the error rates increase as the number of taxa increases. The birth-death trees that we generated using r8s have diameter 2. In order to obtain trees with other diameters, we multiplied the edge lengths by factors of 0.01, 0.1, and 0.5, thus obtaining trees of diameters 0.02, 0.2, and 1.0, respectively. Then, to deviate these trees from ultrametricity, we modified the edge lengths using deviation factor 4. The resulting trees have diameters bounded from above by 4 times the original diameter, but have expected diameters of approximately twice the original diameters. Thus, the final model trees have expected diameters that are 0.04, 0.4, and 2.0. In this way we generated random model trees with 10, 25, 50, 100, 200, 400, and 800 leaves. For each number of taxa and diameter, we generated 30 random birth-death trees (using r8s).

5.4   Experimental Design

For each model tree we generated sequences of length 500 using seq-gen, computed trees using NJ and DCM-NJ+MP. We then computed the Robinson-Foulds error rate for each of the inferred trees, by comparing it to the model tree that generated the data.

5.5   Results and Discussion

In order to obtain statistically robust results, we followed the advice of McGeoch [12] and Moret [13] and used a number of *runs*, each composed of a number of *trials* (a trial is a single comparison), computed the mean and standard deviation over the runs of these events. This approach is preferable to using the same total number of samples in a single run, because each of the runs is an independent pseudorandom stream. With this method, one can obtain estimates of the mean that are closely clustered around the true value, even if the pseudorandom generator is not perfect.

The standard deviation of the mean outcomes in our studies varied depending on the number of taxa. The standard deviation of the mean on 10-taxon trees is 0.2 (which is 20 percent, since the possible values of the outcomes range from 0 to 1), on 25-taxon trees is 0.1 (which is 10 percent), whereas on 200, 400 and 800-taxon trees the standard deviation ranged from 0.02 to 0.04 (which is between 2 and 4 percent). We graph the average of the mean outcomes for the runs, but omit the standard deviations from the graphs.

In Figure 3, we show how neighbor-joining and DCM-NJ+MP are affected by increasing the rate of evolution (i.e., the height). The x-axis is the maximum expected number of changes of a random site across the tree, and the y-axis is the RF rate. We provide a curve for each number of taxa we explored, from 10 up to 800. The sequence length is fixed in this experiment to 500. Note that both neighbor-joining and DCM-NJ+MP have high errors for the lowest rates of evolution, and that at these low rates of evolution the error rates increase as n increases. This is because for these low rates of evolution, increasing the number of taxa makes the smallest edge length (i.e., f) decrease, and thus increases the sequence length needed to have enough changes on the short edges for them to be recoverable. As the rate of evolution increases, the error rates initially decrease for both methods, but eventually the error rates begin to increase again. This increase in error occurs where the exponential portion of the convergence rate (i.e., where the sequence length depends exponentially on $\max \lambda_{ij}$) becomes significant. Note that where this happens is essentially the same for both methods– and that they perform equally well until that point. However, after this point, neighbor-joining's performance is worse, compared to DCM-NJ+MP; furthermore, the error rate increases for neighbor-joining at each of the "large" diameters, as n increases, while DCM-NJ+MP's error rate does not reflect the number of taxa nearly as much.

In Figure 4, we present a different way of looking at the data. In this figure, the x-axis is the number of taxa, the y-axis is the RF rate, and there is a curve for each of the methods. We show thus how increasing n (the number of taxa) while fixing the diameter of the tree affects the accuracy of the trees reconstructed. Note that at low rates of evolution (the left figure), the error rates for both methods increase with the number of taxa. At moderate rates of evolution (the middle figure), error rates increase for both methods but more so for neighbor-joining than for DCM-NJ+MP. Finally, at the higher rate of evolution (the right figure), this trend continues, but the gap is even larger – in fact, DCM-NJ+MP's error increase looks almost flat.

These experiments suggest strongly that except for low diameter situations, the DCM-NJ+MP method (and probably the other "fast-converging" methods) will outperform the neighbor-joining method, especially for large numbers of taxa and high evolutionary rates.

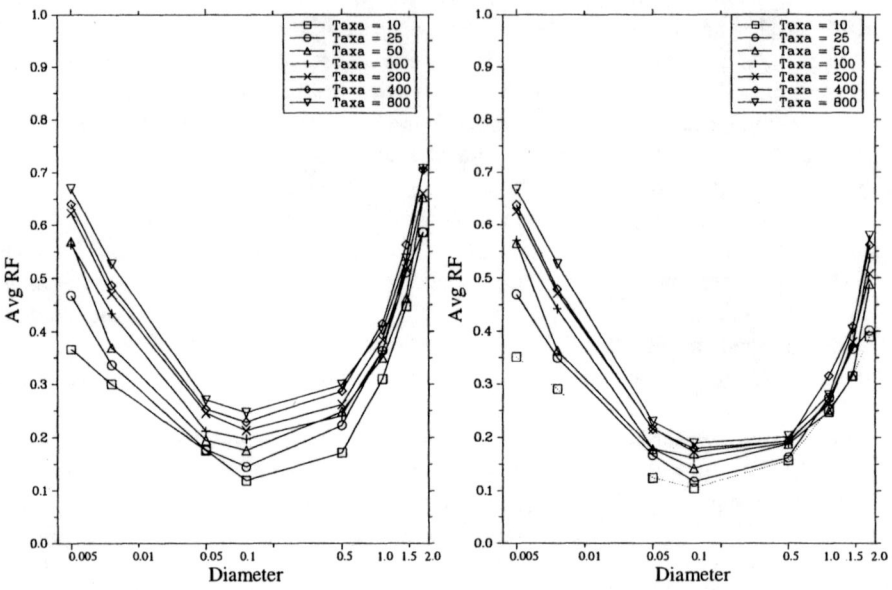

Fig. 3. NJ (left graph) and DCM-NJ+MP (right graph) error rates on random birth-death trees as the diameter (x-axis) grows. Sequence length fixed at 500, and deviation factor fixed at 4.

Table 1 shows the average running times of neighbor-joining and DCM-NJ+MP on the trees that we used in the experiments. The DCM-NJ+MP version that we ran looked at 10 thresholds in Phase 1 instead of looking at all the $\binom{n}{2}$ thresholds.

6   Conclusion

In an earlier study we presented the DCM-NJ+MP method and showed that it outperformed the NJ method for random trees drawn from the uniform distribution on tree topologies and branch lengths. In this study we show that this improvement extends to the case where the trees are drawn from a more biologically realistic distribution, in which the trees are birth-death trees with a moderate deviation from ultrametricity. This study has consequences for large phylogenetic analyses, because it shows that the accuracy of the NJ method may suffer significantly on large datasets. Furthermore, since the

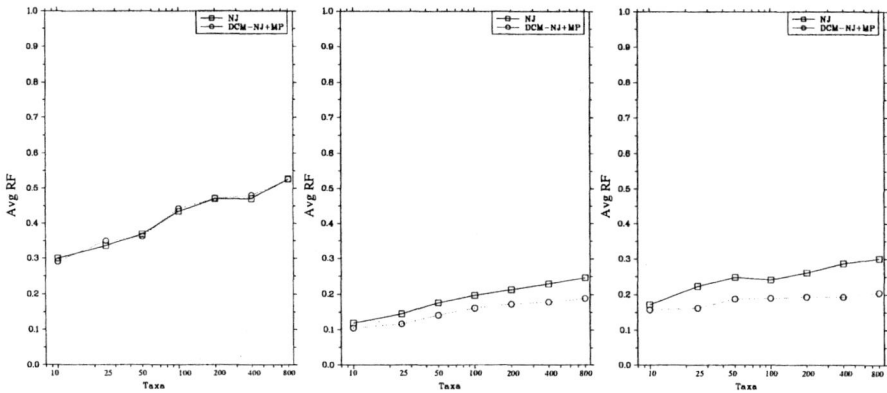

Fig. 4. NJ and DCM-NJ+MP: Error rates on random birth-death trees as the number of taxa (x-axis) grows. Sequence length fixed at 500 and the deviation factor at 4. The expected diameter of the resultant trees are 0.02 (for the left graph), 0.2 (for the middle graph), and 1.0 (for the right graph).

Table 1. The Running Times of NJ and DCM-NJ+MP in Seconds.

Taxa	NJ	DCM-NJ+MP
10	0.01	1.94
25	0.02	9.12
50	0.06	24.99
100	0.35	132.46
200	2.5	653.27
400	20.08	4991.11
800	160.4	62279.3

DCM-NJ+MP method has good accuracy, even on large datasets, our study suggests that other polynomial time methods may be able to handle the large dataset problem without significant error.

Acknowledgments

We would like to thank the David and Lucile Packard Foundation (for a fellowship to Tandy Warnow), the National Science Foundation (for a POWRE grant to Katherine St. John), the Texas Institute for Computational and Applied Mathematics and the Center for Computational Biology at UT-Austin (for support of Katherine St. John), Doug Burger and Steve Keckler for the use of the SCOUT cluster at UT-Austin, and Patti Spencer and her staff for their help.

References

1. K. Atteson. The performance of the neighbor-joining methods of phylogenetic reconstruction. *Algorithmica*, 25:251–278, 1999.
2. V. Berry and O. Gascuel. Inferring evolutionary trees with strong combinatorial evidence. In *Proc. 3rd Ann. Int'l Conf. Computing and Combinatorics (COCOON 97)*, pages 111–123. Springer Verlag, 1997. in *LNCS 1276*.
3. M. Csűrös. Fast recovery of evolutionary trees with thousands of nodes. To appear in RECOMB 01, 2001.
4. M. Csűrös and M. Y. Kao. Recovering evolutionary trees through harmonic greedy triplets. *Proceedings of ACM-SIAM Symposium on Discrete Algorithms (SODA 99)*, pages 261–270, 1999.
5. P. L. Erdos, M. Steel, L. Székély, and T. Warnow. A few logs suffice to build almost all trees–I. *Random Structures and Algorithms*, 14:153–184, 1997.
6. P. L. Erdos, M. Steel, L. Székély, and T. Warnow. A few logs suffice to build almost all trees–II. *Theor. Comp. Sci.*, 221:77–118, 1999.
7. J. Huelsenbeck and D. Hillis. Success of phylogenetic methods in the four-taxon case. *Syst. Biol.*, 42:247–264, 1993.
8. D. Huson, S. Nettles, and T. Warnow. Disk-covering, a fast-converging method for phylogenetic tree reconstruction. *Comput. Biol.*, 6:369–386, 1999.
9. D. Huson, K. A. Smith, and T. Warnow. Correcting large distances for phylogenetic reconstruction. In *Proceedings of the 3rd Workshop on Algorithms Engineering (WAE)*, 1999. London, England.
10. M. Kimura. A simple method for estimating evolutionary rates of base substitutions through comparative studies of nucleotide sequences. *J. Mol. Evol.*, 16:111–120, 1980.
11. W. H. Li. *Molecular Evolution*. Sinauer, Massachuesetts, 1997.
12. C. McGeoch. Analyzing algorithms by simulation: variance reduction techniques and simulation speedups. *ACM Comp. Surveys*, 24:195–212, 1992.
13. B. Moret. Towards a discipline of experimental algorithmics, 2001. To appear in Monograph in Discrete Mathematics and Theoretical Computer Science; Also see http://www.cs.unm.edu/~moret/dimacs.ps.
14. L. Nakhleh, U. Roshan, K. St. John, J. Sun, and T. Warnow. Designing fast converging phylogenetic methods. Oxford U. Press, 2001. To appear in *Bioinformatics*: Proc. 9th Int'l Conf. on Intelligent Systems for Mol. Biol. (ISMB 01).
15. A. Rambaut and N. C. Grassly. Seq-gen: An application for the Monte Carlo simulation of dna sequence evolution along phylogenetic trees. *Comp. Appl. Biosci.*, 13:235–238, 1997.
16. D. F. Robinson and L. R. Foulds. Comparison of phylogenetic trees. *Mathematical Biosciences*, 53:131–147, 1981.
17. M. Sanderson. r8s software package. Available from http://loco.ucdavis.edu/r8s/r8s.html.
18. N. Sautou and M. Nei. The neighbor-joining method: A new method for reconstructing phylogenetic trees. *Mol. Biol. Evol.*, 4:406–425, 1987.
19. D. L. Swofford. PAUP*: Phylogenetic analysis using parsimony (and other methods), 1996. Sinauer Associates, Underland, Massachusetts, Version 4.0.
20. Condor Development Team. Condor high throughput computing program, Copyright 1990-2001. Developed at the Computer Sciences Department of the University of Wisconsin; http://www.cs.wisc.edu/condor.
21. T. Warnow, B. Moret, and K. St. John. Absolute convergence: true trees from short sequences. *Proceedings of ACM-SIAM Symposium on Discrete Algorithms (SODA 01)*, pages 186–195, 2001.

(1+ε)-Approximation of Sorting by Reversals and Transpositions

Niklas Eriksen

Dept. of Mathematics, Royal Institute of Technology
SE-100 44 Stockholm, Sweden
niklas@math.kth.se
http://www.math.kth.se/~niklas

Abstract. Gu *et al.* gave a 2-approximation for computing the minimal number of inversions and transpositions needed to sort a permutation. There is evidence that, from the point of view of computational molecular biology, a more adequate objective function is obtained, if transpositions are given double weight. We present a $(1 + \varepsilon)$-approximation for this problem, based on the exact algorithm of Hannenhalli and Pevzner, for sorting by reversals only.

1 Introduction

This paper is concerned with the problem of sorting permutations using long range operations like inversions (reversing a segment) and transpositions (moving a segment). The problem comes from computational molecular biology, where the aim is to find a parsimonious rearrangement scenario that explains the difference in gene order between two genomes. In the late eighties, Palmer and Herbon [9] found that the number of such operations needed to transform the gene order of one genome into the other could be used as a measure of the evolutionary distance between two species.

The kinds of operations we consider are inversions, transpositions and inverted transpositions. Hannenhalli and Pevzner [7] showed that the problem of finding the minimal number of inversions needed to sort a signed permutation is solvable in polynomial time, and an improved algorithm was subsequently given by Kaplan *et al.* [8]. Caprara, on the other hand, showed that the corresponding problem for unsigned permutations is NP-hard [4]. For transpositions no such sharp results are known, but the (3/2)-approximation algorithms of Bafna and Pevzner [2] and Christie [5] are worth mentioning.

Moving on to the combined problem, Gu *et al.* [6] gave a 2-approximation algorithm for the minimal number of operations needed to sort a signed permutation by inversions, transpositions and inverted transpositions. However, an algorithm looking for the minimal number of operations will produce a solution heavily biased towards transpositions. Instead, we propose the following problem: find the π-sorting scenario s (i.e., transforming π to the identity) that minimizes $inv(s) + 2trp(s)$, where $inv(s)$ and $trp(s)$ are the numbers of inversions and transpositions in s, respectively.

We give a closed formula for this minimal weighted distance. Our formula is similar to the exact formula for the inversion case, given by Hannenhalli and Pevzner [7].

O. Gascuel and B.M.E. Moret (Eds.): WABI 2001, LNCS 2149, pp. 227–237, 2001.

We also show how to obtain a polynomial time algorithm for computing this formula with an accuracy of $(1 + \varepsilon)$, for any $\varepsilon > 0$. As an example, we explicitly state a 7/6-approximation. We also argue that for most applications the algorithm performs much better than guaranteed.

2   Preliminaries

Here we present some useful definitions from Bafna and Pevzner [2] and Hannenhalli and Pevzner [7], as well as a couple of new ones.

In this paper, we work with **signed, circular permutations**. We adopt the convention of reading the circular permutations counterclockwise. We will sometimes linearize the permutation by inverting both signs and reading direction if it contains -1, then making a cut in front of 1 and finally adding $n + 1$ last, where n is the length of the permutation. An example is shown in Figure 1. A **breakpoint** in a permutation is a pair

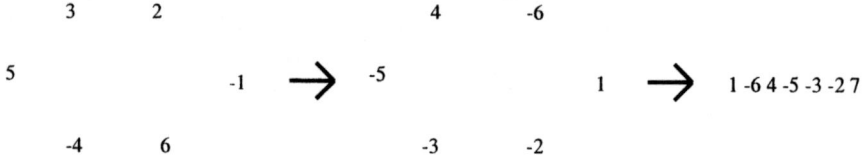

Fig. 1. Transforming a Circular Permutation to a Linear Form.

of adjacent genes that are not adjacent in a given reference permutation. For instance, if we compare a genome to the identity permutation and consider the linearized version of the permutation, the pair (π_i, π_{i+1}) is a breakpoint if and only if $\pi_{i+1} - \pi_i \neq 1$. For unsigned permutations, this would be written $|\pi_{i+1} - \pi_i| \neq 1$

The three operations we consider are inversions, transpositions and inverted transpositions These are defined in Figure 2. Following [2, 7, 8], we transform a signed, circular

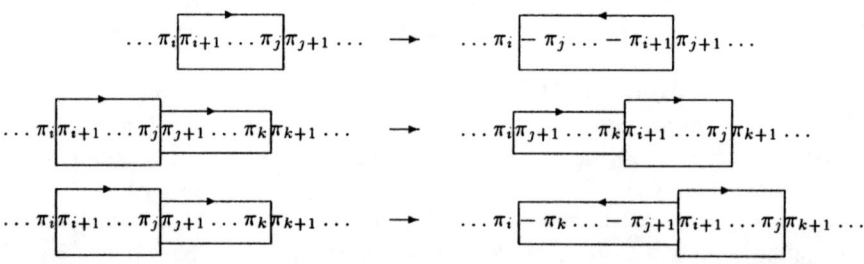

Fig. 2. Definitions of inversion, transposition and inverted transposition on *signed* genomes. If we remove all signs, the definition holds for *unsigned* genomes.

permutation π on n elements to an unsigned, circular permutation π' on $2n$ elements as follows. Replace each element x in π by the pair $(2x-1, 2x)$ if x is positive and by the pair $(-2x, -2x-1)$ otherwise. An example can be viewed in Figure 3. Then, to each operation in π there is a corresponding operation in π', where the cuts are placed after even positions. We also see that the number of breakpoints in π equals the number of breakpoints in π'. Define the **breakpoint graph** on π' by adding a black edge between

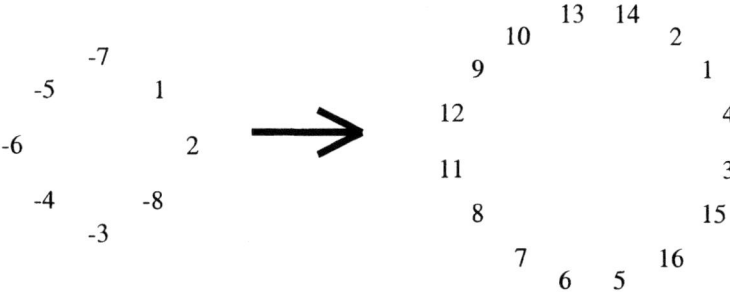

Fig. 3. Transforming a signed permutation of length 8 to an unsigned permutation of length 16.

π'_i and π'_{i+1} if there is a breakpoint between them and a grey edge between $2i$ and $2i+1$, unless these are adjacent (Figure 4). These edges will then form alternating **cycles**. The **length** of a cycle is the number of black edges in it. Sometimes we will also draw black and grey edges between $2i$ and $2i+1$, even though these are adjacent. We will then get a cycle of length one at the places where we do not have a breakpoint in π (these cycles will be referred to as **short cycles**). A cycle is **oriented** if, when we traverse it, at least one black edge is traversed clockwise and at least one black edge is traversed counterclockwise. Otherwise, the cycle is unoriented.

Consider two cycles c_1 and c_2. If we can not draw a straight line through the circle such that the elements of c_1 are on one side of the line and the elements of c_2 are on the other side, then these two cycles are **inseparable**. This relation is extended to an equivalence relation by saying that c_1 and c_j are in the same **component** if there is a sequence of cycles $c_1, c_2, \ldots, c_j$ such that, for all $1 \le i \le j-1$, c_i and c_{i+1} are inseparable.

A component is **oriented** if at least one of its cycles is oriented and **unoriented** otherwise. If there is an interval on the circle, which contains an unoriented component, but no other unoriented components, then this component is a **hurdle**. If we cannot remove a hurdle without creating a second hurdle upon its removal (this is the case if there is an unoriented component, which is not a hurdle, that stretches over an interval that contains the previously mentioned hurdle, but no other hurdles), then the hurdle is called a **super hurdle**. If we have an odd number of super hurdles and no other hurdles, the permutation is known as a **fortress**.

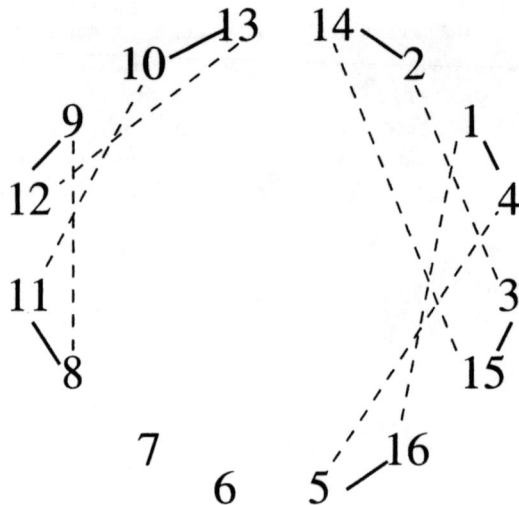

Fig. 4. The breakpoint graph of a transformed permutation. It contains three cycles, two of length 2 and one of length 3. The latter constitutes one component, and the first two constitute another component. The first component is unoriented and the second is oriented. Both components have size 2.

We should observe that for components in the breakpoint graph, the operations needed to remove them do not depend on the actual numbers on the vertices. We could therefore treat the components as separate objects, disregarding the particular permutation they are part of. If we wish, we can also regard the components as permutations, by identifying them with (one of) the shortest permutations whose breakpoint graphs consist of this component only. For example, the 2-cycle component in Figure 4 can be identified with the permutation 1 −3 −4 2 5.

We say that an unoriented component is of **odd length** if all cycles in the component are of odd length. We let $b(\pi)$, $c(\pi)$ and $h(\pi)$ denote the number of breakpoints, cycles (not counting cycles of length one) and hurdles in the breakpoint graph of a permutation π, respectively. For components t, $b(t)$ and $c(t)$ are defined similarly. The **size** s of a component t is given by $b(t) - c(t)$. We also let $c_s(\pi)$ denote the number of cycles in π, including the short ones, and $f(\pi)$ is a function that is 1 is π is a fortress, and 0 otherwise.

3   Expanding the Inversion Formula

Let S_π^I denote the set of all scenarios transforming π into id using inversions only and let $inv(s)$ denote the number of inversions in a scenario s. The inversion distance is defined as $d_{Inv}(\pi) = \min_{s \in S_\pi^I}\{inv(s)\}$. It has been shown in [7, 8] that

$$d_{Inv}(\pi) = b(\pi) - c(\pi) + h(\pi) + f(\pi),$$

where $b(\pi), c(\pi), h(\pi)$ and $f(\pi)$ have been defined in the previous paragraph. In this paper, we define the distance between π and id by

$$d(\pi) = \min_{s \in S_\pi} \{inv(s) + 2\,trp(s)\},$$

where S_π is the set of all scenarios transforming π into id, allowing both inversions and transposition, and $inv(s)$ and $trp(s)$ is the number of inversions and transpositions in scenario s, respectively. Here, transpositions refer to both ordinary and inverted transpositions.

In order to give a formula for this distance, we need a few definitions.

Definition 1. *Regard all components t as permutations and let $d(t)$ be the distance between t and id as defined above for permutations. Consider the set S of components t such that $d(t) > b(t) - c(t)$ (when using inversions only, this is the set of unoriented components). We call this set the set of **strongly unoriented components**. If there is an interval on the circle that contains the component $t \in S$, but no other member of S, then t is a **strong hurdle**. **Strong super hurdles** and **strong fortresses** are defined in the same way as super hurdles and fortresses (just replace hurdle with strong hurdle).*

Observation 1. *In any scenario, each inverted transposition can be replaced by two inversions, without affecting the objective function. This means that in calculating $d(\pi)$, we need not bother with inverted transpositions. Therefore, we will henceforth consider only inversions and ordinary transpositions.*

Lemma 1. *Each strongly unoriented component is unoriented (in the inversion sense).*

Proof. We know that for oriented components t, $d_{inv}(t) = b(t) - c(t)$ and for any permutation π, we have $d(\pi) \leq d_{Inv}(\pi)$. Regarding the component t as a permutation gives $d(t) \leq d_{Inv}(t)$. Thus, for strongly unoriented components we have $d_{inv}(t) \geq d(t) > b(t) - c(t)$ and we can conclude that a strongly unoriented component can not be oriented.

Theorem 2. *The distance $d(\pi)$ defined above is given by*

$$d(\pi) = b(\pi) - c(\pi) + h_t(\pi) + f_t(\pi),$$

or, equivalently (counting short cycles as well),

$$d(\pi) = n - c_s(\pi) + h_t(\pi) + f_t(\pi),$$

where $h_t(\pi)$ is the number of strong hurdles in π, $f_t(\pi)$ is 1 if π is a strong fortress (and 0 otherwise) and n is the length of the permutation π.

Proof. It is easy to see that $d(\pi) \leq n - c_s(\pi) + h_t(\pi) + f_t(\pi)$. If we treat the strong hurdles as in the inversion case, we need only $h_t(\pi) + f_t(\pi)$ inversions to make all strongly unoriented components oriented. All oriented components can be removed efficiently using inversions, and the unoriented components which are not strongly unoriented can, by definition, be removed efficiently.

We now need to show that we can not do better than the formula above. From Hannenhalli and Pevzner we know that we can not decrease $n - c_s(\pi)$ by more than 1 using an inversion. Similarly, a transposition will never decrease $n - c_s(\pi)$ by more than 2, which is obtained by splitting a cycle in three cycles. The question is whether transpositions can help us to remove strong hurdles more efficiently than inversions.

Bafna and Pevzner have shown that applying a transposition can only change the number of cycles by 0 or ± 2. There are thus three possible ways of applying a transposition. First, we can split a cycle into three parts ($\Delta c_s = 2$). If we do this to a strong hurdle, at least one of the components we get must by definition remain a strong hurdle, since otherwise the original component could be removed efficiently. This gives $\Delta h_t = 0$. Second, we can let the transposition cut two cycles ($\Delta c_s = 0$). To decrease the distance by three, we would have to decrease the number of strong hurdles by three which is clearly out of reach (only two strong hurdles may be affected by a transposition on two cycles). Finally, if we merge three cycles ($\Delta c_s = -2$), we would need to remove five strong hurdles. This clearly is impossible.

It is conceivable that the fortress property could be removed by a transposition that reduce $n - c_s(\pi) + h_t(\pi)$ by two and at the same time removes an odd number of strong super hurdles or adds a strong hurdle that is not a strong super hurdle. However, from the analysis above, we know the transpositions that decrease $n - c_s(\pi) + h_t(\pi)$ by two must decrease $h_t(\pi)$ by an even number. We also found that when this was achieved, no other hurdles apart from those removed were affected. Hence, there are no transpositions that reduce $n - c_s(\pi) + h_t(\pi) + f_t(\pi)$ by three.

We find that $d(\pi) \geq n - c_s(\pi) + h_t(\pi) + f_t(\pi)$, and in combination with the first inequality, $d(\pi) = n - c_s(\pi) + h_t(\pi) + f_t(\pi)$.

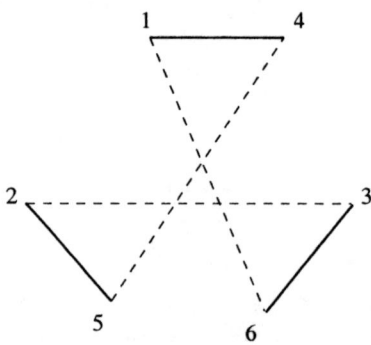

Fig. 5. The breakpoint graph of a cycle of length three which can be removed by a single transposition.

3.1   The Strong Hurdles

Once we have identified all strongly unoriented components in a breakpoint graph, we are able to calculate the number of strong hurdles. We thus need to look into the question of determining which components are strongly unoriented.

From the lemma above, we found that all strongly unoriented components are unoriented. The converse is not true. One example of this is the unoriented cycle in Figure 5, which can be removed with a single transposition. However, many unoriented components are also strongly unoriented. Most of them are characterized by the following lemma.

Lemma 2. *If an unoriented component contains a cycle of even length, then it is strongly unoriented.*

Proof. Since the component is unoriented, applying an inversion to it will not increase the number of cycles. If we apply a transposition to it, it will remain unoriented. Thus, the only way to remove it efficiently would be to apply a series of transpositions, all increasing the number of cycles by two,

Consider what happens if we split a cycle of even length into three cycles. The sum of the length of these three new cycles must equal the length of the original cycle, in particular it must be even. Three odd numbers never add to an even number, so we must still have at least one cycle of even length, which is shorter than the original cycle.

Eventually, the component must contain a cycle of length 2. There are no transpositions reducing $b(t) - c(t)$ by 2 that can be applied to this cycle, and hence the component is strongly unoriented.

Concentrating on the unoriented components with cycles of odd lengths only, we find that some of these are strongly unoriented and some are not. For instance, there are two unoriented cycles of length three. One of them is the cycle in which we may remove three breakpoints (Figure 5) and the other one can be seen in Figure 6 (a). Note that this cycle can not be a component. This is, however, not true for the components in Figure 6 (b) and (c), which are the two smallest strongly unoriented components of odd length.

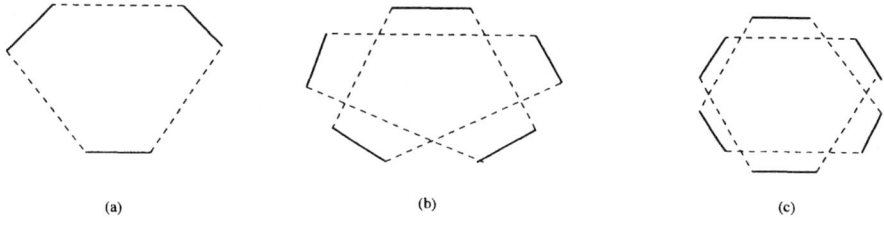

(a)   (b)   (c)

Fig. 6. A cycle of length three which can not be removed by a transposition (a), the smallest strongly unoriented component of odd length (b) and the second smallest strongly unoriented component of odd length (c).

4   The (7/6)-Approximation and the $(1 + \varepsilon)$-Approximation

Even though we at this stage are unable to recognize all strongly unoriented components in an efficient manner, we are still able to approximate the distance reasonably well. We will first show that our identification of the two strongly oriented components of size less than 6 that contain odd cycles exclusively will give us a 7/6-approximation (remember that the size of a component t was defined as $b(t) - c(t)$). We will then show that if we have identified all odd strongly unoriented components of size less than k, we can make a $(1 + \varepsilon)$-approximation for $\varepsilon = 1/k$.

First we look at the case when we know for sure that π is not a fortress, and then we look at the case when π may be a fortress.

4.1   If π Is not a Fortress, We Have a 7/6-Approximation

For all odd unoriented components with size less than 6, we are able to distinguish between those that are strongly unoriented and those that are not. In fact, the only strongly unoriented components in this set can be found in Figure 6 (b) and (c). Thus, the smallest components that may be wrongly deemed as strong hurdles are those of size 6.

Let $h_u(\pi)$, $c_u(\pi)$ and $b_u(\pi)$ be the number of components, the number of cycles and the number of breakpoints among the odd unoriented components of size 6 or larger, respectively. It is clear that $h_u(\pi) \leq c_u(\pi)$ and $h_u(\pi) \leq \frac{b_u(\pi) - c_u(\pi)}{6}$. Let $b_o(\pi)$ and $c_o(\pi)$ denote the number of breakpoints and cycles, respectively, among all other components (that is, the components that we know whether they are strongly unoriented or not). Also, let $h_{none}(\pi)$ denote the number of hurdles we would have if none of the large odd unoriented components are strongly unoriented and let $h_{all}(\pi)$ denote the number of hurdles we would have, if all of these are strongly unoriented. It follows that $h_{none}(\pi) \leq h_t(\pi) \leq h_{all}(\pi) \leq h_{none}(\pi) + h_u(\pi)$. This gives

$$\begin{aligned}
d(\pi) &= b(\pi) - c(\pi) + h_t(\pi) \\
&\leq b_o(\pi) + b_u(\pi) - c_o(\pi) - c_u(\pi) + h_{all}(\pi) \\
&\leq b_o(\pi) - c_o(\pi) + b_u(\pi) - c_u(\pi) + h_{none}(\pi) + h_u(\pi) \\
&\leq b_o(\pi) - c_o(\pi) + b_u(\pi) - c_u(\pi) + h_{none}(\pi) + \frac{b_u(\pi) - c_u(\pi)}{6}
\end{aligned}$$

and

$$\begin{aligned}
d(\pi) &= b(\pi) - c(\pi) + h_t(\pi) \\
&\geq b_o(\pi) + b_u(\pi) - c_o(\pi) - c_u(\pi) + h_{none}(\pi)
\end{aligned}$$

and hence (putting $d_o(\pi) = b_o(\pi) - c_o(\pi) + h_{none}(\pi)$)

$$d(\pi) \in \left[d_o(\pi) + b_u(\pi) - c_u(\pi) \,,\, d_o(\pi) + \frac{7(b_u(\pi) - c_u(\pi))}{6} \right].$$

In most situations, $d_o(\pi)$ will be quite large compared to $b_u(\pi) - c_u(\pi)$ and then the approximation is much better than 7/6. Thus, in practice we may use this algorithm to get a reliable value for $d(\pi)$.

4.2   If π May Be a Fortress, We Still Have a 7/6-Approximation

The analysis is similar to the one in the previous case. To simplify things a bit, we look at the worst case. The effect of π being a fortress is most significant if $d(\pi)$ is small. We need an odd number of strong super hurdles, and no other strong hurdles, to make a fortress. It takes two strong hurdles to form a strong super hurdle, and one strong super hurdle can not exist by itself. Thus, we need at least six strongly unoriented components, arranged in pairs, covering disjoint intervals of the circle.

We consider the case where we have six components, arranged such that we have three possible strong super hurdles. For each of these three pairs, there are three possible cases. If we know that we have a strong super hurdle, then we know that, for each component, $b - c \geq 2$ (there are no components with $b - c < 2$). Thus, for the pair we have $b - c + h \geq 5$. If we know that one of the two components is strongly unoriented, but we are not sure about the other, then we know that we have a strong hurdle and for the second component, we know that $b - c \geq 6$. Together this gives $b - c + h \geq 9$. Finally, if we are ignorant to whether any of the two components are strongly unoriented, we do not even know whether the pair constitutes a strong hurdle. Since both components fulfill $b - c \geq 6$, we get $b - c + h \in [r, r + 1]$, where $r \geq 12$. The worst cases is when are totally ignorant in each of the three pairs and $r = 12$. In that case, we get

$$d(\pi) \in [3 \cdot 12, \; 3 \cdot 13 + 1] = [36, 40],$$

and since this is the worst case, we have a (10/9)-approximation, which is better than 7/6. Again, this ratio will be significantly smaller in most applications.

4.3   The $(1 + \varepsilon)$-Approximation

In order to improve on the (7/6)-approximation, we need to be able to identify strong hurdles among larger components. Since we have not yet found an easy way to do this, we content ourselves with creating a table of all unoriented components of a certain size, which are not strongly unoriented. The table could be created using, for instance, an exhaustive search.

Given a table of all such components of size less than k, and a component t of size less than k, we will be able to tell if t is strongly unoriented or not. Thus, applying the same calculations as in the 7/6 case above, we find that (if π is not a fortress)

$$d(\pi) \in \left[d_o(\pi) + b_u(\pi) - c_u(\pi), \; d_o(\pi) + \frac{(k+1)(b_u(\pi) - c_u(\pi))}{k} \right],$$

or (if π may be a fortress, worst case (for $k > 10$, the worst case is different from the worst case for $k = 6$))

$$d(\pi) \in [2k + 2 \cdot 5, \; 2k + 1 + 2 \cdot 5 + 1],$$

We clearly have

$$\lim_{k \to \infty} \frac{k+1}{k} = \lim_{k \to \infty} 1 + \frac{1}{k} = 1$$

and

$$\lim_{k \to \infty} \frac{2k + 12}{2k + 10} = \lim_{k \to \infty} 1 + \frac{1}{k + 5} = 1.$$

5   The Algorithm

We will now describe the (7/6)-approximation algorithm, which is easily generalized to the $1 + \varepsilon$ case. First remove, by applying a sequence of optimal transpositions, all odd unoriented components of size less than six, that are not strongly unoriented. This can be done as follows: Find a black edge such that its two adjacent grey edges are crossing. For these small components, this can always be done. Cut this black edge and the two black edges that are adjacent to the mentioned grey edges. This transposition will always reduce the distance by two , and for these components, we can always continue afterwards in a similar fashion. In the $1 + \varepsilon$ case, we would have to use the table to find out which transposition to use.

After removing these components, we can apply the inversion algorithm of Hannenhalli and Pevzner. The complexity of this algorithm is polynomial in the length of the original permutation, as is the first step of our algorithm, since identifying the unoriented components that are not strongly unoriented and removing them can be done in linear time. Computing the inversion distance can be done in linear time [1] and thus an approximation of the combined distance can also be computed in linear time.

To get a $(1 + \varepsilon)$-approximation, all we have to do is to tabulate all odd oriented components of size $\frac{1}{\varepsilon}$, that are not strongly unoriented. We also need to tabulate, for each such component, a sequence of transpositions that will remove the component efficiently. It is clear that the algorithm is still polynomial, since looking up a component in the table is done in constant time (for each ε).

6   Discussion

The algorithm presented here relies on the creation of a table of components that can be removed efficiently. Could this technique be used to find an algorithm for **any** similar sorting problem such as sorting by transpositions? In general, the answer is no. In this case, as for sorting with inversions, we know that if a component can not be removed efficiently, we need only one extra inversion. We also know that for components that can be removed efficiently, we can never improve on such a sorting by combining components. For sorting by transpositions, no such results are known and until they are, the table will need to include not only some of the components up to a certain size, but every permutation of every size.

The next step is obviously to examine if there is an easy way to distinguish all strongly unoriented components. For odd unoriented components, this property seems very elusive. It also seems hard to discover a useful sequence of transpositions that removes odd oriented components that are not strongly unoriented. However, investigations on small components have given very promising results. For cycles of length 7, we have the following result: If the cycle is not a strongly unoriented component, then no transposition that increase the number of cycles by two will give a strongly unoriented component. This appears to be the case for cycles of length 9 as well, but no fully exhaustive search has been conducted, due to limited computational resources.

If this pattern would hold, we could apply any sequence of breakpoint removing transpositions to a component, until we either have removed the component, or are unable to find any useful transpositions. In the first case, the component is clearly not strongly unoriented, and in the second case it would be strongly unoriented.

Acknowledgment

I wish to thank my advisor Kimmo Eriksson for valuable comments during the preparation of this paper. Niklas Eriksen was supported by the Swedish Natural Science Research Council and the Swedish Foundation for Strategic Research.

References

1. Bader, D. A., Moret, B. M. E., Yan, M.: A Linear-Time Algorithm for Computing Inversion Distance Between Signed Permutations with an Experimental Study. Preprint
2. Bafna, V., Pevzner, P.: Genome rearrangements and sorting by reversals. Proceedings of the 34th IEEE symposium of the foundations of computer science (1994), 148–157
3. Bafna, V., Pevzner, P.: Sorting by Transpositions. SIAM Journal of Discrete Mathematics **11** (1998), 224–240
4. Caprara, A.: Sorting permutations by reversals and Eulerian cycle decompositions. SIAM Journal of Discrete Mathematics **12** (1999), 91–110
5. Christie, D: A.: Genome rearrangement problems. Ph. D. thesis (1998)
6. Gu, Q.-P., Peng, S., Sudborough, H.: A 2-approximation algorithm for genome rearrangements by reversals and transpositions. Theoretical Computer Science **210** (1999), 327–339
7. Hannenhalli, S., Pevzner, P.: Transforming cabbage into turnip (polynomial algorithm for sorting signed permutations with reversals). Proceedings of the 27th Annual ACM Symposium on the Theory of Computing (1995), 178-189
8. Kaplan, H., Shamir, R., Tarjan, R. E.: Faster and Simpler Algorithm for Sorting Signed Permutations by Reversals. Proceedings of the Eighth Annual ACM-SIAM Symposium on Discrete Mathematics (1997), 344-351
9. Palmer, J. D., Herbon, L. A.: Plant mitochondrial DNA evolves rapidly in structure, but slowly in sequence. Journal of Molecular Evolution **28** (1988), 87–97

On the Practical Solution of
the Reversal Median Problem

Alberto Caprara

DEIS, Università di Bologna
Viale Risorgimento 2, 40136 Bologna, Italy
acaprara@deis.unibo.it

Abstract. In this paper, we study the *Reversal Median Problem* (RMP), which arises in computational biology and is a basic model for the reconstruction of evolutionary trees. Given q genomes, RMP calls for another genome such that the sum of the reversal distances between this genome and the given ones is minimized. So far, the problem was considered too complex to derive mathematical models useful for its practical solution. We use the graph theoretic relaxation of RMP that we developed in a previous paper [6], essentially calling for a perfect matching in a graph that forms the maximum number of cycles jointly with q given perfect matchings, to design effective algorithms for its exact and heuristic solution. We report the solution of a few hundred instances associated with real-world genomes.

1 Introduction

The problem of reconstructing evolutionary trees from genome sequences is of great interest in computational biology [19]. Formally, given a set of *genomes*, each representing a species and defined by a sequence of *genes*, and a notion of *distance* between genome pairs, the problem calls for a *Steiner tree* in the genome space, with the given genomes defining the Steiner node set (the other nodes in the tree are intended to represent species from which the given species have evolved). Distance is aimed at modeling the number of evolutionary changes (also called *elementary operations*) that led from one genome to another. There are several types of elementary operations to be considered [19], such as reversals (also called *inversions*), transpositions, translocations, deletions, insertions, etc. Defining suitable notions of distance and finding algorithms to compute distances has been a challenging topic in the 90s. In particular, much work has been done in the simplified model in which all genomes contain the same set of genes, all genes appearing within a genome are pairwise different (i.e., there are no gene duplications) and elementary operations are of a unique type. For this case, the most realistic notion of distance appears to be the *reversal* (or *inversion*) *distance*, corresponding to the minimum number of reversals that transform one genome into another [10, 26].

A breakthrough result in computational biology states that, if the orientation of the genes within the genomes is known, the reversal distance can be computed in polynomial time [12]. Actually, the time is even *linear* in the genome size by using the method of [1]. Therefore, being more realistic than other distances and efficient to compute,

O. Gascuel and B.M.E. Moret (Eds.): WABI 2001, LNCS 2149, pp. 238–251, 2001.

one would expect the reversal distance to be employed when evolutionary trees are re-constructed (still under the simplifying assumptions of genomes with the same genes, no gene duplication and elementary operations of a unique type). This is not the case, probably because the mathematical modeling of the problem of finding the "best" tree w.r.t. the reversal distance is considered to be too complex, giving as motivation the fact that, though efficient, computation of the distance between two permutations is not at all trivial (actually its complexity was open for a while until the result of [12]). The only attempts so far to use the reversal distance within evolutionary tree reconstruction can be found in [11, 25, 17].

For the reasons above, the use of a simpler (though less realistic) notion of distance, called *breakpoint distance*, which is trivial to compute, has been proposed in [22] and extensively studied afterwards [3, 20, 23, 18, 8, 21, 5]. In particular, much work has been done on the so-called *Breakpoint Median Problem* (BMP), which is the problem of find-ing a genome which is closest to a given set of genomes w.r.t. the breakpoint distance, i.e., the sum of the breakpoint distances between the genome to be found and each given genome is minimized. All the methods to reconstruct evolutionary trees [3, 23, 18] use as a subroutine a procedure to solve BMP, either exactly or heuristically, in order to find the "best" genome associated with a given tree node once the genomes associated with the neighbors of the node are fixed. It is easy to show [22] that BMP is a special case of the *Traveling Salesman Problem* (TSP).

This paper is aimed at considering the use of the reversal distance in the reconstruc-tion of evolutionary trees, mainly focusing on the *Reversal Median Problem* (RMP), the counterpart of BMP in which the reversal distance is used instead of the breakpoint distance. In [6] we described a graph theoretic relaxation of RMP which allowed us to prove that the problem is $\mathcal{NP}$-hard (actually also $\mathcal{APX}$-hard). Essentially all pa-pers dealing with BMP mention RMP as a more realistic model, say that it is $\mathcal{NP}$-hard citing [6], and motivate the use of BMP by sentences like "For RMP, there are no al-gorithms available, aside from rough heuristics, for handling even three relatively short genomes" [3, 22], or "Even heuristic approaches for RMP work well only for small instances" [23, 24]. Our ultimate goal is to show that, although RMP is $\mathcal{NP}$-hard and nontrivial to model, instances of the problem associated with real-world genomes can be solved to (near-)optimality within short computing time.

In this paper we first recall the graph theoretic relaxation of RMP given in [6], in Section 2. In [6] we also presented an *Integer Linear Programming* (ILP) formulation for this relaxation, hoping that this ILP would have allowed us to solve to proven op-timality real-world RMP instances. Actually, this was not the case, as we discuss in the present paper—the exact algorithm that we will propose, presented in Section 3, is not based on this ILP. However, a careful use of the LP relaxation of the ILP is the key part of the best heuristic algorithm that we have developed, illustrated in Section 4. Experimental results are given in Section 5, where we show that our methods can solve to proven optimality many instances associated with real-world genomes, even of relatively large size, and find provably good solutions for the remaining instances.

2 Preliminaries

A genome without gene duplications for which the orientation of the genes is known can be represented by a *signed permutation* π on $N := \{1, \ldots, n\}$, obtained by *signing* a permutation $\tau = (\tau_1 \ldots \tau_n)$ on N, i.e., by replacing each element τ_i by either $\pi_i = +\tau_i$ or $\pi_i = -\tau_i$. In particular, signs model the relative orientation of the genes within the genome. We denote by Σ_n the set of the $2^n n!$ signed permutations on N. A *reversal* of the interval (i, j), $1 \leq i \leq j \leq n$, applied to a signed permutation π, is an operation which both inverts the subsequence $\pi_i \pi_{i+1} \ldots \pi_{j-1} \pi_j$ and switches the signs of the elements in the subsequence, replacing $\pi_1 \ldots \pi_{i-1} \pi_i \pi_{i+1} \ldots \pi_{j-1} \pi_j \pi_{j+1} \ldots \pi_n$ by $\pi_1 \ldots \pi_{i-1} - \pi_j - \pi_{j-1} \ldots - \pi_{i+1} - \pi_i \pi_{j+1} \ldots \pi_n$. The minimum number of reversals needed to transform a signed permutation π^1 into a signed permutation π^2 (or vice versa) is called the *reversal distance* between π^1 and π^2, here denoted by $d(\pi^1, \pi^2)$. Given two signed permutations, the problem of *Sorting By Reversals* (SBR) calls for finding the reversal distance between the two permutations and an associated shortest sequence of reversals. SBR was shown to be polynomially solvable in [12]. At present, the best asymptotic running time is $O(n^2)$, achieved by the algorithm of [14]. Nevertheless, as mentioned above, the simple computation of the reversal distance (without an associated sequence of reversals) can be carried out in $O(n)$ time [1]. Throughout the paper we will only work with signed permutations, therefore for convenience we use the term permutation to indicate a signed permutation.

Given q permutations $\pi^1, \ldots, \pi^q \in \Sigma_n$, $q \geq 3$, representing genomes with the same set of genes, RMP calls for a permutation $\sigma \in \Sigma_n$ such that $\delta(\sigma) := \sum_{k=1}^{q} d(\sigma, \pi^k)$ is minimized. The remainder of the section gives the essential notions from the previous works on SBR and RMP.

The graphs we consider are in general *multigraphs*, possibly having *parallel* edges with common endpoints. More generally, we consider *multisets*, which may contain distinct copies of identical elements. Given a node set V we call an edge set $M \subset \{(i, j) : i, j \in V, i \neq j\}$ a *matching* of V if each node in V is incident to at most one edge in M. If each node in V is incident to exactly one edge in M, the matching is called *perfect*. In general, we work with a graph $G = (V, E)$ and an associated (perfect) matching M of V without requiring $M \subseteq E$. *Cycles* and *paths* are thought of as subsets of $\{(i, j) : i, j \in V, i \neq j\}$, by implicitly assuming that they are *simple*, i.e., they do not visit a same node twice or more. The *length* of a cycle or path is given by the number of its edges, and we consider "degenerate" cycles of length 2 formed by pairs of parallel edges. A *Hamiltonian cycle* of V is a cycle visiting all the nodes in V. Given two matchings M, L of V, the graph $(V, M \cup L)$ corresponds to a set of node-disjoint paths and cycles; we call the latter the cycles *defined* by $M \cup L$. If both M and L are perfect, the graph $(V, M \cup L)$ corresponds to a set of node-disjoint cycles visiting all the nodes in V.

Consider the node set $V := \{0, 1, \ldots, 2n, 2n+1\}$, and the associated perfect matching $H := \{(2i - 1, 2i) : i = 1, \ldots, n\} \cup \{(0, 2n + 1)\}$, called the *base matching* of V. For convenience, let $\tilde{n} := n + 1$ denote the cardinality of any perfect matching of V. There is a natural correspondence between signed permutations in Σ_n and perfect matchings M of V such that $M \cup H$ defines a Hamiltonian cycle of V. These matchings are called *permutation matchings*. In particular, the permutation matching $M(\pi)$

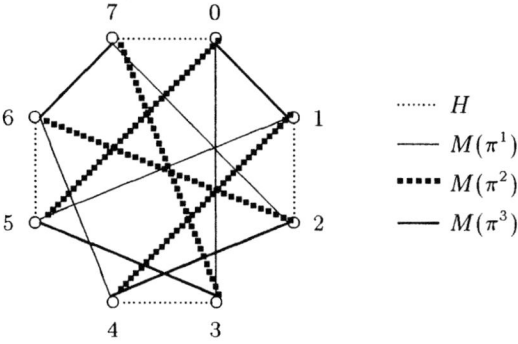

Fig. 1. The MB Graph $G(\pi^1, \pi^2, \pi^3)$ Associated with $\pi^1 = (2\ -3\ 1)$, $\pi^2 = (3\ -1\ -2)$ and $\pi^3 = (1\ -2\ 3)$.

associated with a permutation $\pi \in \Sigma_n$ is defined by

$$M(\pi) := \{(2|\pi_i| - \nu(\pi_i), 2|\pi_{i+1}| - 1 + \nu(\pi_{i+1})) : i \in \{0, \ldots, n\}\}, \qquad (1)$$

where for convenience we have defined $\pi_0 := 0$, $\pi_{n+1} := n + 1$ and $\nu(\pi_i) := 0$ if $\pi_i \geq 0$, $\nu(\pi_i) := 1$ if $\pi_i < 0$. Figure 1 illustrates the permutation matchings associated with $\pi^1 = (2\ -3\ 1)$, $\pi^2 = (3\ -1\ -2)$ and $\pi^3 = (1\ -2\ 3)$. The following result is well known and used in almost all papers on SBR.

Proposition 1. ([2]) $M(\cdot)$ defines a one-to-one mapping between signed permutations in Σ_n and permutation matchings of V.

Given two permutations π^1 and π^2, the set of edges $M(\pi^1) \cup M(\pi^2)$ defines a set of cycles whose edges are alternately in $M(\pi^1)$ and in $M(\pi^2)$ (possibly containing cycles of length 2). Let $c(\pi^1, \pi^2)$ denote the number of these cycles. For permutations π^1, π^2, π^3 in Figure 1, $d(\pi^1, \pi^3) = 2$, $c(\pi^1, \pi^3) = 2$, $d(\pi^1, \pi^2) = d(\pi^2, \pi^3) = 3$, $c(\pi^1, \pi^2) = c(\pi^2, \pi^3) = 1$. A fundamental result proved in [16,2], restated according to our notation, is the following

Proposition 2. ([16, 2]) For two permutations $\pi^1, \pi^2 \in \Sigma_n$, $d(\pi^1, \pi^2) \geq \tilde{n} - c(\pi^1, \pi^2)$.

We will call $\tilde{n} - c(\pi^1, \pi^2)$ the *cycle distance* between π^1 and π^2 or between $M(\pi^1)$ and $M(\pi^2)$. In practice, the lower bound on the reversal distance given by the cycle distance turns out to be very strong. It is convenient to extend the notion of cycle distance also to general perfect matchings. Given two perfect matchings T, S of V (not necessarily permutation matchings), let $c(T, S)$ denote the number of cycles defined by $T \cup S$. The *cycle distance* between T and S is given by $\tilde{n} - c(T, S)$. It is easy to verify that this is indeed a distance, for it satisfies the triangle inequality.

Let $Q := \{1, \ldots, q\}$. We define the MB graph associated with permutations $\pi^1, \ldots,$ $\pi^q \in \Sigma_n$ as the graph $G(\pi^1, \ldots, \pi^q)$ with node set V and edge multiset $M(\pi^1) \cup \ldots \cup M(\pi^q)$. Note that possible common edges in $M(\pi^j)$ and $M(\pi^k)$, $j \neq k$, are

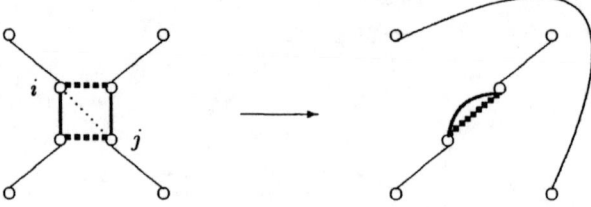

Fig. 2. The Contraction of Edge (i, j).

considered distinct parallel edges in $G(\pi^1, \ldots, \pi^q)$. Figure 1 represents the MB graph $G(\pi^1, \pi^2, \pi^3)$ associated with $\pi^1 = (2 - 3\ 1)$, $\pi^2 = (3 - 1 - 2)$ and $\pi^3 = (1 - 2\ 3)$.

For a given RMP instance defined by permutations $\pi^1, \ldots, \pi^q \in \Sigma_n$, and a permutation $\sigma \in \Sigma_n$, let $\gamma(\sigma) := \sum_{k \in Q} c(\sigma, \pi^k)$. Theorem 2 immediately implies that $\delta(\sigma) = \sum_{k \in Q} d(\sigma, \pi^k) \geq q\tilde{n} - \gamma(\sigma)$. Together with Theorem 1, this suggests to study the following *Cycle Median Problem* (CMP): Given permutations $\pi^1, \ldots, \pi^q \in \Sigma_n$, find a permutation $\tau \in \Sigma_n$ such that $q\tilde{n} - \gamma(\tau)$ is minimized. In terms of $G(\pi^1, \ldots, \pi^q)$, the problem calls for a permutation matching T such that $q\tilde{n} - \sum_{k \in Q} c(T, M(\pi^k))$ is minimized. The correspondence between solutions τ and T is given by $M(\tau) = T$. The immediate generalizations of Theorem 2 for RMP is the following. Let δ^* and $q\tilde{n} - \gamma^*$ denote the optimal solution values of RMP and CMP, respectively.

Proposition 3. *([6]) Given an RMP instance and the associated CMP instance, $\delta^* \geq q\tilde{n} - \gamma^*$.*

For the MB graph in Figure 1, an optimal solution of both RMP and CMP is given by π^1, for which $\delta(\pi^1) = d(\pi^1, \pi^1) + d(\pi^1, \pi^2) + d(\pi^1, \pi^3) = 0 + 3 + 2 = 5$ and $\gamma(\pi^1) = c(\pi^1, \pi^1) + c(\pi^1, \pi^2) + c(\pi^1, \pi^3) = 4 + 1 + 2 = 7$. Hence, $\delta^* = 5$ and $\gamma^* = 7$.

The strength of the cycle distance lower bound on the reversal distance suggests that the solution of CMP should yield a strong lower bound on the optimal solution value of RMP. CMP is indeed the key problem addressed in this paper to derive effective algorithms for RMP.

We conclude this section with an important notion that will be used in the next sections. Given a perfect matching M on node set V and an edge $e = (i, j) \in \{(i, j) : i, j \in V, i \neq j\}$, we let M/e be defined as follows. If $e \in M$, $M/e := M \setminus \{e\}$. Otherwise, letting $(i, a), (j, b)$ be the two edges in M incident to i and j, $M/e := M \setminus \{(i, a), (j, b)\} \cup \{(a, b)\}$. The following obvious lemma will be used later in the paper.

Lemma 1. *Given two perfect matchings M, L of V and an edge $e = (i, j) \in M$, $M \cup L$ defines a Hamiltonian cycle of V if and only if $(M/e) \cup (L/e)$ defines a Hamiltonian cycle of $V \setminus \{i, j\}$.*

Given an MB graph $G(\pi^1, \ldots, \pi^q)$, the *contraction* of an edge $e = (i, j) \in \{(i, j) : i, j \in V, i \neq j\}$ is the operation that modifies $G(\pi^1, \ldots, \pi^q)$ as follows. Edge (i, j) is removed along with nodes i and j. For $k = 1, \ldots, q$, $M(\pi^k)$ is replaced by $M(\pi^k)/e$, and the base matching H is replaced by H/e. Figure 2 illustrates the contraction of edge (i, j).

Note that the graph obtained after an edge contraction is not necessarily an MB graph, in particular there may be some $k \in Q$ such that $M(\pi^k) \cup H$ defines more than one cycle after contraction. This is apparently a drawback, since our methods deal with instances obtained by contracting edges. To overcome this, our algorithms are suited for the following generalization of CMP. Consider a graph G on node set V, with $|V| = 2\tilde{n}$ for some integer $\tilde{n} > 1$, along with a perfect matching H, called base matching, and let $E := \{(i, j) : i, j \in V, i \neq j\} \setminus H$. Given q perfect matchings $M_1, \ldots, M_q \subset E$ (which do not necessarily define Hamiltonian cycles with H), find a perfect matching $T \subset E$ such that $T \cup H$ is a Hamiltonian cycle and $q\tilde{n} - \sum_{k \in Q} c(T, M_k)$ is minimized. In the rest of the paper, we will use the term CMP to denote this more general version.

3 A Combinatorial Branch-and-Bound Algorithm

We next describe a simple branch-and-bound algorithm for CMP (and RMP) based on a straightforward lower bound (as opposed to the sophisticate LP relaxation of the next section) that can be computed in time linear in n. The key issue of our approach is that branching corresponds to edge contraction, and yields another problem with the same structure for which the above lower bound can be computed as well. We describe the method for CMP and then mention the slight modification required to solve RMP.

We start by illustrating the combinatorial lower bound, which is only based on the property that the cycle distance satisfies the triangle inequality and is well known for other median problems in metric spaces (see e.g. [10]). As in Section 2, let $q\tilde{n} - \gamma^*$ denote the optimal solution value of CMP.

Lemma 2. *Given a CMP instance associated with matchings $M_1, \ldots, M_q$,*

$$\gamma^* \leq \frac{q\tilde{n}}{2} + \sum_{k=1}^{q-1} \sum_{l=k+1}^{q} \frac{c(M_k, M_l)}{q-1}. \tag{2}$$

The lower bound on the optimal CMP (and RMP) value given by $q\tilde{n}$ minus the left-hand side of (2), called lb_C, can be computed in $O(nq^2)$ time since computation of $c(M_k, M_l)$ for all $k, l = 1, \ldots, q, k \neq l$, takes $O(n)$ time. In the next section we give an LP interpretation of this bound.

Recall the definition of edge contraction in Section 2. The branching scheme of our algorithm is inspired by the following

Lemma 3. *Given a CMP instance and an edge $e \in E$, the best CMP solution containing e is given by $\tilde{T} \cup \{e\}$, where $\tilde{T}$ is the optimal solution of the CMP instance obtained by contracting edge e.*

Proof. Let $T := \tilde{T} \cup \{e\}$. First of all, note that T is a feasible CMP solution by Lemma 1. Now suppose there is another solution T^* with $e \in T^*$ and

$$\sum_{k \in Q} c(T^*, M_k) > \sum_{k \in Q} c(T, M_k). \tag{3}$$

For all k such that $e \in M_k$, a cycle of length 2 formed by two copies of e is defined by both $T^* \cup M_k$ and $T \cup M_k$. Furthermore, contraction of edge e ensures a one-to-one correspondence between cycles defined by $T \cup M_k$ (resp. $T^* \cup M_k$) which are not two copies of e and cycles in $\tilde{T} \cup M_k$ (resp. $T^* \setminus \{e\} \cup M_k$) after the contraction of e. Hence, (3) would contradict the optimality of $\tilde{T}$. □

According to Lemma 3, if we fix edge e in the CMP solution, an upper bound on the number of cycles is given by $|\{k : M_k \ni e\}|$ (i.e., the number of cycles of length 2 defined by two copies of e) plus the upper bound (2) computed after the contraction of e. This allows us to design a branch-and-bound algorithm where, starting from node $0 \in V$, we enumerate all permutation matchings by fixing, in turn, either edge $(0, 1)$, or $(0, 2), \ldots$, or $(0, 2n)$ in the solution. Recursively, if the last edge fixed in the current partial solution is (i, j) and the edge in H incident to j is (j, k), we proceed with the enumeration by fixing in the solution, in turn, edge (k, l), for all l with no incident edge fixed so far. We proceed in a depth first way. With this scheme we can perform the lower bound test after each fixing in $O(nq^2)$ time (in fact, the recomputation of the lower bound after each fixing is done parametrically and, in practice, turns out to be faster than from scratch). In order to have a fast processing of the subproblems, the only operations performed are edge contraction and the bound computation, and the incumbent solution is updated only when the current partial solution is a complete solution.

The main drawback with the above scheme is that good solutions are found after considerable computing time. To overcome this, we start from the lower bound lb_C computed for the original problem and call the branch-and-bound first with a *target value* $\tilde{t} := lb_C$, searching for a CMP solution of value $\tilde{t}$ and backtracking as soon as the lower bound for the current partial solution is $> \tilde{t}$. If a solution of value $\tilde{t}$ is found, it is optimal and we stop, otherwise no solution of value better than $lb_C + 1$ exists. Accordingly, we call the branch-and-bound with target value $\tilde{t} := lb_C + 1$, and so on, stopping as soon as we find a solution with the target value. Even if this has the effect of reconsidering some subproblem more than once, every call of branch-and-bound with a new value of $\tilde{t}$ takes typically much longer than the previous one, therefore the increase in running time due to many calls is negligible. On the other hand, the scheme allows for the fast derivation of good lower bounds, noting that $\tilde{t}$ is a lower bound on the optimal CMP value and is increased by 1 after each call, with the minimal core memory requirements of a depth-first branch-and-bound (we often examine several million subproblems so the explicit storage of the subproblems would be impossible).

The above algorithm can easily be modified to find optimal RMP solutions. In particular, each time the current partial solution is a complete solution, of value (say) $q\tilde{n} - \gamma$ (w.r.t. the CMP objective function), the above algorithm tests if $q\tilde{n} = \gamma = \tilde{t}$. If this is the case, the algorithm stops as the current solution is optimal. In the modified version, we compute the value δ of the current solution w.r.t. the RMP objective function.

If $\delta = \tilde{t}$, again we stop. Otherwise, we possibly update the incumbent RMP solution value δ^* (initially set to ∞). In any case, the algorithm stops when the target value $\tilde{t}$ (increased at each iteration) satisfies $\delta^* \leq \tilde{t}$.

The branch-and-bound algorithm is effective for many instances, providing a provably optimal solution within short time, especially in the relevant special case of $q = 3$, for which the lower bound is reasonably close to the optimal value. In particular, the processing of each subproblem within the branch-and-bound enumeration is very fast (for our instances, a few hundred thousand subproblems per second on a PC). For the remaining instances, the method is good at finding lower bounds on the optimal CMP (and RMP) value but does not provide good heuristic solutions (even if various heuristics are applied within the enumeration scheme). The following section describes a heuristic based on a natural LP relaxation that performs well in practice.

4   An LP-Based Heuristic

In [6], we proposed a natural ILP formulation of CMP with one binary variable x_e for each edge $e \in E$, equal to 1 if e is in the CMP solution and 0 otherwise, and one binary variable y_C for each cycle that the CMP solution may define with M_k, $k \in Q$. For details, we refer to [6] or to the full paper.

The ILP formulation contains an exponential (in n) number of variables and constraints. Nevertheless, due to a fundamental result of Grötschel, Lovász and Schrijver [9], the associated LP relaxation can be solved in polynomial time provided the *separation* of the constraints and the *generation* of the y variables can be done efficiently. It is shown in [6] that this is indeed the case. However, in practice, this LP relaxation turns out to be very difficult to solve with the present state-of-the-art LP solvers. In the full paper, we provide experimental results showing that, even for $q = 3$, the largest LPs solvable in a few hours correspond to instances with $n = 30$, that can be solved within seconds by the combinatorial branch-and-bound algorithm of the previous section, whereas LPs for $n \geq 40$ may take days or even weeks to be solved. Hence, an exact algorithm based on the ILP formulation seems (at present) useless in practice. In this section, we show how to use the LP relaxation within an effective heuristic.

The heuristic starts from the (in most cases, fractional) vectors x^* produced at each iteration within the iterative solution of the LP relaxation by column generation and separation. For a given x^*, we apply a *nearest neighbor* algorithm, which starts from some node $i_0 \in V$, selects the edge $(i_0, j) \in E$ such that $x^*_{(i_0,j)}$ is maximum, considers the node l such that $(j, l) \in H$, selects the edge $(l, p) \in E$ such that $p \neq i_0$ and $x^*_{(l,p)}$ is maximum, considers the node r such that $(p, r) \in H$, and so on, until the edges selected form a permutation matching. We then apply to the final solution a *2-exchange* procedure, where we check if the removal of two edges in the current solution and their replacement with the two other edges that yield a permutation matching yields an improvement in the CMP value. If this is the case, the replacement is performed and the procedure is iterated. Otherwise, i.e., if no 2-exchange in the current solution improves the CMP value, the procedure terminates. Each time a new solution is considered, namely the initial solution produced by nearest neighbor and the solution after each improvement, we compute the corresponding RMP value to possibly update the best RMP solution so far.

It is easy to see that the complexity of nearest neighbor is $O(n^2)$, plus $O(qn)$ for the evaluation of the value of the CMP (and RMP) solution found. Also easy is to verify that checking if a 2-exchange yields an improvement in the CMP value can be done in $O(q)$ time if one has a data structure that, for each $k \in Q$ and $i \in V$, tells which is the cycle in $T \cup M_k$ that visits node i, where T is the current solution (cycles are simply identified by numbers). This data structure can be reconstructed in $O(qn)$ time every time the current solution is improved. Hence, each iteration of the 2-exchange procedure, that either yields an improvement or terminates the procedure, takes $O(qn^2)$ time since $O(n^2)$ 2-exchanges are considered.

For every vector x^*, we try each node $i \in V$ as starting node in nearest neighbor. This is because starting from different nodes may yield solutions of considerably different quality. Now the point is how to produce each x^* within reasonable time, as the solution of each LP within the iterative procedure is quite time consuming, as mentioned above. A natural choice is to work with a "sparsified" graph, where only a subset of the edges $E' \subseteq E$ is considered in the LP (the variables associated with the other edges other being fixed to 0). To this aim, the algorithm is organized in *rounds*. In the first round, we let $E' := \bigcup_{k \in Q} M_k$. For the other rounds, we consider the other edges by increasing value of lower bound lb_C. In particular, for each edge $e \in E$, we compute $lb_C(e)$, the value of lower bound lb_C if edge e is fixed in the solution. In the second round we let $E' := \bigcup_{k \in Q} M_k \cup \{e \in E : lb_C(e) \leq lb_C\}$, in the third round $E' := \bigcup_{k \in Q} M_k \cup \{e \in E : lb_C(e) \leq lb_C + 1\}$, and so on. We stress that in the nearest neighbor heuristic and in the 2-exchange procedure the *whole* set E of edges is considered (the definition of E' is only meant to speed-up the solution of the LPs).

To further drive the heuristic with the LP solutions, every time in a round the value of the current LP $\bar{\gamma}$ is such that $\lceil q\tilde{n} - \bar{\gamma} \rceil < \delta^*$, where δ^* is the best RMP solution value so far, we *fix* to 1 the $\lceil \tilde{n}/10 \rceil$ x variables whose value is highest, imposing the associated edges in the solution. We make sure that the partial solution is contained into a CMP solution, and perform the fixing only if at least 3 iterations (of column generation or separation followed by an LP) were performed since the last fixing. The round terminates when the optimal value of the LP (containing only the edges in E' and with some edges fixed in the solution), say $\bar{\gamma}$, satisfies $\lceil q\tilde{n} - \bar{\gamma} \rceil \geq \delta^*$.

The overall heuristic terminates either after a time limit or when the round corresponding to $E' = E$ terminates.

5   Experimental Results

Our algorithms were implemented in C and ran on a Digital Ultimate Workstation 533MHz. As already mentioned, the LP solver used was CPLEX 6.5. Moreover, we computed the reversal distance between permutations using the implementation of the linear time algorithm of [1], which is available at http://www.cs.unm.edu/~moret/GRAPPA (in particular, we used the function *invdist_noncircular_nomem()*).

The main test instance in the early development of our code is associated with three real-world genomes each with $n = 33$ genes and reported in [25]. This instance already gives some clear indication about the solution methods proposed in this paper, as our branch-and-bound algorithm can solve it within less than 0.2 seconds (the optimal value being 41), the LP heuristic finds an optimal solution in less than 0.05 seconds, whereas the associated (complete) LP relaxation was not solved after $500,000$ seconds!

In our experiments, we mainly solved instances with $q \in \{3, 4, 5\}$. The main reason is that the RMP instances solved within the reconstruction of evolutionary trees are associated with small values of q (typically, $q = 3$). Moreover, as shown by the tables, the effectiveness of the lower bounds proposed in this paper quickly decreases as q increases; therefore not much can be said about the optimality or near-optimality of the solutions provided for instances with higher values of q (with some exceptions, see below).

All tables report the values of q and n, the time limit imposed on the branch-and-bound algorithm and on the LP heuristic for all instances (within the caption), and the following information:

> *# inst*: the number of instances considered for each value of q and n (average and maximum values refer to this number of instances);
> *# opt*: the number of instances solved to optimality by the branch-and-bound algorithm within the time limit;
> δ^*: the average value of the best RMP solution found—the optimal one if the branch-and-bound algorithm terminates before the time limit, otherwise the best solution produced by the LP heuristic;
> lb_C: the average value of lower bound lb_C;
> lb_{BB}: the average value of the lower bound produced by the branch-and-bound algorithm (equal to the optimum if the algorithm terminates before the time limit);
> *B&B subpr.*: the average number of subproblems considered within the branch-and-bound algorithm;
> *B&B time*: the average (maximum) time required by the branch-and-bound algorithm (possibly equal to the time limit);
> δ^H: the average value of the best RMP solution produced by the LP heuristic;
> *H time*: the average (maximum) time required to find the best solution by the LP heuristic;
> *gap*: the average (maximum) difference between the best RMP solution found and the lower bound produced by the branch-and-bound algorithm.

In Table 1 we present results on instances associated with uniformly random permutations. The table shows that, for $q = 3$, the maximum size of such instances solvable to proven optimality within reasonable time is between 30 and 40 (with a time limit of 1 hour instead of 10 minutes, 6 instances out of 10 can be solved for $n = 40$). For $q = 4$ the threshold is just above 20, whereas it is below 20 for $q = 5$. This reflects the bad quality of the combinatorial lower bound for $q > 3$. The LP heuristic in some cases requires some time to find the best solution, even if solutions whose value is 1 or at most 2 units larger are typically found in fractions of a second. Our feeling is that the solutions found by the LP heuristic are near optimal even for the cases in which the lower bound cannot certify this. Table 2 refers to the randomly generated instances mentioned

Table 1. Results on Uniformly Random Instances, Time Limit of 10 Minutes.

q	n	# inst.	# opt.	δ*	lb_C	lb_{BB}	B&B subpr.	B&B time	$δ^H$	H time	gap
3	10	10	10	14.2	13.5	14.2	328.6	0.0 (0.0)	14.2	0.0 (0.2)	0.0 (0)
3	20	10	10	29.2	27.5	29.2	43312.0	0.1 (0.3)	29.2	0.2 (0.8)	0.0 (0)
3	30	10	10	45.0	42.8	45.0	6504469.6	10.2 (45.6)	45.3	34.0 (232.4)	0.0 (0)
3	40	10	2	61.7	57.2	60.3	317477043.2	524.9 (600.0)	61.8	58.0 (427.7)	1.4 (4)
3	50	10	0	79.2	72.3	75.1	335299968.0	600.0 (600.0)	79.2	184.7 (584.9)	4.1 (5)
4	10	10	10	21.1	17.9	21.1	6026.2	0.0 (0.0)	21.1	0.1 (0.4)	0.0 (0)
4	20	10	10	44.1	37.2	44.1	70430739.2	171.2 (536.8)	44.1	2.8 (8.6)	0.0 (0)
4	30	10	0	69.2	57.0	63.4	214200012.8	600.0 (600.0)	69.2	29.8 (183.1)	5.8 (7)
4	40	10	0	93.7	76.3	82.2	192300006.4	600.0 (600.0)	93.7	80.2 (254.2)	11.5 (13)
4	50	10	0	121.2	96.6	101.9	173600000.0	600.0 (600.0)	121.2	96.9 (450.2)	19.3 (20)
5	10	10	10	28.7	22.4	28.7	69344.3	0.2 (0.3)	28.7	0.1 (0.2)	0.0 (0)
5	20	10	0	59.3	46.3	56.6	175400012.8	600.0 (600.0)	59.3	63.4 (588.1)	2.7 (3)
5	30	10	0	91.8	71.2	79.7	148699993.6	600.0 (600.0)	91.8	121.7 (423.8)	12.1 (14)
5	40	10	0	125.7	95.3	103.3	129500006.4	600.0 (600.0)	125.7	141.8 (359.5)	22.4 (25)
5	50	10	0	161.5	120.6	127.7	113400000.0	600.0 (600.0)	161.5	275.2 (507.2)	33.8 (35)

Table 2. Results on Random Instances from set1 in [17], Time Limit of 10 Minutes.

r	q	n	# inst.	# opt.	δ*	lb_C	lb_{BB}	B&B subpr.	B&B time	$δ^H$	H time	gap
2	3	20	10	10	14.0	13.5	14.0	364.7	0.0 (0.0)	14.2	0.0 (0.0)	0.0 (0)
2	3	40	10	10	14.7	14.0	14.7	873.4	0.0 (0.0)	14.9	0.0 (0.0)	0.0 (0)
2	3	80	10	10	15.1	14.2	15.1	5341.7	0.0 (0.0)	15.1	0.1 (0.2)	0.0 (0)
2	3	160	10	10	15.1	14.2	15.1	21075.8	0.0 (0.1)	15.1	0.7 (0.7)	0.0 (0)
2	3	320	10	10	15.0	14.2	15.0	56076.9	0.1 (0.2)	15.0	5.2 (5.2)	0.0 (0)
4	3	10	10	10	14.5	13.6	14.5	492.9	0.0 (0.0)	14.8	0.0 (0.0)	0.0 (0)
4	3	20	10	10	27.5	26.5	27.5	31017.6	0.0 (0.3)	27.9	0.0 (0.1)	0.0 (0)
4	3	40	10	9	47.1	45.0	46.9	39167574.4	63.0 (600.0)	47.4	0.3 (2.7)	0.2 (2)
4	3	80	10	10	56.5	55.1	56.5	4342064.8	7.0 (70.2)	56.8	0.4 (1.1)	0.0 (0)
4	3	160	10	10	57.5	56.5	57.5	16382.0	0.0 (0.1)	57.6	1.0 (1.6)	0.0 (0)
4	3	320	10	10	57.6	56.6	57.6	45550.9	0.1 (0.1)	57.6	6.0 (10.2)	0.0 (0)
8	3	10	10	10	14.3	13.8	14.3	267.7	0.0 (0.0)	14.6	0.0 (0.0)	0.0 (0)
8	3	20	10	10	29.5	27.9	29.5	42045.8	0.1 (0.2)	30.1	0.1 (0.2)	0.0 (0)
8	3	40	10	4	61.6	57.3	60.3	258996275.2	433.1 (600.0)	62.1	0.9 (6.1)	1.3 (3)
8	3	80	10	0	125.5	112.7	115.0	292400000.0	600.0 (600.0)	125.5	9.8 (18.3)	10.5 (13)
8	3	160	10	0	190.6	177.5	179.7	272000025.6	600.0 (600.0)	190.6	46.2 (276.5)	10.9 (24)
8	3	320	10	9	205.0	203.4	204.8	37536243.2	66.7 (600.0)	205.0	29.2 (73.1)	0.2 (2)
16	3	10	10	10	14.2	13.4	14.2	247.3	0.0 (0.0)	14.4	0.0 (0.0)	0.0 (0)
16	3	20	10	10	29.5	28.0	29.5	96419.2	0.1 (0.7)	29.9	0.0 (0.1)	0.0 (0)
16	3	40	10	3	62.0	57.8	60.6	299220787.2	501.5 (600.0)	62.5	0.4 (0.6)	1.4 (3)
16	3	80	10	0	130.8	116.0	118.3	288000000.0	600.0 (600.0)	130.8	10.8 (23.8)	12.5 (14)
16	3	160	10	0	276.6	236.2	237.3	204600000.0	600.0 (600.0)	276.6	248.8 (508.1)	39.3 (42)
16	3	320	10	0	537.1	451.4	451.8	146000000.0	600.0 (600.0)	537.1	244.5 (600.0)	85.3 (98)

in [17] and publicly available at http://www.cs.unm.edu/~moret/GRAPPA. We report only results for the instances set1—the results for those in set2 are analogous and can be found in the full paper. All these instances have $q = 3$ permutations, each generated starting from the identity permutation and simulating a number of evolutionary changes. Parameter r increases with the number of changes simulated, therefore for larger values of r the RMP solution value is larger. The two tables show that, at least for $q = 3$, RMP is typically easy to solve if the solution value (i.e., the overall reversal distance) is considerably smaller w.r.t. the case of uniformly random permutations, for which the average solution value seems to be roughly $1.5n$ (see Table 1). Specifically, instances with $δ^* < n$ can be solved quickly, even for $n = 320$. We stress that these are the instances for which the reversal distance gives more reliable indication about the actual evolutionary distance, see also [19]. Note that in both tables, for $r = 8$, instances with $n = 320$ (for which $\delta^* < n$) are easier than instances with $n = 80$ and $n = 160$ (for which $\delta^* > n$).

Table 3. Results on Instances from 13 Chloroplast Genomes, Time Limit of 1 Minute.

q	n	# inst.	# opt.	δ^*	lb_C	lb_{BB}	B&B subpr.	B&B time	δ^H	H time	gap
3	105	286	286	19.4	18.7	19.4	48697.0	0.1 (1.1)	19.4	– (–)	0.0 (0)
4	105	715	651	28.5	25.0	28.3	6257725.6	12.1 (60.0)	28.5	0.4 (5.1)	0.1 (4)
5	105	1287	1073	36.5	31.2	36.3	9265956.4	24.1 (60.0)	36.5	0.8 (7.3)	0.2 (4)

1 Trachelium 2 Campanula 3 Adenophora 4 Symphyandra 5 Legousia
6 Asyneuma 7 Triodanus 8 Wahlenbergia 9 Merciera   10 Codonopsis
11 Cyananthus 12 Platycodon   13 Tobacco

Table 4. Results on Instances from 15 Mitochondrial Genomes, Time Limit of 1 Minute.

q	n	# inst.	# opt.	δ^*	lb_C	lb_{BB}	B&B subpr.	B&B time	δ^H	H time	gap
3	37	455	394	45.0	42.8	44.7	8338102.3	13.3 (60.0)	45.0	0.4 (1.7)	0.2 (5)
4	37	1365	59	69.2	57.1	62.9	21866065.4	58.0 (60.0)	69.2	0.8 (10.8)	6.3 (16)
5	37	3003	19	90.8	71.3	79.1	15178139.2	59.8 (60.0)	90.8	2.6 (43.8)	11.6 (23)

1 Homo Sapiens   2 Albinaria Coerulea   3 Arbacia Lixula
4 Artemia Franciscana   5 Ascaris Suum   6 Asterina Pectinifera
7 Balanoglossus Carnosus 8 Cepaea Nemoralis   9 Cyprinus Carpio
10 Drosophila Yakuba   11 Florometra Serratissima 12 Katharina Tunicata
13 Lumbricus Terrestris   14 Onchocerca Volvulus   15 Struthio Camelus

Table 5. Results on Instances from 13 Chloroplast Genomes for Large Values of q, Time Limit of 1 Hour.

q	n	# inst.	# opt.	δ^*	lb_C	lb_{BB}	B&B subpr.	B&B time	δ^H	H time	gap
6	105	13	13	42.1	34.8	42.1	135100475.1	465.7 (3149.2)	42.1	– (–)	0.0 (0)
7	105	13	12	51.0	41.5	50.9	150696379.1	673.4 (3600.0)	51.0	4.2 (4.2)	0.1 (1)
8	105	13	8	60.6	47.9	59.8	272373385.8	1548.1 (3600.0)	60.6	0.8 (0.9)	0.8 (4)
9	105	13	10	69.0	54.6	68.5	263701838.8	1828.0 (3600.0)	69.0	1.3 (1.9)	0.5 (4)
10	105	13	7	78.3	61.2	76.9	234546156.3	1943.5 (3600.0)	78.3	2.1 (6.2)	1.4 (5)
11	105	13	8	86.7	68.2	85.6	204401368.6	1989.2 (3600.0)	86.7	1.6 (2.7)	1.1 (4)
12	105	13	8	95.1	74.5	94.2	185584896.0	2121.0 (3600.0)	95.1	1.4 (1.5)	0.8 (3)
13	105	1	1	103.0	81.0	103.0	179367376.0	2361.0 (2361.0)	103.0	– (–)	0.0 (0)

We conclude presenting results for instances associated with real-world genomes. We considered two sets of genomes, one corresponding to chloroplast genomes mentioned in [17] and available at http://www.cs.unm.edu/~moret/GRAPPA, each with $n = 105$ genes, and the other to mitochondrial genomes that we obtained from Mathieu Blanchette [4], each with $n = 37$ genes. For each genome set and $q = 3, 4, 5$, we considered, respectively, the instances associated with all triples, four-tuples and fivetuples in the set. The results are reported in Table 3 and 4. As the number of instances is quite large, we imposed one minute time limit both on the branch-and-bound and the heuristic. Moreover, we applied the LP heuristic only to the instances that were not solved to optimality by the branch-and-bound.

The behavior for the mitochondrial instances is analogous to the case of random permutations, namely almost all instances can be solved to proven optimality for $q = 3$ whereas for $q = 4$ and 5 the gap is 0 only for very few instances, being (on average) equal to 9% and 13%, respectively.

On the other hand, chloroplast instances turn out to be relatively easy to solve, the solution value being much smaller than for random permutations. With very few exceptions, all instances are solved within one unit of the optimum in very short time. For this reason, we tried also to solve instances for higher values of q. For $q \in \{6, \ldots, 12\}$ we solved the 13 instances obtained by taking q consecutive (in a circular sense) genomes starting from the first, the second, ..., the thirteenth. For $q = 13$ we solved the instance with the whole genome set. Table 5 gives the corresponding results, with a time limit of one hour, motivated by the fact that the number of instances considered is small and that, for $n = 105$, 1 minute or one hour often makes a big difference, which is not the case for "small" n (say, $n < 50$), probably because of the exponential nature of the algorithm. The table shows that, for these genomes, even an instance with $q = 13$ can be solved optimally within reasonable time, and that the average gap over all the instances is about 1 unit.

Acknowledgments

This work was partially supported by MURST and CNR, Italy. Moreover, I would like to thank Dan Gusfield, David Sankoff and Nick Tran for helpful discussions on the subject; Mathieu Blanchette, David Bryant, Nadia El-Mabrouk and Stacia Wyman for having provided me with the real-world instances mentioned in the paper; and David Bader, Bernard Moret, Tandy Warnow, Stacia Wyman and Mi Yan for having made their code for SBR publicly available.

References

1. D.A. Bader, B.M.E. Moret, M. Yan, "A Linear-Time Algorithm for Computing Inversion Distance Between Signed Permutations with an Experimental Study", *Proceedings of the Seventh Workshop on Algorithms and Data Structgures (WADS'01)* (2001), to appear in Lecture Notes in Computer Science; available at http://www.cs.unm.edu/.
2. V. Bafna and P.A. Pevzner, "Genome Rearrangements and Sorting by Reversals", *SIAM Journal on Computing* 25 (1996) 272–289.
3. M. Blanchette, G. Bourque and D. Sankoff, "Breakpoint Phylogenies", in S. Miyano and T. Takagi (eds.), *Proceedings of Genome Informatics 1997*, (1997) 25–34, Universal Academy Press.
4. M. Blanchette, personal communication.

5. D. Bryant, "A Lower Bound for the Breakpoint Phylogeny Problem", to appear in *Journal of Discrete Algorithms* (2001).

6. A. Caprara, "Formulations and Hardness of Multiple Sorting by Reversals", *Proceedings of the Third Annual International Conference on Computational Molecular Biology (RE-COMB'99)* (1999) 84–93, ACM Press.

7. A. Caprara, G. Lancia and S.K. Ng, "Sorting Permutations by Reversals through Branch-and-Price", to appear in *INFORMS Journal on Computing* (2001).

8. M.E. Cosner, R.K. Jansen, B.M.E. Moret, L.A. Rauberson, L.-S. Wang, T. Warnow and S. Wyman, "An Empirical Comparison of Phylogenetic Methods on Chloroplast Gene Order Data in Campanulaceae", in [19], 99-121.

9. M. Grötschel, L. Lovász and A. Schrijver, "The Ellipsoid Method and its Consequences in Combinatorial Optimization", *Combinatorica* 1 (1981), 169–197.

10. D. Gusfield, *Algorithms on Strings, Trees and Sequences: Computer Science and Computational Biology*, (1997), Cambridge University Press.

11. S. Hannenhalli, C. Chappey, E.V. Koonin and P.A. Pevzner, "Genome Sequence Comparison and Scenarios for Gene Rearrangements: A Test Case", *Genomics* 30 (1995) 299–311.

12. S. Hannenhalli and P.A. Pevzner, "Transforming Cabbage into Turnip (Polynomial Algorithm for Sorting Signed Permutations by Reversals)", *Journal of the ACM* 48 (1999) 1–27.

13. M. Jünger, G. Reinelt and G. Rinaldi, "The traveling salesman problem", in M. Ball, T. Magnanti, C. Monma, G. Nemhauser (eds.), *Network Models*, Handbooks in Operations Research and Management Science, 7 (1995) 225–330, Elsevier.

14. H. Kaplan, R. Shamir and R.E. Tarjan, "Faster and Simpler Algorithm for Sorting Signed Permutations by Reversals", *SIAM Journal on Computing* 29 (2000) 880–892.

15. J. Kececioglu and D. Sankoff, "Efficient Bounds for Oriented Chromosome Inversion Distance", *Proceedings of 5th Annual Symposium on Combinatorial Pattern Matching*, Lecture Notes in Computer Science 807 (1994) 307–325, Springer Verlag.

16. J. Kececioglu and D. Sankoff, "Exact and Approximation Algorithms for Sorting by Reversals, with Application to Genome Rearrangement", *Algorithmica* 13 (1995) 180–210.

17. B.M.E. Moret, L.-S. Wang, T. Warnow and S.K. Wyman. "Highly accurate reconstruction of phylogenies from gene order data", *Tech. Report TR-CS-2000-51* Dept. of Computer Science, University of New Mexico (2000), available at http://www.cs.unm.edu/.

18. B.M.E. Moret, S.K. Wyman, D.A. Bader, T. Warnow and M. Yan, "A New Implementation and Detailed Study of Breakpoint Analysis", *Proceedings of the Sixth Pacific Symposium on Biocomputing (PSB 2001)* (2001) 583–594, World Scientific Pub.

19. D. Sankoff and J.H. Nadeau (eds.) *Comparative Genomics: Empirical and Analytical Approaches to Gene Order Dynamics*, (2000) Kluwer Academic Publishers.

20. I. Pe'er and R. Shamir, "The Median Problems for Breakpoints are $\mathcal{NP}$-Complete", ECCC Report No. 71 (1998), University of Trier, 1998, available at http://www.eccc.uni-trier.de/.

21. I. Pe'er and R. Shamir, "Approximation Algorithms for the Median Problem in the Breakpoint Model", in [19], 225–241.

22. D. Sankoff and M. Blanchette, "Multiple Genome Rearrangement and Breakpoint Phylogenies", *Journal of Computational Biology* 5 (2000) 555–570.

23. D. Sankoff, D. Bryant, M. Denault, B.F. Lang, G. Burger, "Early Eukaryote Evolution Based on Mitochondrial Gene Order Breakpoints", to appear in *Journal of Computational Biology* (2001).

24. D. Sankoff and N. El-Mabrouk, "Duplication, Rearrangement and Reconciliation", in [19], 537–550.

25. D. Sankoff, G. Sundaram and J. Kececioglu, "Steiner Points in the Space of Genome Rearrangements", *International Journal of Foundations of Computer Science* 7 (1996) 1–9.

26. J. Setubal and J. Meidanis, *Introduction to Computational Molecular Biology*, (1997), PWS Publishing.

Algorithms for Finding Gene Clusters

Steffen Heber[1] and Jens Stoye[2]

[1] Department of Computer Science & Engineering
University of California, San Diego
sheber@ucsd.edu
[2] Max Planck Institute for Molecular Genetics
Berlin, Germany
stoye@molgen.mpg.de

Abstract. Comparing gene orders in completely sequenced genomes is a standard approach to locate clusters of functionally associated genes. Often, gene orders are modeled as permutations. Given k permutations of n elements, a k-tuple of intervals of these permutations consisting of the same set of elements is called a *common interval*. We consider several problems related to common intervals in multiple genomes. We present an algorithm that finds all common intervals in a family of genomes, each of which might consist of several chromosomes. We present another algorithm that finds all common intervals in a family of circular permutations. A third algorithm finds all common intervals in signed permutations. We also investigate how to combine these approaches. All algorithms have optimal worst-case time complexity and use linear space.

1 Introduction

The conservation of gene order has been extensively studied so far [25, 19, 16, 12]. There is strong evidence that genes clustering together in phylogenetically distant genomes frequently encode functionally associated proteins [23, 4, 24] or indicate recent horizontal gene transfer [11, 5]. Due to the increasing amount of completely sequenced genomes, the comparison of gene orders to find conserved gene clusters is becoming a standard approach for protein function prediction [20, 17, 22, 6].

In this paper we describe efficient algorithms for finding gene clusters for various types of genomic data. We represent gene orders by permutations (re-orderings) of integers. Hence gene clusters correspond to intervals (contiguous subsets) in permutations, and the problem of finding conserved gene clusters in different genomes translates to the problem of finding *common intervals* in multiple permutations.

In addition to this bioinformatic application, common intervals also relate to the *consecutive arrangement problem* [2, 7, 8] and to cross-over operators for genetic algorithms solving sequencing problems such as the traveling salesman problem or the single machine scheduling problem [3, 15, 18].

Recently, Uno and Yagiura [26] presented an optimal $O(n + K)$ time and $O(n)$ space algorithm for finding all $K \leq \binom{n}{2}$ common intervals of two permutations π_1 and π_2 of n elements. We generalized this algorithm to a family $\Pi = (\pi_1, \ldots, \pi_k)$ of $k \geq 2$ permutations in optimal $O(kn + K)$ time and $O(n)$ space [10] by restricting the

O. Gascuel and B.M.E. Moret (Eds.): WABI 2001, LNCS 2149, pp. 252–263, 2001.

set of common intervals to a smaller, generating subset. To apply common intervals to the bioinformatic problem of finding conserved clusters of genes in data derived from completely sequenced genomes we further extended the above algorithm to additional types of permutations.

Genomes of higher organisms generally consist of several linear chromosomes while bacterial, archaeal, and mitochondrial DNA is organized in one to several circular pieces. While in the first case the algorithm from [10] might report too many gene clusters if the multiple chromosomes are simply concatenated, in the latter case gene clusters might be missed if the circular pieces are cut at some arbitrary point. We handle this problem by adapting the original algorithm to *multichromosomal permutations* as well as *circular permutations*.

For prokaryotes, it is also known that, in the vast majority of cases, functionally associated genes of a gene cluster lie on the same DNA strand [20, 12]. We take this into account by constructing *signed permutations* where the sign of a gene indicates the strand it lies on. We then determine all common intervals with the additional restriction that within each permutation, the elements of a common interval must have the same sign, while between permutations the sign might vary. This allows us to restrict the set of common intervals to biologically meaningful candidates.

The paper is organized as follows. In Section 2 we formally define common intervals and related terminology. We briefly describe the algorithms of Uno and Yagiura [26] and of Heber and Stoye [10] to find all common intervals of 2 (respectively $k \geq 2$) permutations. Then we present time- and space-optimal algorithms for the problem of finding all common intervals in multichromosomal permutations (Section 3), in signed permutations (Section 4), and in circular permutations (Section 5). In Section 6 we show how the various approaches can be combined without sacrificing the optimal time complexity. Section 7 concludes with few final remarks.

2   Common and Irreducible Intervals

2.1   Basic Definitions

A *permutation* π of (the elements of) the set $N := \{1, 2, \ldots, n\}$ is a re-ordering of the elements of N. We denote by $\pi(i) = j$ that the ith element in this re-ordering is j. For $1 \leq x \leq y \leq n$, we set $[x, y] := \{x, x + 1, \ldots, y\}$ and call $\pi([x, y]) := \{\pi(i) \mid i \in [x, y]\}$ an *interval* of π.

Let $\Pi = (\pi_1, \ldots, \pi_k)$ be a family of k permutations of N. Without loss of generality we assume in this section that $\pi_1 = id_n := (1, \ldots, n)$. A subset $c \subseteq N$ is called a *common interval* of Π if and only if there exist $1 \leq l_j < u_j \leq n$ for all $1 \leq j \leq k$ such that

$$c = \pi_1([l_1, u_1]) = \pi_2([l_2, u_2]) = \ldots = \pi_k([l_k, u_k]).$$

Note that this definition excludes common intervals of size one.

In the following we represent a common interval c either by specifying its elements or by the shorter notation $\pi_j([l_j, u_j])$ for a $j \in \{1, \ldots, n\}$. (For $\pi_j = id_n$ this notation further simplifies to $[l_j, u_j]$.) The set of all common intervals of $\Pi = (\pi_1, \ldots, \pi_k)$ is denoted C_Π.

Example 1. Let $N = \{1, \ldots, 9\}$ and $\Pi = (\pi_1, \pi_2, \pi_3)$ with $\pi_1 = id_9$, $\pi_2 = (3, 2, 1, 9, 7, 8, 6, 5, 4)$, and $\pi_3 = (4, 5, 6, 8, 7, 1, 2, 3, 9)$. With respect to π_1 we have

$$C_\Pi = \{[1, 2], [1, 3], [1, 9], [2, 3], [4, 5], [4, 6], [4, 8], [5, 6], [5, 8], [6, 8], [7, 8]\}.$$

$\square$

In order to keep this paper self-contained, in the remainder of this section we recall the algorithms of Uno and Yagiura [26] and of Heber and Stoye [10] that find all common intervals of 2 (respectively $k \geq 2$) permutations. We will restrict our description to basic ideas and only give details where they are necessary for an understanding of the new algorithms described in Sections 3–6 of this paper.

2.2   Finding All Common Intervals of Two Permutations

Here we consider the problem of finding all common intervals of $k = 2$ permutations $\pi_1 = id_n$ and π_2 of N.

An easy test if an interval $\pi_2([x, y])$, $1 \leq x < y \leq n$, is a common interval of $\Pi = (\pi_1, \pi_2)$ is based on the following functions:

$$
\begin{aligned}
l(x, y) &:= \min \pi_2([x, y]) \\
u(x, y) &:= \max \pi_2([x, y]) \\
f(x, y) &:= u(x, y) - l(x, y) - (y - x).
\end{aligned}
$$

Since $f(x, y)$ counts the number of elements in $[l(x, y), u(x, y)] \setminus \pi_2([x, y])$, an interval $\pi_2([x, y])$ is a common interval of Π if and only if $f(x, y) = 0$. A simple algorithm to find C_Π is to test for each pair of indices (x, y) with $1 \leq x < y \leq n$ if $f(x, y) = 0$, yielding a naive $O(n^3)$ time or, using running minima and maxima, a slightly more involved $O(n^2)$ time algorithm.

In order to save the time to test $f(x, y) = 0$ for some pairs (x, y), Uno and Yagiura [26] introduce the notion of *wasteful* candidates for y.

Definition 1. *For a fixed x, a right interval end $y > x$ is called* wasteful *if it satisfies $f(x', y) > 0$ for all $x' \leq x$.*

Based on this notion, Uno and Yagiura give an algorithm called RC (short for *Reduce Candidate*) that has as its essential part a data structure Y consisting of a doubly-linked list $ylist$ for the indices of non-wasteful right interval end candidates and, storing intervals of $ylist$, two further doubly-linked lists $llist$ and $ulist$ that implement the functions l and u in order to compute f efficiently. An outline of Algorithm RC is shown in Algorithm 1 where $L.succ(e)$ denotes the successor of element e in a doubly linked list L. After initializing the lists of Y, a counter x (corresponding to the currently investigated left interval end) runs from $n - 1$ down to 1. In each iteration step, during the update of Y, $ylist$ is trimmed such that afterwards the function $f(x, y)$ is monotonically increasing for the elements y remaining in $ylist$. In lines 5–7, this allows us to efficiently find all common intervals with left end x by evaluating $f(x, y)$ running left-to-right through the elements $y > x$ of $ylist$ until an index y is encountered with $f(x, y) > 0$ when the reporting procedure stops.

Algorithm 1 (Reduce Candidate, RC)

Input: A family $\Pi = (\pi_1 = id_n, \pi_2)$ of two permutations of $N = \{1, \ldots, n\}$.
Output: The set of all common intervals C_Π.
1:   initialize Y
2: **for** $x = n - 1, \ldots, 1$ **do**
3:      update Y // trim $ylist$, update $llist$ and $ulist$
4:      $y \leftarrow x$
5:      **while** $(y \leftarrow ylist.succ(y))$ defined **and** $f(x, y) = 0$ **do**
6:         output $[l(x, y), u(x, y)]$
7:      **end while**
8: **end for**

For details of the data structure Y and the update procedure in line 3, see [26, 10]. The analysis shows that the update of data structure Y in line 3 can be performed in amortized $O(1)$ time, such that the complete algorithm takes $O(n + K)$ time to find the K common intervals of π_1 and π_2.

2.3 Irreducible Intervals

Before we show how to generalize Algorithm RC to find all common intervals of $k \geq 2$ permutations, we first present a useful generating subset of the set of common intervals, the set of *irreducible intervals* [10] and report a few of their properties.

We say that two common intervals $c_1, c_2 \in C_\Pi$ have a *non-trivial overlap* if $c_1 \cap c_2 \neq \emptyset$ and neither includes the other. A list $p = (c_1, \ldots, c_{\ell(p)})$ of common intervals $c_1, \ldots, c_{\ell(p)} \in C_\Pi$ is a *chain* (of length $\ell(p)$) if every two successive intervals in p have a non-trivial overlap. A chain of length one is called a *trivial chain*, all other chains are called *non-trivial chains*. A chain that can not be extended to its left or right is a *maximal chain*. It is easy to see that for a chain p of common intervals, the interval $\tau(p) := \bigcup_{c' \in p} c'$ is a common interval as well. We say that p *generates* $\tau(p)$.

Definition 2. *A common interval c is called* reducible *if there is a non-trivial chain that generates c, otherwise it is called* irreducible.

This definition partitions the set of common intervals C_Π into the set of reducible intervals and the set of irreducible intervals, denoted I_Π. Obviously, $1 \leq |I_\Pi| \leq |C_\Pi| \leq \binom{n}{2}$.

Example 1 (cont'd). For $\Pi = (\pi_1, \pi_2, \pi_3)$ as above, the irreducible intervals (with respect to $\pi_1 = id_9$) are

$$I_\Pi = \{[1, 2], [1, 9], [2, 3], [4, 5], [5, 6], [6, 8], [7, 8]\}.$$

The reducible intervals are generated as follows:

$$
\begin{aligned}
[1, 3] &= [1, 2] \cup [2, 3], \\
[4, 6] &= [4, 5] \cup [5, 6], \\
[4, 8] &= [4, 5] \cup [5, 6] \cup [6, 8], \\
[5, 8] &= [5, 6] \cup [6, 8].
\end{aligned}
$$

We cite the following two results from [10] (without proofs) which indicate the great value of the concept of irreducible intervals.

Lemma 1. *Given a family $\Pi = (\pi_1, \ldots, \pi_k)$ of permutations of $N = \{1, 2, \ldots, n\}$, the set of irreducible intervals I_Π allows us to reconstruct the set of all common intervals C_Π in optimal $O(|C_\Pi|)$ time.* □

Lemma 2. *Given a family $\Pi = (\pi_1, \ldots, \pi_k)$ of permutations of $N = \{1, 2, \ldots, n\}$, we have $1 \le |I_\Pi| \le n - 1$.* □

2.4   Finding All Common Intervals of k Permutations

Now we can describe the algorithm from [10] that finds all K common intervals of a family of $k \ge 2$ permutations of N in $O(kn + K)$ time.

For $1 \le i \le k$, set $\Pi_i := (\pi_1, \ldots, \pi_i)$. Starting with $I_{\Pi_1} = \{[j, j + 1] \mid 1 \le j \le n - 1\}$, the algorithm successively computes I_{Π_i} from $I_{\Pi_{i-1}}$ for $i = 2, \ldots, k$ (see Algorithm 2). The algorithm employs a mapping

$$\varphi_i : I_{\Pi_{i-1}} \to I_{\Pi_i}$$

that maps each element $c \in I_{\Pi_{i-1}}$ to the smallest common interval $c' \in C_{\Pi_i}$ that contains c. It is shown in [10] that this mapping exists and is surjective, i.e., $\varphi_i(I_{\Pi_{i-1}}) := \{\varphi_i(c) \mid c \in I_{\Pi_{i-1}}\} = I_{\Pi_i}$. Furthermore, it is shown that $\varphi_i(I_{\Pi_{i-1}})$ can be effi-

---

Algorithm 2 (Finding All Common Intervals of k Permutations)

Input: A family $\Pi = (\pi_1 = id_n, \pi_2, \ldots, \pi_k)$ of k permutations of $N = \{1, \ldots, n\}$.
Output: The set of all common intervals C_Π.
1: $I_{\Pi_1} \leftarrow ([1, 2], [2, 3], \ldots, [n - 1, n])$
2: **for** $i = 2, \ldots, k$ **do**
3:   $I_{\Pi_i} \leftarrow \{\varphi_i(c) \mid c \in I_{\Pi_{i-1}}\}$ // (see Algorithm 3)
4: **end for**
5: generate C_Π from $I_\Pi = I_{\Pi_k}$ using Lemma 1
6: output C_Π

---

ciently computed by a modified version of Algorithm RC where the data structure Y is supplemented by a data structure S that is derived from $I_{\Pi_{i-1}}$. S consists of several doubly-linked *clists* containing intervals of *ylist*, one for each maximal chain of the intervals in $I_{\Pi_{i-1}}$.

Using π_1 and π_i, as in Algorithm RC, the *ylist* of Y allows for a given x to access all non-wasteful right interval end candidates y of $C_{(\pi_1, \pi_i)}$. The aim of S is to further reduce these candidates to only those indices y for which simultaneously $[x, y] \in C_{\Pi_{i-1}}$ (ensuring $[x, y] \in C_{\Pi_i}$) and $[x, y]$ contains an interval $c \in I_{\Pi_{i-1}}$ that is not contained in any smaller interval from C_{Π_i}. Together this ensures that exactly the irreducible intervals $[x, y] \in \varphi_i(I_{\Pi_{i-1}})$ are reported.

An outline of the modified version of Algorithm RC is shown in Algorithm 3. Essentially, S keeps a list of *active* intervals, i.e., intervals from $I_{\Pi_{i-1}}$ for which the

---

Algorithm 3 (Extended Algorithm RC)

Input: A family $\Pi = (\pi_1 = id_n, \pi_i)$ of two permutations of $N = \{1, \ldots, n\}$; a set of irreducible intervals $I_{\Pi_{i-1}}$.
Output: The set of irreducible intervals I_{Π_i}.
 1: initialize Y and S
 2: **for** $x = n - 1, \ldots, 1$ **do**
 3:    update Y and S // trim $ylist$, update $l/ulist$; activate elements of the $clists$
 4:    **while** $([x', y] \leftarrow S.first\_active\_interval(x))$ defined **and** $f(x, y) = 0$ **do**
 5:      output $[l(x, y), u(x, y)]$
 6:      deactivate $[x', y]$
 7:    **end while**
 8: **end for**

---

image under mapping φ_i has not yet been determined. In the reporting loop of lines 4–7, rather than testing if $f(x, y) = 0$ running from left to right through all indices $y > x$ of $ylist$, only right ends of active intervals are tested. Therefore, function $S.first\_active\_interval(x)$ returns the active intervals in left-to-right order with respect to their right end y. If an active interval $[x', y]$ gives rise to a common interval, i.e., if $f(x, y) = 0$, then an element of $\varphi_i(I_{\Pi_{i-1}})$ is encountered and the active interval is deactivated. Similar to Algorithm RC, reporting stops whenever the first active interval with right end y is encountered such that $f(x, y) > 0$.

Again, for details of the data structure and the update procedure in line 3 we refer to the original description [10]. There it is also shown that updating the data structure S takes amortized $O(1)$ time. Hence, due to the reduced output size (see Lemma 2), the Extended Algorithm RC takes only $O(n)$ time. Together with Lemma 1 this implies the overall time complexity $O(kn + K)$ for Algorithm 2. The additional space usage is $O(n)$.

3   Common Intervals of Multichromosomal Permutations

In view of biological reality, in the following we consider variants of the common intervals problem that have to be addressed when dealing with real genomic data. Our first variant that we consider is the scenario where the genome consists of multiple chromosomes.

As above, let $N := \{1, 2, \ldots, n\}$ represent a set of n genes. A *chromosome c* of N is defined as a linearly ordered subset of N and will be represented as a linear list. A *multichromosomal permutation* π of N is defined as a set of chromosomes, containing each element of N exactly once, i.e.,

$$\pi = \{c_1, \ldots, c_l\} \quad \text{with} \quad N = \dot{\bigcup}_{1 \leq i \leq l} c_i.$$

Given a family $\Pi = (\pi_1, \ldots, \pi_k)$ of k multichromosomal permutations of N, a subset $s \subseteq N$ is called a *common interval* of Π if and only if for each multichromosomal permutation π_i, $i = 1, \ldots, k$, there exists a chromosome with s as an interval.

Example 2. Let $N = \{1, \ldots, 6\}$ and $\Pi = (\pi_1, \pi_2, \pi_3)$ with $\pi_1 = \{(1, 2, 3), (4, 5, 6)\}$, $\pi_2 = \{(1, 5, 6, 4), (3, 2)\}$, and $\pi_3 = \{(1, 6, 4, 5), (3), (2)\}$. Here chromosome ends are indicated by parentheses. The only common interval is $\{4, 5, 6\}$.   □

A modification of Algorithm 2 can be used for finding all common intervals of k multichromosomal permutations. We start by arranging the chromosomes of each multichromosomal permutation in arbitrary order. This way we obtain a family $\Pi' = (\pi'_1, \pi'_2, \ldots, \pi'_k)$ of k (standard) permutations π'_i, $i = 1, \ldots, k$. Without loss of generality we assume that $\pi'_1 = id_n$. Now, as above, set $\Pi'_i := (\pi'_1, \pi'_2, \ldots, \pi'_i)$. Then, starting with

$$I_{\Pi'_1} := \{[j, j + 1] \mid j, j + 1 \text{ on the same chromosome in } \pi_1 \text{ and } 1 \leq j < n\},$$

the algorithm successively computes $I_{\Pi'_i}$ from $I_{\Pi'_{i-1}}$ for $i = 2, \ldots, k$ using a modification of Algorithm 2, where in the extended Algorithm RC (Algorithm 3) the reporting procedure is not only stopped whenever $f(x, y) > 0$, but also as soon as the genes at indices x and y belong to different chromosomes of π_i.

By the definition of $I_{\Pi'_1}$, this algorithm will never place two genes from different chromosomes in π_1 together in a common interval. Moreover, by the modification of Algorithm 3, two genes from different chromosomes of the other genomes $\pi_2, \ldots, \pi_k$ will never be placed together in a common interval. Nevertheless, the location of common intervals that lie on the same chromosome in all genomes is not affected by the modification of the algorithm. Since the additional test if x and y belong to the same chromosome is a constant time operation, and the output can only be smaller than that of the original Algorithm 3, the new algorithm also takes $O(n)$ time to generate $I_{\Pi'_i}$ from $I_{\Pi'_{i-1}}$. The outer loop (Algorithm 2) and the final generation of the common intervals from the irreducible intervals (Lemma 1) are unchanged, so that we have the following

Theorem 1. *Given k multichromosomal permutations of $N = \{1, \ldots, n\}$, all K common intervals can be found in optimal $O(kn + K)$ time using $O(n)$ additional space.*
                    □

4   Common Intervals of Signed Permutations

In this section we consider the problem of finding all common intervals in a family of signed permutations. It is common practice when considering genome rearrangement problems, to denote the direction of a gene in the genome by a plus $(+)$ or minus $(-)$ sign depending on the DNA strand it is located on [21]. In the context of sorting signed permutations by reversals [1, 9, 13, 14], the sign of a gene tells the direction of the gene in the final (sorted) permutation and changes with each reversal. In our context, it has been observed that for prokaryotes, functionally coupled genes, e.g. in operons, virtually always lie on the same DNA strand [20, 12]. Hence, when given signed permutations, we require that the sign does not change within an interval. Between the different permutations, the sign of the intervals might vary, though. This restricts the (original) set of all common intervals to the biologically more meaningful candidates.

Example 3. Let $N = \{1, \ldots, 6\}$ and $\Pi = (\pi_1, \pi_2, \pi_3)$ with $\pi_1 = (+1, +2, +3, +4, +5, +6)$, $\pi_2 = (-3, -1, -2, +5, +4, +6)$, and $\pi_3 = (-4, +5, +6, -2, -3, -1)$. With respect to π_1 the interval $[1, 3]$ is a common interval, but $[4, 5]$ and $[4, 6]$ are not. □

Obviously, the number of common intervals in signed permutations can be considerably smaller than the number of common intervals in unsigned permutations. Hence, applying Algorithm 2 followed by a filtering step will not yield our desired time-optimal result.

However, the problem can be solved easily by applying the algorithm for multichromosomal permutations from the previous section. Since a common interval in signed permutations can never contain two genes with different sign, we break the signed permutations into pieces ("chromosomes") wherever the sign changes. This is clearly possible in linear time. Then we apply the algorithm from the previous section to the obtained family of multichromosomal permutations, the result being exactly the common intervals of the original signed permutations. Hence, we have the following

Theorem 2. *Given k signed permutations of $N = \{1, \ldots, n\}$, all K common intervals can be found in optimal $O(kn + K)$ time using $O(n)$ additional space.* □

5   Common Intervals of Circular Permutations

As discussed in the Introduction, much of the DNA in nature is circular. Consequently, by representing genomes as (possibly multichromosomal) *linear* permutations of genes and then looking for common gene clusters, one might miss clusters that span across the (mostly arbitrary) dissection point where the circular genome is linearized.

In this section we consider an arrangement of the set of genes $N = \{1, 2, \ldots, n\}$ along a circle and call this a *circular permutation*. Given a family $\Pi = (\pi_1, \ldots, \pi_k)$ of k circular permutations of N, a (sub)set $c \subseteq N$ of genes is called a *common interval* if and only if the elements of c occur uninterruptedly in each circular permutation.

Example 4. Let $N = \{1, \ldots, 6\}$ and $\Pi = (\pi_1, \pi_2, \pi_3)$ with $\pi_1 = (1, 2, 3, 4, 5, 6)$, $\pi_2 = (2, 4, 5, 6, 1, 3)$, and $\pi_3 = (6, 4, 1, 3, 2, 5)$. Apart from the trivial intervals (N, the singletons, and N minus each singleton), the common intervals of Π are $\{1, 2, 3\}$, $\{1, 2, 3, 4\}, \{1, 4, 5, 6\}, \{2, 3\}, \{4, 5, 6\}, \{5, 6\}$. □

In the following we will show how to find all K common intervals in a family of circular permutations in optimal $O(kn + K)$ time. Again, this can be done by an easy modification of the original algorithm from Section 2, in combination with the following observation.

Lemma 3. *Let c be a common interval of a family Π of circular permutations of N. Then its complement $\bar{c} := N \setminus c$ is also a common interval of Π.*

Proof. This follows immediately from the definition of common intervals of circular permutations. □

Note that Lemma 3 does not hold for irreducible intervals.

Algorithm 4 (Finding All Common Intervals of k Circular Permutations)

Input: A family $\Pi = (\pi_1 = id_n, \pi_2, \ldots, \pi_k)$ of k circular permutations of $N = \{1, \ldots, n\}$.
Output: The set of all common intervals C_Π.
  1: $I_{\Pi_1}^* \leftarrow (\{1, 2\}, \{2, 3\}, \ldots, \{n-1, n\}, \{n, 1\})$
  2: **for** $i = 2, \ldots, k$ **do**
  3:    $I_{\Pi_i}^* \leftarrow \{\varphi_i^*(c) \mid c \in I_{\Pi_{i-1}}^*\}$ // (see text)
  4: **end for**
  5: generate C_Π^* from $I_\Pi^* = I_{\Pi_k}^*$ using Lemma 1
  6: $\overline{C}_\Pi^* \leftarrow \{\bar{c} \mid c \in C_\Pi^*\}$
  7: output $C_\Pi^* \cup \overline{C}_\Pi^*$

The general idea is now to first find only the common intervals of size $\leq \lfloor \frac{n}{2} \rfloor$, and then find the remaining common intervals by complementing these. The procedure is outlined in Algorithm 4. The main difference to Algorithm 2 is that function φ_i is replaced by a variant, denoted φ_i^*, that works on circular permutations and only generates irreducible intervals of size $\leq \lfloor \frac{n}{2} \rfloor$. This function is implemented by multiple calls to the original function φ_i. The two circular permutations π_1 and π_i are therefore linearized in two different ways each, namely by once cutting them between positions n and 1, and once cutting between positions $\lfloor \frac{n}{2} \rfloor$ and $\lfloor \frac{n}{2} \rfloor + 1$. Then φ_i is applied to each of the four resulting pairs of linearized permutations. For convenience, the output of common intervals of length $> \lfloor \frac{n}{2} \rfloor$ is suppressed. Finally, the resulting intervals of the four runs of φ_i are merged, sorted according their start and end positions using counting sort, and duplicates are removed. Clearly, φ_i^* generates all irreducible intervals of size $\leq \lfloor \frac{n}{2} \rfloor$ in $O(n)$ time. Hence, we have the following

Theorem 3. *Given k circular permutations of $N = \{1, \ldots, n\}$, all K common intervals can be found in optimal $O(kn + K)$ time using $O(n)$ additional space.* $\square$

6   Combination of the Algorithms

In this section we show how to handle arbitrary combinations of multichromosomal, signed, and circular permutations.

Combining multichromosomal and signed permutations is straightforward, but it is not obvious how to handle combinations which involve circular chromosomes without loosing the optimal running time. Circular chromosomes of different genomes might now have incompatible gene contents and Lemma 3 no longer holds as the following example shows.

Example 5. Let $N = \{1, \ldots, 8\}$ and $\Pi = (\pi_1, \pi_2)$ with $\pi_1 = \{(1, 2, 3, 4), (5, 6, 7, 8)\}$ and $\pi_2 = \{(1, 3, 5, 6, 7), (2, 4, 8)\}$ where all chromosomes are circular. While $c = \{5, 6\}$ is a common interval, its complement $N \setminus c = \{1, 2, 3, 4, 7, 8\}$ is not. $\square$

We overcome these problems by a preprocessing step where we include artificial *breakpoints* into the genomes. The breakpoints do not affect common intervals but refine the permutations so that they can be handled by our algorithms. The preprocessing is performed as follows.

We compare permutation π_1 successively to each of the other permutations π_i, $2 \leq i \leq k$ and test for each pair of neighboring genes in π_1 (i.e., for each chromosome $c = (\pi_1(l), \pi_1(l+1), \ldots, \pi_1(r))$ the pairs $\{\pi_1(j), \pi_1(j+1)\}$ for $l \leq j \leq r-1$, plus the pair $\{\pi_1(l), \pi_1(r)\}$ for circular c) if they lie on the same chromosome in π_i or not. If not, they can not be elements of the same common interval and we introduce a new artificial breakpoint between the two genes in π_1. Then we do the same comparison in the opposite direction, i.e., we introduce breakpoints between neighboring genes of π_i, $2 \leq i \leq k$ whenever they do not lie on the same chromosome of π_1. At the first time a breakpoint is inserted in a circular chromosome, the chromosome is linearized by cutting at the breakpoint and replacing it in the genome by the appropriately circularly shifted linear chromosome. Breakpoints in a linear chromosome dissect the chromosome. This preprocessing can be performed in $O(kn)$ time.

After the preprocessing, the genes that do not occur in any circular chromosome can be handled by the algorithm for multichromosomal permutations (Section 3) in a straightforward way. The genes that occur in at least one circular chromosome are partitioned into sets of genes which correspond to a single circular chromosome. This partition is well defined, since the set of genes of each remaining circular chromosome corresponds, in the other genomes, either to one circular or to one or several linear chromosomes. Each element of this partition is now treated separately. We start by restricting all genomes to the selected gene set. If each of the restricted genomes is circular we can apply the algorithm for circular permutations (Section 5) directly. Otherwise we choose a restricted genome that consists of one or several linear chromosomes and arrange the chromosomes in an arbitrary order. Denote l (r) the first (last) gene in this order. We proceed as in the multichromosomal case (Section 3) except we encounter a circular genome π_c. If l and r are neighboring genes in π_c we linearize π_c by cutting between them and proceed as for a linear genome. Otherwise, similar to the case of circular permutations (Section 5), we copy π_c four times and linearize the copies by cutting one copy on the left of l, one copy on the right of l, one copy on the left of r, and one copy on the right of r. For each of these genomes we compute all irreducible intervals. The resulting intervals are merged, sorted according their start and end positions using counting sort, and duplicates are removed. This procedure guarantees that we determine all irreducible intervals except for those, which contain l and r simultaneously. But due to our choice of l and r there is at most one such interval, the trivial one, which contains all genes. We test this interval separately.

Since the above described preprocessing and the modifications of the algorithms for multichromosomal and circular permutations do not affect the optimal asymptotic running time, we have

Theorem 4. *Given k multichromosomal, signed, circular or linear (or mixed) permutations of $N = \{1, \ldots, n\}$, all K common intervals can be found in optimal $O(kn+K)$ time using $O(n)$ additional space.* □

7 Conclusion

In this paper we have presented time and space optimal algorithms for variants of the common intervals problem for k permutations. The variants we considered, multichromosomal permutations, signed permutations, circular permutations, and their combinations, were motivated by the requirements imposed by real data we were confronted with in our experiments. While in preliminary testing we have applied our algorithms to bacterial genomes, it is obvious that in a realistic setting, one should further relax the problem definition. In particular, one should allow for missing or additional genes in a common interval while imposing a penalty whenever this occurs. Such relaxations seem to make the problem much harder, though.

Acknowledgments

We thank Richard Desper and Zufar Mulyukov for carefully reading this manuscript as well as Pavel Pevzner for a helpful discussion on the ideas contained in the manuscript.

References

1. V. Bafna and P. Pevzner. Genome rearrangements and sorting by reversals. *SIAM J. Computing*, 25(2):272–289, 1996.
2. K. S. Booth and G. S. Lueker. Testing for the consecutive ones property, interval graphs and graph planarity using *PQ*-tree algorithms. *J. Comput. Syst. Sci.*, 13(3):335–379, 1976.
3. R. M. Brady. Optimization strategies gleaned from biological evolution. *Nature*, 317:804–806, 1985.
4. T. Dandekar, B. Snel, M. Huynen, and P. Bork. Conservation of gene order: A fingerprint of proteins that physically interact. *Trends Biochem. Sci.*, 23(9):324–328, 1998.
5. J. A. Eisen. Horizontal gene transfer among microbial genomes: new insights from complete genome analysis. *Curr. Opin. Genet. Dev.*, 10(6):606–611, 2000.
6. W. Fujibuchi, H. Ogata, H. Matsuda, and M. Kanehisa. Automatic detection of conserved gene clusters in multiple genomes by graph comparison and P-quasi grouping. *Nucleic Acids Res.*, 28(20):4029–4036, 2000.
7. D. Fulkerson and O. Gross. Incidence matrices with the consecutive 1s property. *Bull. Am. Math. Soc.*, 70:681–684, 1964.
8. M. C. Golumbic. *Algorithmic Graph Theory and Perfect Graphs*. Academic Press, New York, 1980.
9. S. Hannenhalli and P. A. Pevzner. Transforming cabbage into turnip: Polynomial algorithm for sorting signed permutations by reversals. *J. ACM*, 46(1):1–27, 1999.
10. S. Heber and J. Stoye. Finding all common intervals of *k* permutations. In *Proceedings of the 12th Annual Symposium on Combinatorial Pattern Matching, CPM 2001*, volume 2089 of *Lecture Notes in Computer Science*, pages 207–219. Springer Verlag, 2001. To appear.
11. M. A. Huynen and P. Bork. Measuring genome evolution. *Proc. Natl. Acad. Sci. USA*, 95(11):5849–5856, 1998.
12. M. A. Huynen, B. Snel, and P. Bork. Inversions and the dynamics of eukaryotic gene order. *Trends Genet.*, 17(6):304–306, 2001.
13. H. Kaplan, R. Shamir, and R. E. Tarjan. A faster and simpler algorithm for sorting signed permutations by reversals. *SIAM J. Computing*, 29(3):880–892, 1999.
14. J. D. Kececioglu and D. Sankoff. Efficient bounds for oriented chromosome inversion distance. In M. Crochemore and D. Gusfield, editors, *Proceedings of the 5th Annual Symposium on Combinatorial Pattern Matching, CPM 94*, volume 807 of *Lecture Notes in Computer Science*, pages 307–325. Springer Verlag, 1994.

15. S. Kobayashi, I. Ono, and M. Yamamura. An efficient genetic algorithm for job shop scheduling problems. In *Proc. of the 6th International Conference on Genetic Algorithms*, pages 506–511. Morgan Kaufmann, 1995.

16. W. C. III Lathe, B. Snel, and P. Bork. Gene context conservation of a higher order than operons. *Trends Biochem. Sci.*, 25(10):474–479, 2000.

17. E. M. Marcotte, M. Pellegrini, H. L. Ng, D. W. Rice, T. O. Yeates, and D. Eisenberg. Detecting protein function and protein-protein interactions from genome sequences. *Science*, 285:751–753, 1999.

18. H. Mühlenbein, M. Gorges-Schleuter, and O. Krämer. Evolution algorithms in combinatorial optimization. *Parallel Comput.*, 7:65–85, 1988.

19. A. R. Mushegian and E. V. Koonin. Gene order is not conserved in bacterial evolution. *Trends Genet.*, 12(8):289–290, 1996.

20. R. Overbeek, M. Fonstein, M. D'Souza, G. D. Pusch, and N. Maltsev. The use of gene clusters to infer functional coupling. *Proc. Natl. Acad. Sci. USA*, 96(6):2896–2901, 1999.

21. P. A. Pevzner. *Computational Molecular Biology: An Algorithmic Approach*. MIT Press, Cambridge, MA, 2000.

22. B. Snel, G. Lehmann, P. Bork, and M. A. Huynen. STRING: A web-server to retrieve and display the repeatedly occurring neigbourhood of a gene. *Nucleic Acids Res.*, 28(18):3443–3444, 2000.

23. J. Tamames, G. Casari, C. Ouzounis, and A. Valencia. Conserved clusters of functionally related genes in two bacterial genomes. *J. Mol. Evol.*, 44(1):66–73, 1997.

24. J. Tamames, M. Gonzalez-Moreno, J. Mingorance, A. Valencia, and M. Vicente. Bringing gene order into bacterial shape. *Trends Genet.*, 17(3):124–126, 2001.

25. R. L. Tatusov, A. R. Mushegian, P. Bork, N.P. Brown, W. S. Hayes, M. Borodovsky, K. E. Rudd, and E. V. Koonin. Metabolism and evolution of *Haemophilus influenzae* deduced from a whole-genome comparison with *Escherichia coli*. *Curr. Biol.*, 6:279–291, 1996.

26. T. Uno and M. Yagiura. Fast algorithms to enumerate all common intervals of two permutations. *Algorithmica*, 26(2):290–309, 2000.

Determination of Binding Amino Acids Based on Random Peptide Array Screening Data

Peter J. van der Veen[1], L.F.A. Wessels[1,2], J.W. Slootstra[3], R.H. Meloen[3], M.J.T. Reinders[2], and J. Hellendoorn[1]

[1] Faculty of Information Technology and Systems
Control Engineering Laboratory
Delft University of Technology
Mekelweg 4, P.O. Box 5031, 2600 GA Delft, The Netherlands
{P.J.vanderVeen,L.Wessels,J.Hellendoorn}@ITS.TUDelft.NL
[2] Faculty of Information Technology and Systems
Information and Communication Theory Group
Delft University of Technology
Mekelweg 4, P.O. Box 5031, 2600 GA Delft, The Netherlands
M.J.T.Reinders@ITS.TUDelft.NL
[3] Pepscan Systems BV, Lelystad, The Netherlands

Abstract. An algorithmic aid is presented which identifies those amino acids in a peptide that are essential to bind against a specific monoclonal antibody. The input data comes from random peptide array screening experiments, which results in a binding strength for all these different peptides. On the basis of this data, the proposed algorithm generates a rule, which describes the amino acids that are necessary to ensure strong binding and consequently separates the best binding peptides from the worst binding ones. The generation of this rule is performed using only information about the occurrence of an amino acid in a peptide, i.e., it doesn't use the relative position of the amino acids in the peptide. Results obtained from experimental data show that for several different monoclonal antibodies, the amino acids which are important according to the generated rules, coincide with amino acids included in motifs which are known to be important for binding. The information gained from this algorithm is useful for the design of subsequent experiments aimed at further optimization of the best binding peptides found during the peptide screening experiment.

1 Introduction

Synthetic peptides can be designed to bind against the same antibody as complete proteins do [5]. For this, the peptide should mimic the area of the protein which is recognized by the antibody. Although proteins normally consist of a large number of amino acids, the contact area between an antibody and a protein generally consists of 15-22 amino acids on each side [6]. Several experiments suggest that only three to five of these amino acids are responsible for the major binding contribution between a protein and an antibody [4,9]. These amino

O. Gascuel and B.M.E. Moret (Eds.): WABI 2001, LNCS 2149, pp. 264–277, 2001.
© Springer-Verlag Berlin Heidelberg 2001

acids define the so called 'energetic epitope' [6]. In this paper, we focus on the determination of those amino acids that constitute this 'energetic epitope'.

The most obvious way to construct a synthetic peptide is by determining the amino acids comprising the epitope of the protein. However, these amino acids can be hard to identify and if they are identified, the usage of these amino acids will not always yield a well binding peptide. This can for example be due to a conformational difference between the amino acids at the protein surface and the amino acids in the peptide. When no well binding peptide can be constructed from a part of the original protein, a random search is generally performed to find a well binding peptide. Such a search can e.g. be done by a phage display analysis [1,7,10] or a peptide screening analysis [3]. In a peptide screening analysis, thousands of randomly generated peptides can be measured against the same antibody. This results in a relative binding strength for every measured peptide. In general, the best binding peptide found during such an analysis is, however, far worse than required. Therefore, several lead variants are usually screened to improve the performance of the best binding peptides.

Although thousands of different measurements are preformed during a single run of a peptide screening analysis, the number of tested peptides is relatively small compared to the number of possible different peptides. The peptides under study consist of 12 successive amino acids, implying that one can generate 20^{12} unique peptides. In principle, one should test all these possibilities to find the best binding peptide. Naturally, this is not feasible and hence a feasible search procedure is required.

One logical method to search for such a well binding peptide is by selecting several of the best known binding peptides and test a large number of mutations of these peptides. But, again due to the large search space, only a limited number of these mutations can be tested during one experiment. Still the question remains which (combination of) amino acids of these peptides should be replaced. In this paper we propose an algorithm that is meant to help in this decision by identifying those amino acids which have a higher chance of being important for binding. Effectively this reduces the search space and makes it possible to spend more effort in mutating the other amino acids of these peptides.

The method generates a rule containing a combination of amino acids which is over-represented in the best binding and underrepresented in the worst binding peptides. Because the amino acids in the rule are over-represented amongst the best binding peptides, these amino acids will most probably improve the binding strength and are most probably contained in the 'energetic epitope'.

2   Data

The data used in this paper is obtained from four distinct peptide screening experiments, where one experiment has been performed for every monoclonal antibody described. In each experiment the binding strength of either 3640 or 4550 randomly generated peptides is measured against an antibody. Each of these peptides consists of 12 amino acids, where each amino acid has an equal probability of occurring at a given position in the peptide. The resulting values represent the binding strength between the different peptides and the monoclonal antibody at a fixed dilution.

The peptides are sorted on their binding strengths to place peptides with a similar binding strength close together, several (scaled) binding curves are shown in Section 4.

3   Method

The actual binding process between a peptide and an antibody is very complex and depends on the interplay between several amino acids. In this paper, we do not consider the full complexity of the problem, rather, our approach is based on the principle that some amino acids are typically needed in a peptide to ensure a high affinity binding against an antibody.

To find these amino acids, we need a representation that does not include the positional information of the peptide's amino acids. To this end, each peptide is represented by a 20 dimensional feature vector. Each feature represents the occurrence of a specific amino acid in the peptide (only the 20 naturally occurring L-amino acids are used to generate the peptides). Thus, if an amino acid is included in the peptide (even if it occurs more than once) the corresponding feature is set to 1 otherwise to 0. An example of this representation is given in Table Table 1. Our aim is to design an algorithm that constructs, for a given

Table 1. Splitting a Peptide up in 20 Features.

	Peptide	A	C	D	E	F	G	H	I	K	L	...
1	QGWFFMQINTQY	0	0	0	0	1	1	0	1	0	0	
2	LMWNPNIKTCER	0	1	0	1	0	0	0	1	1	1	
3	CDSADVSGDHLI	1	1	1	0	0	1	1	1	0	1	
4	MHGVIAQGQQDV	1	0	1	0	0	1	1	1	0	0	
5	FHNQLYYSPDYV	0	0	1	0	1	0	1	0	0	1	
6	HEEWFQLFYYMQ	0	0	0	1	1	0	1	0	0	1	

antibody, based on the measured data, a logical rule which describes the amino acids that are needed in a well binding peptide. This rule maps each of the measured peptides to one of two classes: 1) 'well binding' peptides that contain the necessary amino acids and thus satisfy the rule and 2) 'bad binding' peptides which do not contain the required amino acids and do not satisfy the rule. Since the resulting rule will be used as a decision support in the analysis of the measurements, it is imperative that the generated rule is interpretable. Therefore, we chose to construct rules using logical expressions of amino acids (allowing only AND and OR constructions). An example is given in equation

(1).

$$\text{IF } (\textsf{X OR X OR } \cdots) \text{ AND } (\textsf{X OR X OR } \cdots) \text{ AND } \cdots \qquad (1)$$
$$\text{THEN } \textsf{well binding} \text{ ELSE } \textsf{bad binding}$$

Each X in equation (1) can be replaced by any of the 20 different amino acids.

A classification result of a possible rule is shown in Figure Figure 1, where for every peptide a dot is drawn, indicating whether the peptide is classified as well binding (a dot at 100) or as bad binding (a dot at zero). On the left side of the figure, the number of well binding peptides is larger than the number of bad binding ones. Clearly, we now have a situation where the rule associates

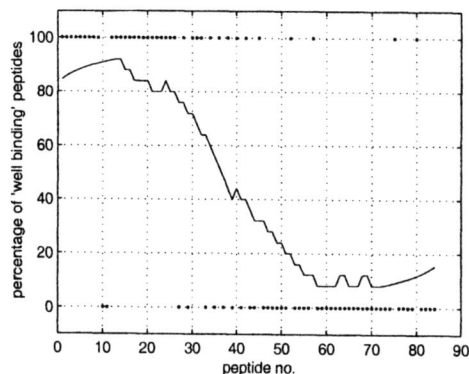

Fig. 1. Classification of Several Peptides. A dot at 100 means that the corresponding peptide is classified as well binding. The continuous line shows the estimated well binding distribution.

every peptide with a binary value (well binding/bad binding) while the data set associates every peptide with a continuous valued binding strength. Since the binding strength *gradually* transitions from the largest to the smallest value it is undesirable to define a crisp boundary between the two categories.

Therefore, to evaluate the rule, we need an additional requirement. This requirement is built into the algorithm as follows: It is assumed that the measured binding strength of a given peptide is proportional to the similarity between this peptide and the best n binding peptides. According to this assumption, we may interpret the measured binding strength as a measure of the *probability density* of the well binding peptides (over the sorted peptides). By requiring that the distribution of the well binding peptides equals the binding energy curve, the classification rule can be optimized. Clearly this will have the desired result that peptides having a high binding energy will be more often classified as well binding and vice versa.

That leaves us with the question how to find the distribution of the well binding peptides given the (binary) output of the classification rule, as in Figure Figure 1. Here we convolve the output of the classification rule with a sliding window that calculates the percentage of **well binding** peptides in that window (resembling a Parzen estimation). Figure Figure 1shows such a resulting distribution when using an averaging window of 25 peptides wide. At the border we clip the averaging window, such that the filtered line has the same size as the number of measured peptides. The average value for the first peptide is calculated using only half of the window size (13 peptides). The next one is calculated using 14 peptides, etc. Clearly, the number of peptides used in this figure is smaller than the number of peptides used in the experiment. On the real data sets a window size of 100 was employed.

The estimated distribution can now be compared with the measured binding energy. Figure Figure 2 shows an example of such a comparison. In this figure, the

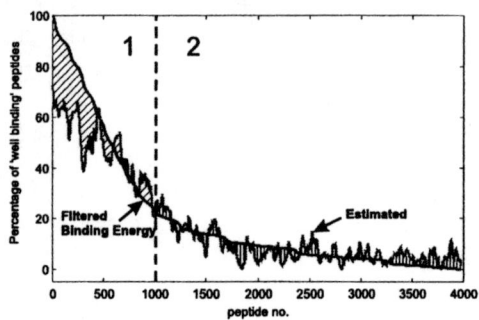

Fig. 2. Comparison between the filtered measured binding energy and the estimated **well binding** distribution. The line separating Area 1 and Area 2 is automatically placed at the 'knee' of the measured binding curve. The position of the knee is derived from the intersection point of two straight lines fitted to the binding curve.

measured binding energy curve is rescaled to fit between 0 and 100. Additionally, we filtered the binding energy in the same way as the classified peptides were filtered. The latter was done to improve the comparison between the two curves by diminishing the effect of the different preprocessing steps of the data (e.g. the value of the best binding peptides is considerably reduced by the averaging).

The classification rule is now optimized by minimizing the difference between the binding energy curve and the **well binding** *distribution*. This is done by minimizing the root mean square error between the two curves (i.e., the dashed area in Figure Figure 2).

Inspection of the binding energy curve reveals that this curve has a sharp peak on the extreme left side. In other words the **well binding** peptides are underrepresented in the data set in comparison to the **bad binding** peptides. This asymmetric distribution has a severe consequence in the calculation of the root mean square error difference between the two curves: i.e., errors made on the bad binding peptides accumulate to constitute a large percentage of the overall error. This may have the undesirable effect that after minimization of the error, the binding curve is well approximated in the area where bad binding peptides occur, while the peak, i.e., the section containing the best binding peptides (in which we are interested) is poorly approximated.

To overcome this undesired behavior, the calculation of the root mean square error (RMS) between the measured binding energy and the estimated 'well binding' distribution is split in two terms: 1) one for the peptides above the background binding strength (Region 1 in Figure Figure 2) and 2) one for the other elements (Region 2 in Figure Figure 2). The final error measure is the weighted sum of the RMS error in each of these areas, where each of the terms is weighted by the number of samples involved:

$$\text{Cost} = \frac{RMS_1}{n_1} + \frac{RMS_2}{n_2} \tag{2}$$

where n_1 and n_2 represent the number of samples associated with the two terms.

The optimization algorithms applied, try to minimize this measure by generating a rule of the form given in Equation (1), i.e., a rule consisting of the logical AND composition of a set of terms. Each term is defined here as the logical OR composition of a set of amino acids. In this paper, a genetic algorithm and a greedy algorithm are employed to find this rule.

For the genetic algorithm, the rule is coded in a chromosome which contains 20 elements for each OR term (each element codes for the usage of one of the 20 amino acids in the OR term). This is necessary to code for every possible combination of amino acids in one OR term. A result of this algorithm is shown in Section 4.

Most results presented in this paper are generated by a greedy algorithm. This algorithm starts with an empty rule. During the first step the algorithm selects that amino acid which results in the lowest cost. This results in a rule with a single term consisting of only one amino acid. In subsequent steps, the rule is modified by either 1) addition of a new amino acid to one of the existing terms, 2) deletion of one amino acid from one of the terms 3), addition of a new (single amino acid) term or 4) deletion of a term consisting of a single amino acid. A greedy choice is made between all these possibilities.

At every iteration step, the algorithm is forced to add or remove one amino acid to the rule. Even when the resulting rule has a higher cost than the current one. If the current solution is already the optimal one, the algorithm is forced to generate a rule with a lower performance. The optimal solution, however, will be found back in the next iteration step. The advantage of this enforced addition/removal of one amino acid at every iteration step is that it gives the algorithm the possibility to escape from a small local optimum.

Eventually, when the algorithm is unable to find a better solution, it will iterate between the best rule and the second best rule found. If this happens, the algorithm terminates and the best rule is returned.

4   Results

The proposed algorithm is applied to four different data sets. The results are discussed in this section. Except when stated otherwise, the distribution of well binding peptides have been derived with an averaging window of width 100. The percentage shown on the vertical axes of the figures represents the value of the distribution and can be interpreted as the percentage of peptides in the vicinity of a given peptide which are classified as `well binding`.

mAb A[1]. Figure 3 shows the result for monoclonal antibody A. The dashed

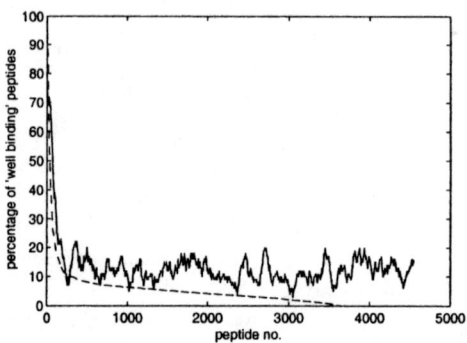

Fig. 3. Results of Monoclonal Antibody A. The dashed line is the rescaled version of the measured relative binding strength. The solid line gives the estimated distribution of well binding peptides when applying the rule in Equation (3).

line in the figure is the rescaled measured binding curve (the values are rescaled to fit in the same window as the results). The solid line is the distribution of `well binding` peptides when using the rule in Equation (3).

$$\text{Well binding} \Rightarrow (\text{G OR V}) \text{ AND F AND T} \tag{3}$$

The figure shows that both curves resemble each other quite well, which gives a good indication that it is possible to classify the data in this way. On the left side of the figure, the estimated distribution starts at approximately 70% implying that the algorithm could not identify a pattern in some of the best binding peptides.

For this antibody it is known that the motif **FT** improves the relative binding strength considerably. This kind of information is not known for the amino acids **G** or **V**, but a large number of the well binding peptides containing **F** and **T** also contain a **G** or a **V**. Including these elements in the rule makes it possible to reduce the number of false positives. This can be understood as follows: increasing the size of the rule without increasing the number of misclassified well binding peptides reduces the number of false positives. The larger the number of amino acids in a peptide that need to fulfill the rule, the smaller the number of peptides that will be classified as `well binding`.

---

[1] mAb A and mAb B are fictitious names.

The genetic algorithm is also applied on this data. According to our implementation of this algorithm, the number of OR terms have to be predefined. The best result found using two, three and four terms are shown in Equations (4)-(6).

$$\text{Well binding} \Rightarrow (\text{G OR H OR V}) \text{AND F} \tag{4}$$

$$\text{Well binding} \Rightarrow (\text{Q OR S OR V}) \text{AND F AND T} \tag{5}$$

$$\text{Well binding} \Rightarrow (\text{C OR G OR V}) \text{AND F AND T AND}$$
$$(\text{I OR L OR S OR V}) \tag{6}$$

The result of (6) has a lower cost than the one proposed by the greedy algorithm. But it does not deviate from the greedy result in stating that F and T are the two most important amino acids. So, the information gained from the greedy algorithm is comparable to the result of the genetic network.

The greedy algorithm and the genetic algorithm also performed comparable on the other data sets. We prefer the greedy algorithm, while this algorithm is much faster than the genetic algorithm. Therefore, only results obtained with the greedy algorithm will be presented in the rest of this paper.

Using the rule generated by the greedy algorithm (3), peptide no. 3 in Table Table 2 is classified as a **bad binding** peptide. This is partly due to the lack

Table 2. The Five Best Binding Peptides of Monoclonal Antibody A.

no.	Peptide	Result
1	C M E Q I M R **G** **T** P **F** **T**	Well binding
2	K L S **T** P I R H **G** A **F** **T**	Well binding
3	L I H Y D N **T** N M W Y **T**	**Bad binding**
4	**V** S **F** M **F** L A P **T** **G** **F** **T**	Well binding
5	M W P L P W **V** A I P **F** **T**	Well binding

of the amino acid F in the peptide. In this case the quite similar amino acid Y probably performs the same contribution to the binding as F normally does. This substitution effect is obviously not included in the algorithm.

Equation (3) is created using all the available peptides. To check the stability of the algorithm the results are also calculated after repeatedly removing 10% of the measured peptides from the training set. The resulting 10 rules are shown in equation (7):

$$1: \text{Well binding} \Rightarrow (\text{G OR V}) \text{AND F AND T}$$
$$2: \text{Well binding} \Rightarrow (\text{G OR V}) \text{AND F AND T}$$
$$3: \text{Well binding} \Rightarrow (\text{G OR V}) \text{AND F AND T}$$
$$4: \text{Well binding} \Rightarrow (\text{G OR K}) \text{AND F AND T}$$
$$5: \text{Well binding} \Rightarrow (\text{G OR V}) \text{AND F AND T}$$
$$6: \text{Well binding} \Rightarrow (\text{G OR M}) \text{AND F AND T}$$

7: Well binding $\Rightarrow$ (G OR V) AND F AND T

8: Well binding $\Rightarrow$ (G OR V) AND F AND T

9: Well binding $\Rightarrow$ (G OR V) AND F AND T

10: Well binding $\Rightarrow$ (G OR V) AND F AND T AND

(A OR D OR K OR R OR S)   (7)

The term (A OR D OR K OR R OR S) in the last rule is true for 97% of the peptides. This term is probably due to an error made in an early greedy step in the algorithm. When an OR group doesn't improve the result of the algorithm and it has two or more elements, the algorithm can only increase the number of amino acids in the group to reduce its effect. The V seams to be less important because it is replaced once by a K and once by an M. In all ten rules, the F and T have to be included in a peptide to be classified as well binding.

mAb 32F81. The number of well binding peptides for monoclonal antibody 32F81 is much larger than for mAb A, which can be seen in Figure 4. Again the

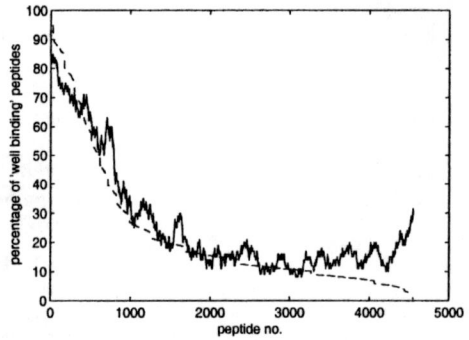

Fig. 4. Results of Monoclonal Antibody 32F81. The dashed line is the rescaled version of the measured relative binding strength. The solid line gives the percentage of elements which fulfill Equation (8).

algorithm is able to find a rule which approximates the binding strength curve quite well. The rule found for this antibody (Equation (8)), describes the data better in the sense that 80% of the best binding peptides are classified as well binding. The right part of the predicted result is remarkable (after peptide 4000). Here the distribution of well binding peptides rises while the measured binding energy shows a rapid decay. Some of the peptides which occur in this area show an exceptionally low binding strength in several different peptide screening experiments. The reason for this exceptionally bad binding behavior of these peptides is not found by the algorithm.

The best five binding peptides of monoclonal antibody 32F81 are shown in Table Table 3.

Table 3. The Five Best Binding Peptides of Monoclonal Antibody 32F81.

no.	Peptide	Result
1	<u>F</u> <u>K</u> C <u>T</u> A D <u>Q</u> Y V <u>T</u> <u>F</u> A	Well binding
2	Q G S W <u>D</u> <u>K</u> <u>F</u> G M N <u>E</u> <u>H</u>	Well binding
3	L P W <u>H</u> <u>E</u> I L <u>K</u> Q N V N	Well binding
4	<u>D</u> <u>K</u> I <u>E</u> N Y I Q W P <u>H</u> <u>D</u>	Well binding
5	<u>E</u> <u>K</u> C <u>D</u> <u>H</u> Q C A Q C V Y	Well binding

Well binding $\Rightarrow$ (E OR T) AND K AND (D OR F OR H)   (8)

All peptides shown fulfill the rule and are classified as well binding peptides. For this peptide it should be noted that the result using only the amino acid K is already quite discriminative: 96% of the 400 best binding peptides contain this amino acid. The comparison between the distribution found by the algorithm and the distribution using a rule containing only K is shown in Figure 5.

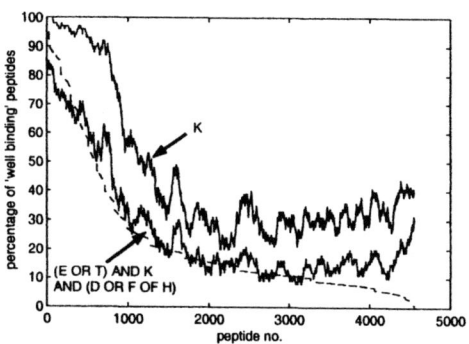

Fig. 5. Comparison of the resulting rule of the algorithm (8) and the rule: K, for monoclonal antibody 32F81.

The result using the rule 'K' has almost a 100% accuracy for the best binding peptides. However the number of false positives is also quite high. The cost function we employ prefers the answer with a lower number of false positives.

mAb B. A part of the distribution of monoclonal antibody B is shown in Figure 6. The number of well binding peptides is extremely small for this experiment. This makes it necessary to reduce the size of the averaging window when estimating the distribution. Namely, when a window size of 100 would be employed here, the maximal value of the estimated distribution would be 15%, effectively removing the peak of the distribution. To avoid this, the averaging window has been reduced to a length of 16. A disadvantage of such a short window is the larger fluctuations of the predicted result (as can be seen in the tail of the curve in Figure Figure 6).

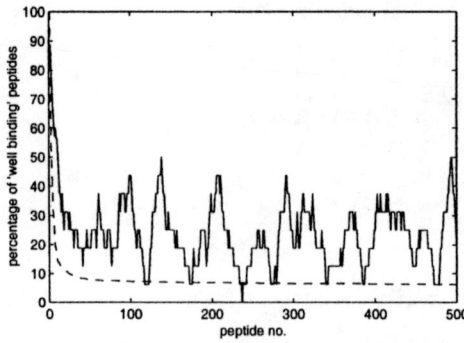

Fig. 6. Part of the Result for Monoclonal Antibody B. The dashed line is the rescaled version of the measured relative binding strength. The solid line gives the percentage of elements which fulfill the rule in Equation (9) (calculated using an averaging window of 16).

The rule used to generate Figure Figure 6 is given in equation (9). A known well binding motif for this peptide is **TEDSAVE**. The amino acids **A**, **V** and **E** of this motif are included in the rule found by the algorithm.

$$\text{Well binding} \;\Rightarrow\; \text{A AND (E OR N) AND (I OR V)} \qquad (9)$$

The five best binding peptides for this antibody are given in Table Table 4. The

Table 4. The Five Best Binding Peptides of Monoclonal Antibody B.

no.	Peptide	Result
1	D **A** **A** **E** D D **V** **N** G R F M	Well binding
2	**V** T **N** C C R S **A** **V** H **E** K	Well binding
3	R **N** M S K T S **A** S **A** **V** **E**	Well binding
4	W **E** L W **I** K K K **A** K **V** **V**	Well binding
5	**E** **A** **I** W K K C K G **V** T F	Well binding

AVE motif is present in the third peptide of this table.

mAb 6A.A6. The last result is for monoclonal antibody 6A.A6 (Figure Figure 7). The distribution of **well binding** peptides is estimated quite well for the best binding peptides. However, for the bad binding peptides, the distribution does not fit the actual measured binding energy well, resulting in a large number of false positives. An interesting element of rule (10) is the OR part (D OR E). These two amino acids are very similar (see e.g. [2]) and can often be exchanged.

$$\text{Well binding} \;\Rightarrow\; \text{S AND (D OR E)} \qquad (10)$$

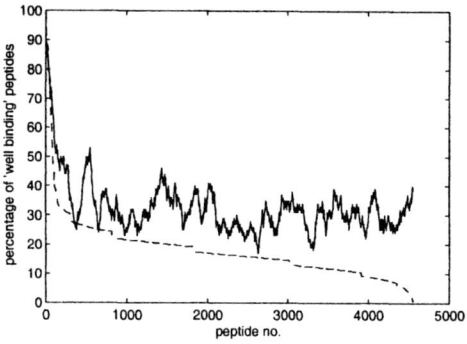

Fig. 7. Results for Monoclonal Antibody 6A.A6. The dashed line is the rescaled version of the measured relative binding strength. The solid line gives the percentage of elements which fulfill the rule in Equation (10).

The best known binding peptide against this antibody has the following sequence: SRLPPNSDVVLG. The amino acids S and D are included in the peptide, thus the best known peptide will be classified as a well binding peptide. Replacement studies of this peptide show that the motif PNSD is very important for a well binding result [8]. This pattern is much longer then the result found by the algorithm. This might be caused by the lack of very well binding peptides that have this motif in the data set. It should be noted that the binding strength of even the best measured peptides of the random data set are much smaller then this best known result. The five best binding peptides are shown in Table Table 5. The known important motif PNSD is not available among these

Table 5. The Five Best Binding Peptides of Monoclonal Antibody 6A.A6.

no.	Peptide	Result
1	R <u>S</u> <u>D</u> <u>S</u> M G I L L L P L	Well binding
2	P Q <u>S</u> <u>D</u> T L H Y C I Q N	Well binding
3	C <u>S</u> <u>D</u> V V <u>E</u> T R C Q I <u>D</u>	Well binding
4	<u>E</u> <u>S</u> N T W V K C Y <u>S</u> Y W	Well binding
5	<u>E</u> F M <u>S</u> <u>D</u> I A L R G T V	Well binding

peptides.

5   Discussion

We have introduced an algorithm that identifies those amino acids in a peptide which are most probably needed to bind against a monoclonal antibody. The relevant amino acids are described by a rule that distinguishes between well and bad binding peptides. The algorithm we have introduced automatically generates such a rule description from peptide screening experimental data.

This is based on the fact that if a certain motif or combination of amino acids improves the binding strength between a peptide and a monoclonal antibody, most of the peptides containing this motif will bind better than average against this antibody. This results in an over-representation of these amino acids in the better binding peptides. Such an over-representation of amino acids is utilized by the proposed algorithm to generate a rule of amino acids that distinguishes between well binding and bad binding peptides.

A greedy optimization algorithm is employed, which means that it does not necessarily find the rule with the minimal cost. Other optimization procedures like genetic algorithms are able to perform better in finding the absolute optimum.

Better solutions than those obtained with the greedy approach were in our experience, only marginally better in terms of the cost function, and were characterized by more OR terms (like the term (A OR D OR K OR R OR S) in the last rule in Equation (7)).

It is important, however, to keep in mind that a large number of peptides will fulfill these long OR terms, i.e., such rules are non-specific. Since these rules are employed to generate new peptides for subsequent analysis, we are more interested in rules with short OR terms which reduce the search space by their specificity.

We could adapt the cost function to achieve this by adding a penalty term for long rules. However, if we look more carefully at the greedy optimization, we notice that it starts by including the most discriminating amino acids in the rule. This results in a reasonable performance after only a few iteration steps. Additional OR terms are only included when this improves the performance of the rule (in fact the performance should improve when normally one, but at most two additional amino are included in the OR term). This greedy property of the algorithm reduces the chance that these unimportant long terms are included in the final rule.

The rules as such can not be directly employed to generate new well binding motifs, since positional information and frequency of occurrence is lost in the feature representation. However, within the group of well binding peptides that satisfy the rule, the amino acids specified in the rule occur at specific positions in these peptides. A logical approach is to construct new candidate peptides by preserving for the N best binding peptides, those amino acids that appear in the rule, while randomly mutating the rest.

Acknowledgments

This work was funded by the DIOC-5 program of the Delft Inter-Faculty Research Center at the TU Delft and by Pepscan Systems BV located in Lelystad.

References

1. D.A. Daniels and D.P. Lane. Phage peptide libraries. *Methods: A Companion to Methods in Enzymology*, 9:494–507, 1996.

2. M.O. Dayhoff, R.M. Schwartz, and B.C. Orcutt. A model of evolutionary change in proteins. *Atlas of Protein Sequence and Structure*, 5, supplement 3:345–352, 1978.

3. H.M. Geysen, R.H. Meloen, and S.J. Barteling. Use of peptide synthesis to probe viral antigens for epitopes to a resolution of a single amino acid. *Proc Natl Acad Sci USA*, 81:3998–4002, 1984.

4. L. Jin and J.A. Wells. Dissecting the energetics of an antibody-antigen interface by alanine shaving and molecular grafting. *Protein Science*, 3:2351–2357, 1994.

5. R.H. Meloen, J.I. Casal, K. Dalsgaard, and J.P.M. Langeveld. Synthetic peptide vaccines: Succes at last. *Vaccine*, 13(10):885–886, 1995.

6. M.H.V. Van Regenmortel and S. Muller. Synthetic peptides as antigens. In S. Pillai and P.C. Van der Vliet, editors, *Laboratory Techniques in Biochemistry and Molecular Biology, Vol. 28*. Elsevier, Amsterdam, The Netherlands, 1999.

7. J.K. Scott. Discovering peptide ligands using epitope libraries. *Trends in biotechnology*, 17:241–245, 1992.

8. J.W. Slootstra, W.C. Puijk, G.J. Ligtvoet, J.P.M. Langeveld, and R.H. Meloen. Structural aspects of antibody-antigen ineteraction revealed through small random peptide libraries. *Molecular Diversity*, 1:87–96, 1995.

9. E. Trifilieff, M.C. Dubs, and M.H.V. Van Regenmortel. Antigenic cross-reactivity potential of synthetic peptides immobilized on polyethylene rods. *Molecular Immunology*, 28(8):889–896, 1991.

10. Y.L. Yip and R.L. Ward. Epitope discovery using monoclonal antibodies and phage peptide libraries. *Combinatorial Chemistry & High Throughput Screening*, 2:125–138, 1999.

A Simple Hyper-Geometric Approach for Discovering Putative Transcription Factor Binding Sites

Yoseph Barash, Gill Bejerano, and Nir Friedman

School of Computer Science & Engineering
The Hebrew University, Jerusalem 91904, Israel
{hoan,jill,nir}@cs.huji.ac.il

Abstract. A central issue in molecular biology is understanding the regulatory mechanisms that control gene expression. The recent flood of genomic and post-genomic data opens the way for computational methods elucidating the key components that play a role in these mechanisms. One important consequence is the ability to recognize groups of genes that are co-expressed using microarray expression data. We then wish to identify *in-silico* putative transcription factor binding sites in the promoter regions of these gene, that might explain the co-regulation, and hint at possible regulators. In this paper we describe a simple and fast, yet powerful, two stages approach to this task. Using a rigorous hyper-geometric statistical analysis and a straightforward computational procedure we find small conserved sequence kernels. These are then stochastically expanded into PSSMs using an EM-like procedure. We demonstrate the utility and speed of our methods by applying them to several data sets from recent literature. We also compare these results with those of MEME when run on the same sets.

1 Introduction

A central issue in molecular biology is understanding the regulatory mechanisms that control gene expression. The recent flood of genomic and post-genomic data, such as microarray expression measurements, opens the way for computational methods elucidating the key components that play a role in these mechanisms.

Much of the specificity in transcription regulation is achieved by *transcription factors*, which are largely responsible for the so called combinatorial aspects of the regulatory process (the number of possible behaviors being much larger than the number of factors). These are proteins that, when in the suitable state, can bind to specific DNA sequences. By binding to the chromosome in a location near the gene, these factors can either activate or repress the transcription of the gene. While there are many potential sites where these factors can bind, it is clear that much of the regulation occurs by factors that bind in the *promoter region* which is located upstream of the transcription start site.

Unlike DNA-DNA hybridization, the dynamics of protein-DNA recognition are not completely understood. Nonetheless, experimental results show that transcription factors have specific preference to particular DNA sequences. Somewhat generalizing, the affinity of most factors is determined to a large extent by one or more relatively short regions of 6–10bp. (One must bear in mind that DNA strands span a complete turn

O. Gascuel and B.M.E. Moret (Eds.): WABI 2001, LNCS 2149, pp. 278–293, 2001.
© Springer-Verlag Berlin Heidelberg 2001

every 10 bases, thus geometric considerations make it unlikely that a single protein binds to a longer region, although counterexamples are known.) A common situation is the formation of *dimers* in which two DNA binding proteins form a complex. Each of the two proteins, binds to a short sequence, and together they bind to a sequence that can be 12–18bp long, with a short spacer separating the two regions. Common protein motifs such as the DNA binding Helix-Turn-Helix (HTH) motif also induce the same preference on the regulatory site.

The recent advances in microarray experiments allow to monitor the expression levels of genes in a genome-wide manner [8, 9, 14, 15, 22, 23]. An important aspect of these experiments is that they allow to find groups of genes that have similar expression patterns across a wide range of conditions [12]. Arguably, the simplest biological explanation of co-expression is co-regulation by the same transcription factors.[1]

This observation sparked several works on *in-silico* identification of putative transcription factor binding sites [4, 17, 19–21]. The general scheme that most of these papers take involves two phases. First, they perform, or assume, some clustering of genes based on gene expression measurements. Second, they search for short DNA patterns that appear in the promoter region of the genes in each particular cluster. These works are based to a large extent on methods that were developed to find common motifs in protein and DNA sequences. These include combinatorial methods [6, 19, 21, 24, 25], parameter optimization methods such as Expectation Maximization (EM) [1], and Markov Chain Monte Carlo (MCMC) simulations [18, 20]. See [19] for a review of these lines of work.

The use of expression profiles helps to select relatively "clean" clusters of genes (i.e., most of them are indeed co-regulated by the same factors). Our interest here lies with the second phase, and is thus not limited to gene expression analysis. Given high quality clusters of genes, suspected for any reason to be co-regulated, we address the hardness of the computational problem of finding putative binding sites in these clusters.

In this paper we describe a fast, simple, yet powerful, approach for finding putative binding sites with respect to a given cluster of genes. Like some of the other works we divide this phase into two stages. In the first stage we scan, in an exhaustive manner, for simple patterns from an enumerable class (such as all 7-mers). We use a straightforward, natural, and well understood statistical model for filtering significant patterns out of this class. Using the hyper-geometric distribution, we compute the probability that a subset of genes of the given size will have these many occurrences of the pattern we examine, when chosen randomly from the group of all known genes. In the second stage, we use the patterns that were chosen as seeds for training a more expressive *position-specific scoring matrix* (PSSM) to model the putative binding site. These models are both more accurate representation of the binding site, and potentially capture much longer conserved regions.

By assuming that most binding sites do contain highly conserved short subsequences and by explicitly using our post-genomic knowledge of all known and putative genes to contrast clusters of genes against the genome background, we acquire quality seeds

---

[1] Clearly this is not always the case. Co-regulation can be achieved by other means, and similar expression patterns can be a result of parallel pathways or a close serial relationship. Nonetheless, this is often the case, and a reasonable hypothesis to test.

for the construction of PSSMs through a simplified hyper-geometric model. The seeds allow us to track down potential binding site locations through a specific relatively conserved region within them. We then use these short seeds to guide the construction of potentially much longer PSSMs encompassing more, or possibly the complete binding site. In particular, they allow us to align multiple sequences without resorting to an expensive search procedure (such as MCMC simulations).

Indeed, an important feature of our approach is the evaluation speed. Once we finish a preprocessing stage, we can evaluate clusters very efficiently. The preprocessing is genome-wide and not cluster specific. It can be done only once and stored for all future reference. This is important both for facilitating interactive analysis, and for serving as computationally-cheap quality starting points for other, more complex analysis tools (such as [2]) on top of our method.

In the next three sections we outline our algorithmic approach, discussing significance of events, seed finding, and seed expansion into PSSMs, respectively. In Section 5 we describe experimental and comparative results, and then conclude with a discussion.

2   Scoring Events for Significance

2.1   Preliminaries

Suppose we are given a set of genes $\mathcal{G}$. Ideally, these are all the known and putative genes in a genome. With each gene $g \in \mathcal{G}$ we associate a promoter sequence[2] s_g. For simplicity we assume that each of these sequences is of the same size, L.

Suppose we are now given a subset of genes $G \subset \mathcal{G}$ suspected to be co-regulated by some transcription factor. (For example, based on clustering of genes by their expression patterns.) Our aim is to find patterns in the promoter region of these genes, that we will consider as putative binding sites. The assumption being that the co-regulation is mediated by factors that are present in most of the genes in group G, but overall rare in $\mathcal{G}$. Thus, a pattern is considered significant if it is characteristic of G compared to the background $\mathcal{G}$.

Before we discuss what constitutes a pattern in our context, we address the basic statistical definition of a characteristic property. Suppose we find a pattern that appears in the promoter sequences of several genes in G. How do we measure the significance of these appearances with respect to $\mathcal{G}$? A related question one may ask, is whether the set G is significantly different, in terms of the composition of its upstream region, from $\mathcal{G}$.

For now, we concentrate on events occurring in the promoter region of a gene. We focus on *binary* events, such as "s_g contains the subsequence ACGTTCG or its reverse complement". Alternatively, one can consider *counting* the number of occurrences of an event in each promoter sequence, e.g., "the number of times the subsequence ACGTTCG appears in s_g". The analysis of such counting events, while attractive in our biological context, is more complex, in particular since multiple occurrences of an event in a sequence are not independent of each other. See [21, 24] for approximate solutions to this problem.

---

[2] Or an upstream region that best approximates it, when the transcription start site is unknown.

Formally, a binary event E is defined by a *characteristic function* I_E : $\{A, C, G, T\}^\star \to \{0, 1\}$, that determines whether that event occurred or not in any given nucleotide sequence. Given a set G, we define $\#_E(G) = \sum_{g \in G} I_E(s_g)$ to be the number of times E occurs in the promoter regions of group G. We want to assess the significance of observing E at least $\#_E(G)$ times in G, when taking the set of genes $\mathcal{G}$ as the background for our decision.

There are two general approaches for testing such significance. In both cases we compute *p-values*: the probability of the observations occurring under the *null-hypothesis*. This value serves as a measure of the significance of the pattern - the lower p-value is, the more plausible it is that an observation is significant, rather than a chance artifact. The two approaches differ, however, in the nature of each null-hypothesis.

2.2   Random Sequence Null Hypothesis

In this approach, the null hypothesis assumes that the sequences s_g for $g \in G$ are generated from a background sequence model $P_0(s)$. This background distribution attempts to model "prototypical" promoter regions, but does not include any group-specific motifs. Thus, if the event E detects such special motifs, then the probability of randomly sampling genes that satisfy E is small.

The background sequence model can be, for example, a Markov process of some order (say 2 or 3) estimated from the sequences in $\mathcal{G}$ (or, preferably, from $\mathcal{G} - G$). Using this background model we need to compute the probability $p_E = P_0(I_E(s) = 1)$ that a random sequence of Length L will match the event of interest. Now, if we also assume under the null hypothesis that the n sequences in G are independent of each other, then the number of matches to E in G is distributed $Bin(n, p_E)$. We can then compute the p-value of finding $\#_E(G)$ or more such random sequences by the tail weight of a Binomial distribution.

The key technical issue in this approach is computing p_E. This, of course, depends on the assumed form of the background distribution, and on the complexity of the event. However, even for the simple definition of a pattern as an exact subsequence (i.e., $I_E(s) = 1$ iff s contains a specific subsequence) and background probability of the form of an order 1 Markov chain, the required computation is not trivial. This forces the development of various approximations to p_E of varying accuracy and complexity [4,7,21].

2.3   Random Selection Null Hypothesis

Alternatively, in the approach we focus on here, one does not make any assumption about the distribution of promoter sequences. Instead, the null hypothesis is that G was selected at random from $\mathcal{G}$, in a manner that is independent of the contents of the genes' promoter regions.

Assume that $K = \#_E(\mathcal{G})$ out of $N = |\mathcal{G}|$ genes satisfy E. Thus, we require the number[3] of genes that satisfy E in $\mathcal{G}$. The probability of an observation under the null hypothesis is the probability of randomly choosing $n = |G|$ genes in such a way that

---

[3] But not the identity, simplifying the implied underlying *in-vitro* measurements.

$k = \#_E(G)$ of them include the event E. This is simply the *hyper-geometric* probability of finding k red-balls among n draws without replacement from an urn containing K red balls and $N - K$ black ones:

$$P_{\text{hyper}}(k \mid n, K, N) = \frac{\binom{K}{k}\binom{N-K}{n-k}}{\binom{N}{n}}$$

The p-value of the observation is the probability of drawing k or more genes that satisfy E in n draws. This requires summing the tail of the hyper-geometric distribution

$$p\text{-}value(E, G) = \sum_{k'=k}^{n} P_{\text{hyper}}(k' \mid n, K, N)$$

The main appeal of this approach lies in its simplicity, both computationally and statistically. This null hypothesis is particularly attractive in the post-genomic era, where nearly all promoter sequences are known. Under this assumption, irrelevant clustering selects genes in a manner that is independent of their promoter region.

2.4 Dealing with Multiple Hypotheses

We have just defined the significance of a single event E with respect to a group of genes G. But when we try many different events $E_1, \ldots, E_M$ over the same group of genes long enough, we will eventually stumble upon a surprising event even in a group of randomly selected sequences, chosen under the null hypothesis.

Judging the significance of findings in such repeated experiments is known as *multiple hypotheses testing*. More formally, in this situation we have computed a set of p-values $p_1, \ldots, p_M$, the smallest corresponding to the most surprising event. We now ask how significant are our findings considering that we have performed M experiments.

One approach is to find a value $q = q(M)$, such that the probability that any of the events (or the smallest one) has a p-value less than q is small. Using the union bound under the null hypothesis we get that

$$P(\min_m p_m \leq t) \leq \sum_m P(p_m \leq q) = M \cdot q$$

Thus, if we want to ensure that this probability of a false recognition is less than 0.01 (i.e., 99% confidence), we need to set the *Bonferroni* threshold $q = \frac{0.01}{M}$ (see, for example, [11]). The Bonfferoni threshold is strict, as it ensures that each and every validated scoring event is not an artifact. Our aim, however, is a bit different. We want to retrieve a set of events, such that *most* of them are not artifacts. We are often willing to tolerate a certain fraction of artifacts among the events we return. A statistical method that addresses this kind of requirement is the *False Discovery Rate* (FDR) method of [3]. Roughly put, the intuition here is as follows. Under the null hypothesis, there is some probability that the best scoring event will have a small p-value. However, if the group was chosen by the null hypothesis, it can be shown that the p-values we compute are distributed uniformly.

Thus, the p-value of the second best event is expected to be roughly twice as large as the p-value of the best event. Given this intuition, we should be less strict in rejecting the null hypothesis for the second best pattern and so on.

To carry out this idea, we sort the events by their observed p-values, so that $p_1 \leq p_2 \leq \ldots \leq p_M$. We then return the events $E_1, \ldots, E_k$ where $k \leq M$ is the maximal index such that $p_k \leq \frac{kq}{M}$ and q is the significance level we want to achieve in selecting. We have replaced a strict validation test of single events, with a more tolerable version validating a group of events. We may now detect significant patterns, weaker than the most prominent one, that were previously below the threshold computed for the later.

3 Finding Promising Seeds

3.1 Simple Events

We want to consider patterns over relatively short subsequences. We fix a parameter ℓ that determines the length of the sequences we are interested in. Events are then defined over the space of 4^ℓ ℓ-mers.

Arguably the simplest ℓ-mer pattern is a specific subsequence (or consensus). Thus, if σ is an ℓ-mer it defines the event "σ is a subsequence of s". A useful aspect of such events, is that they are *exhaustively enumerable* for the range of ℓ we are interested in. This suggests examining all ℓ-mer patterns in G and ranking them according to their significance.

However, known binding sites that are identified by biological assays, display variability in the binding sequence. Thus, we do not expect to see only exact matches to the ℓ-mer consensus. Instead, we want to allow approximate matches when we search G. To formalize, consider a distance measure between two ℓ-mers, $d(\sigma, \sigma')$. The simplest such function is the hamming distance. However, we may consider more realistic functions, such as distances that penalize changes in a position specific manner. (Biology suggests, for example, that central positions in short binding sites are more conserved.) For concreteness, we focus on the hamming distance measure in the reminder of the paper. However, we stress that the following discussion applies directly to any chosen distance measure.

Let σ be an ℓ-mer. We define a δ-ball centered around σ to be the set $\text{Ball}_\delta(\sigma)$ of ℓ-mers that are of distance at most δ from σ. Thus, in the hamming distance, example, $\text{Ball}_1(\text{AAA}) = \{\text{AAA, CAA, GAA, TAA, ACA, AGA, ATA, AAC, AAG, AAT}\}$. We match an event E with $\text{Ball}_\delta(\sigma)$ such that $I_E(s) = 1$ iff s or its reverse complementary contain an ℓ-mer $\in \text{Ball}_\delta(\sigma)$.

Given ℓ and δ we wish to examine all balls that have at least one occurrence in G (the rest will never appear in any sub group). Balls that occur in all genes in G are also discarded (as they occur in all genes of any sub group). We denote this set of non-trivial events with respect to G as $\mathcal{B}_{(\ell,\delta)}$. Note that for $\delta > 0$, it may include balls whose centers do not appear in any promoter region.

Finding the set $\mathcal{B}_{(\ell,\delta)}$ of balls, and annotating for each gene whether it matches each ball can be done in a straightforward manner. The time requirement then is $N \cdot L \cdot 4^\ell$, and the space requirement $N \cdot |\mathcal{B}_{(\ell,\delta)}|$.

This genome-wide preprocessing needs to be done only once. Storing its results we can rapidly compute p-values of all $\mathcal{B}_{(\ell,\delta)}$ events with respect to any proposed subset of genes. We simply look up which events occurred in the genes in the cluster, and then compute the hyper-geometric tail distribution. Furthermore, one may wish to increase, shrink, or shift the regions under consideration (e.g., from 1000bp to 2000bp upstream), or adjust the upstream regions of several genes (say, due to elucidation of exact transcription start site). While in general the preprocessing phase must be repeated, in practice, since it is mainly made up of counting events, we may efficiently subtract, and add, respectively the counts in the symmetrical difference between the old and new sets of strings, avoiding repeating the complete process over again. With many completely sequenced genomes and gene expression data of model organisms in various settings just beginning to accumulate, our division of labour is especially useful.

3.2 Reducing the Event Space

The definition of $\mathcal{B}_{(\ell,\delta)}$, holding all events we wish to examine, may include as many as $\min(4^\ell, LN)$ balls. We note however, that many of these balls overlap. Thus, if σ and σ' are two ℓ-mers that differ, in the hamming distance example[4], in exactly one letter, then the overlap between $\mathrm{Ball}_\delta(\sigma)$ and $\mathrm{Ball}_\delta(\sigma')$ is clearly substantial. Moreover, if we notice that most of the "mass" of these balls (in terms of the number of occurrences in genes in $\mathcal{G}$) lies in the intersection, we expect that the significance of the events defined by both of them will be similar, since they will be highly correlated.

A way to decrease the storage requirements, and thus extend the range of manageable ℓ's can be found by a guided choice of a representative subset of $\mathcal{B}_{(\ell,\delta)}$ during preprocessing. Based on the above intuitions we want a covering set of balls with maximal mass, to minimize the size of the subset, and minimal overlap, to diversify the events themselves. A heuristic solution can be offered in the form of a greedy algorithm. Starting from an empty subset we repeatedly choose balls of maximal mass that do not violate the minimal overlap demand, until we can no longer continue. We now proceed to examine and store the results only for the events corresponding to the chosen balls.

We stress that since this sparsification is done during preprocessing, before we observe any group G, it should not alter the statistical significance of the results we observe when G is later given to us.

4   Learning Finer Representations

4.1   Position Specific Scoring Matrices

Using the methods of the previous section we can collect a set of promising patterns that are significant for G. These patterns are based on the notion of a δ-ball. Biological knowledge about transcription factor binding sites suggests that the definition of a binding site is in fact more subtle. Some positions are highly conserved, while others are less so. In the literature, there are two main representation of such sites. The first

---

[4] Analogous proximity thresholds can be defined for other distance measures.

is the IUPAC consensus sequences. This approach determines the consensus string of the binding site using a 15 letter alphabet that describe which subset of $\{A, C, G, T\}$ is possible at each position.

A *position specific scoring matrix* (PSSM) (see, e.g., [10]) offers a more refined representation. A PSSM of length ℓ is an object $\mathcal{P} = \{p_1, \ldots, p_\ell\}$, composed of ℓ column distributions over the alphabet $\{A, C, G, T\}$. The distribution p_i, specifies the probability of seeing each nucleotide at the i'th position in the pattern.

Once we have a PSSM $\mathcal{P}$, we can score each ℓ-mer σ by computing its combined probability given $\mathcal{P}$. A more common practice is to compute the log-odds between the PSSM probability and a background probability of nucleotides. Thus, if p_0 is assumed to be the nucleotide probability in promoter regions, then the score of an ℓ-mer σ is:

$$Score_{\mathcal{P}}(\sigma) = \sum_i \log \frac{p_i(\sigma[i])}{p_0(\sigma[i])}$$

If this score is positive σ is more probable according to $\mathcal{P}$ than it is according to the background probability. In practice we set a threshold α (replacing zero) for detecting a pattern. Thus, a pair $(\mathcal{P}, \alpha)$ defines an event $I_{(\mathcal{P}, \alpha)}(s)$. This event occurs iff the best matching subsequence of length ℓ in s, or in its reverse complement, has a score higher than α. That is, if

$$\max_i(Score_{\mathcal{P}}(s[i, \ldots, i+\ell-1]), Score_{\mathcal{P}}(\overline{s[i, \ldots, i+\ell-1]}) > \alpha$$

4.2   Selecting a Threshold

Before we discuss how to learn the PSSM, we consider choosing a threshold α for a given PSSM $\mathcal{P}$. It is possible to set $\alpha = 0$, treating the background and the PSSM as equiprobable. However, since the pattern is a rarer event, we want a stricter threshold. Another potential approach tries to reduce the probability of false recognition. That is, to find an α such that the probability that a random background sequence σ will score higher than α is smaller than a prespecified ϵ. Then, if we want to allow on average one false detection every k genes, we would set $\epsilon = \frac{1}{k*L}$. Unfortunately, we are not aware of an efficient computational procedure to find such thresholds.

Here we suggest a simple alternative. We search for a threshold α, such that the induced detections in the group G will be most significant. Thus, given a group G of genes, and a PSSM $\mathcal{P}$, we search for

$$\alpha^* = \arg \min_\alpha p\text{-}value(G, I_{(\mathcal{P}, \alpha)})$$

That is, we adjust the threshold α so that the event defined by $(\mathcal{P}, \alpha)$ has the smallest p-value with respect to G. This *discriminative* choice of a threshold ensures that we adjust it to take into account the amount of "spurious" matches to the PSSM outside of G. Thus, we strive for a threshold that maximizes the number of matches within G and at the same time minimizes the number of matches outside G. The use of p-values provides a principled way of balancing these two requirements.

We can find this threshold quite efficiently. We compute the best score of the PSSM over each gene in $\mathcal{G}$, and sort this list of scores. We then evaluate only thresholds which are, say, half way between any two adjacent values in our list of sorted scores (each succeeding threshold admits another gene into the group of supposedly detected events). Using, for example, radix sort, this procedure takes time $O(NL)$.

4.3 Learning PSSMs

Learning PSSMs is composed of two tasks. Estimating the parameters of the PSSM given a set of training sequences that are examples of the pattern we want to match, and finding these sequences. The latter is clearly a harder problem and requires some care.

We start with the first task. Suppose we are given a collection $\sigma_1, \ldots, \sigma_n$ of ℓ-mers that correspond to *aligned* sites. We can easily estimate a PSSM $\mathcal{P}$ that corresponds to these sequences. For each position i, we count the number of occurrences of each nucleotide in that position. This results in a count $N(i, c) = \sum_j 1\{\sigma_j[i] = c\}$.

Given the counts we estimate the probabilities. To avoid entries with zero probability, we add *pseudo-counts* to each position. Thus, we assign

$$p_i(c) = \frac{N(i, c) + \gamma}{n + 4\gamma} \tag{1}$$

The key question is how to select the training sequences and how to align them. Our approach builds on our ability to find seeds of conserved sequences. Suppose that we find a significant δ-ball using the methods of the previous section. We can then use this as a *seed* for learning a PSSM. The simplest approach takes the ℓ-mers that match the ball within the promoter regions of G as the training sequences for the PSSM. The learned PSSM then quantifies which differences are common among these sequences and which ones are rare. This gives a more refined view of the pattern that was captured by the δ-ball.

This simple approach learns an ℓ-PSSM from the δ-ball events found in the data. However, using PSSMs we can extend the pattern to a much longer one. We start by aligning not only the sequences that match the δ-ball, but also their flanking regions. These are aligned by virtue of the alignment of the core ℓ-mers. We can then learn a PSSM over a much wider region (say 20bp). If there are conserved positions outside the core positions, this approach will find them.[5]

Consider, for example, a HTH DNA binding motif, or a binding factor dimer, where each component matches 6-10bps with several unspecific gap positions between the two specific sites. If we find one of the two sites using the methods of the previous sections, then growing a PSSM on the flanking regions allows us to discover the other conserved positions.

Once we construct such an initial PSSM, we can improve it using a standard EM-like iterative procedure. This procedure consists of the following steps. Given a PSSM $\mathcal{P}_0$, we compute a threshold α_0 as described above. We then consider each position in the training sequences and compute the probability that the pattern appears at that position. Formally, we compute the likelihood ratio $(\mathcal{P}_0, \alpha_0)$ assigns to the appearance of

---

[5] This assume that there are no variable lengths gaps inside the patterns. The structural constraints on transcription factors suggest that these are not common.

the pattern at $s[i, \ldots, i+\ell-1]$. We then convert this ratio to a probability by computing

$$\rho_{s,i} = \text{logit}(Score_{\mathcal{P}_0}(s[i, \ldots, i+\ell-1])) - \alpha_0)$$

where $\text{logit}(x) = 1/(1 + e^{-x})$ is the *logistic function*. We then re-scale these probabilities by dividing by a normalization factor Z_s so that the posterior probability of observing the pattern in s and its reverse complement sums to 1. Once we have computed these posterior probabilities, we can accumulate *expected counts*

$$N(i, c) = \sum_g \sum_j \frac{\rho_{s_g, j}}{Z_{s_g}} 1\{s_g[j + i] = c\}.$$

These represent the expected number of times that the i'th position in the PSSM takes the value c, based on the posterior probabilities.

Once we collected these expected counts, we re-estimate the weights of the PSSM using Eq. 1 to get a new a PSSM. We optimize the threshold of this PSSM, and repeat the process. Although this process does not guarantee improvement in the p-value of the learned PSSM, it is often the case that successive iterations do lead to significant such improvements. Note that our iterations are analogous to EM's hill-climbing behaviour, and differ from Gibbs samplers where one performs a stochastic random walk aimed at a beneficial equilibrium distribution.

5   Experimental Results

We performed several experiments on data from the yeast genome to evaluate the utility and limitations of the methods described above. Thus, we focused on several recent examples from the literature that report binding sites found either using computational tools or by biological verification. To better calibrate the results, we also applied MEME [1], one of the standard tools in this field, on the same examples.

In this first analysis we chose to use the simple hamming distance measure and treat the 1000bp sequence upstream of the ORF starting position as the promoter region. We note that the latter is a somewhat crude approximation, as this region also contains an untranslated region of the transcript.

We ran our method in two stages. In the first stage, we searched for patterns of length 6–8 with δ ranging between 0–2 mismatches, and an allowed ball overlap factor of 0–1. Generally speaking, in these runs the patterns found with no mismatches or ball overlaps had better p-values. This happens because we search for relatively short patterns, allowing for a non-trivial probability of a random match. For this reason we report below only results with exact matches and no overlap. We believe that higher values of both parameters will be useful for longer patterns (say of length 12 or 13). In the second stage we run the EM-like procedure described above on all the patterns that received significant scores. We chose to learn PSSMs of width 20 using 15 iterations of our procedure.

To compare the results of these two stages, we ran MEME (version 3.0.3) in two configurations. The first restricted MEME to retrieve only short patterns of width 6–8, corresponding to our ℓ-mers stage. The second configuration used MEME's own defaults for pattern retrieval resembling our end product PSSMs.

We applied our procedure to several data sets from the recent literature. Selected results are summarized in Table 1. In this table we rank the top results from the different

Table 1. Selected results on binding site regions of several yeast data sets, comparing our findings with those of MEME.

Source/ Cluster	Trans. Factor	Consensus	Seed rank	Seed p-value	PSSM rank	PSSM p-value	MEME ≤ 8 rank	MEME ≤ 8 e-value	MEME ≤ 50 rank	MEME ≤ 50 e-value
Spellman et al. [22]										
CLN2	MBF	ACGCGT	1	4e-26	1	3e-42	1	1e-18	1	7e-31
SIC1	SWI5p	CCAGCA	1	1e-07	1	1e-12	1	8e-00	8	5e+02
Tavazoie et al. [23]										
3	putative	GATGAG	2	9e-07	5	6e-09	4	1e+06	2	1e-14
	putative	GAAAAatT	3	4e-07	2	1e-11	23	8e+07	3	7e-10
8	STRE	aAGGgG	1	6e-07	3	4e-06	20	1e+08	–	–
14	putative	TTCGCGT	1	2e-09	2	7e-11	13	1e+07	–	–
	putative	TGTTTgTT	3	2e-07	–	–	–	–	13	4e+05
30	MET31/32p	gCCACAgT	1	2e-11	1	2e-11	2	5e+02	8	1e+03
Iyer et al. [16]										
MBF	MBF	ACGCGT	1	1e-12	1	3e-18	3	1e+04	19	1e-03
SBF	SBF	CGCGAAA	1	1e-32	1	1e-37	2	1e-17	–	–

runs of each procedure by their p-values (or e-values) reported by the programs after removing repeated patterns. We report the relative rank of the patterns singled out in the literature and their significance scores. We discuss these results in order.

The first data set is by Spellman *et al.* [22]. They report several cell-cycle related clusters of genes. In a recent paper, Sinha and Tompa [21] report results of a systematic search for binding sites in these clusters of IUPAC consensus regions using a random sequence null hypothesis utilizing a Markov chain of order 3. The main technical developments in [21] are methods for approximating the p-value computation with respect to such a null-hypothesis.

We examined two clusters reported on by Sinha and Tompa. In the first one, CLN2, our method identifies the pattern ACGCGT and various expansions of it. This pattern was found using patterns of length 6, 7, and 8 with significant p-values. The PSSMs learned from these patterns were quite similar, all containing the above motif. Figure 1(a) shows an example. In the second cluster, SIC1, the signal appears with a marginal p-value (close to the Bonfferoni cutoff) already at $\ell = 6$. The trained PSSM recovers the longer pattern with a significant p-value. In both cases, the top ranking patterns correspond to the known binding site.

The second data set is by Tavazoie *et al.* [23]. That paper also examines cell-cycle related expression levels that were grouped using k-means clustering. They examined 30 clusters, and applied an MCMC-based procedure for finding PSSM patterns in the promoter regions of genes in each cluster. We examined the clusters they report as

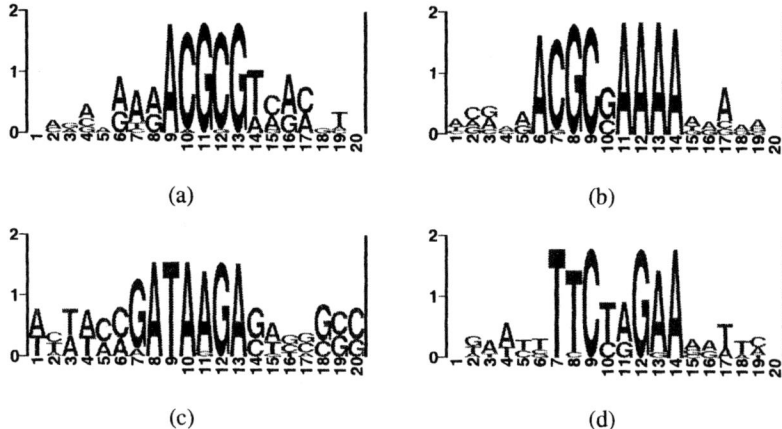

Fig. 1. Examples of PSSMs Learned by Our Procedure. (a) CLN2 cluster. (b) SBF cluster. (c) Gasch *et al.* Cluster M. (d) Gasch *et al.* Cluster I/J.

statistically significant, and were able to reproduce binding sites that are very close to the PSSMs they report; see Table 1.

In a recent paper, Iyer *at al.* [16] identify, using experimental methods, two groups of genes that are regulated by the MBF/SBF transcription factor. Here, again, we managed to recover the binding sites they discuss with high confidence. For example, we show one of our matching PSSMs in Figure 1(b).

Finally, we discuss the recent data set of yeast response to environmental stress by Gasch *et al.* [14]. We report on two clusters of genes "M", and "I/J". In cluster M the string CACGTGA is found in several of the highest scoring patterns. However, when we turned to grow PSSMs out of our seeds, a matrix of a lower ranking seed GATAAGA exceeded the rest, exemplifying that seed ordering is not necessarily maintained when the patterns are extended. The latter, more prominent PSSM is shown in Figure 1(c). In cluster I/J a significant short pattern rising above our threshold is not found. However when we extended the top most seed we obtained the PSSM of Figure 1(d) which both nearly crosses our significance threshold, and holds biological appeal, showing two conserved short regions flanking a less conserved 2-mer.

In general, the scores of the learned PSSMs vary. In some cases, the best seeds yield the best scoring PSSMs. More often, the best scoring PSSM corresponds to a seed lower in the list (we took into account only seeds that have p-value matching the FDR decision threshold). In most cases the PSSM learned to recognize regions flanking the seed sequence. In some cases more conserved regions were discovered. In general our approach manages to identify short patterns that are close to the pattern in the data. Moreover, using our PSSM learning procedure we are able to expand these into more expressive patterns.

We note that in most analysed cases MEME also identified the shorter patterns. However, there are two marked differences. First and foremost is run time. Compared on a 733 MHz Pentium III Linux machine our seed discovery programs ran between half a minute and an hour, exhaustively examining all possible patterns, while the EM-like PSSM growing iterations added a couple of minutes. The shortest MEME run on the same data sets took about an hour, while longer ones ran for days, when asked to return only the top thirty patterns. Second, MEME often gave top scores to spurious patterns that are clear artifacts of the sequence distributions in the promoter regions (such as poly A's). When using MEME one can try to avoid these problems by supplying a more detailed background model. This has the effect of removing most low complexity patterns from the top scoring ones. Our program avoids most of these pitfalls by performing its significance tests with respect to the genome background to begin with.

6   Discussion

In this paper we examined the problem of finding putative transcription factor binding sites with respect to a selected group of genes. We advocate significance calculations with respect to the *random selection* null hypothesis. We claim that this hypothesis is both simple and clear and is more suitable for gene expression experiments than the *random sequence* null hypothesis. We then use a simple hyper-geometric test in a framework for constructing models of binding sites. This framework starts by *systematically* scanning a family of simple "seed" patterns. These seeds are then used for building PSSMs. We describe how to construct statistical tests to select the most surprising threshold value for a PSSM and combine this with an EM-like iterative procedure to improve it. We thus combine a first phase of kernel identification based on a rigorous statistical analysis of word over-representation, with a subsequent phase of optimization, leading to a PSSM, which can be used to scan sequences for new matches of the putative regulatory motif.

We showed that even before performing iterative optimization of the PSSMs, our method recovers highly selective seed patterns very rapidly. We reconstructed results from several recent papers that use more elaborate and computationally intensive tools for finding binding sites, as well as present novel binding sites.

A potential weakness of our model is the fact that we disregard multiple copies of a match in the same sequence (the restriction to binary events). Despite the fact that this phenomenon is known to happen in eukaryotic genes, we recall that a mathematical analysis of counting the number of occurrences in a single string is more elaborate, and computationally intensive. This may indeed lead in such cases to under-estimation, which is problematic mainly for small clusters of co-regulated genes. The recognition of two conserved patterns separated by a relatively long spacer (say of 10bp or more), resulting from a HTH motif or a dimer complex, can however be attacked by looking for proximity relationships between pairs of occurrences of different significant seeds.

As this field is showing an influx of interest, our work resembles several others in different aspects. We highlight only the most relevant ones.

The use of the hyper-geometric distribution in the context of finding binding sites is used by Jensen and Knudsen [17] to find short conserved subsequences of length 4–6 bp. They demonstrate the ability to reconstruct sequences, but suffer statistical problems when they consider longer ℓ-mers, due to the large number of competing hypotheses.

Already in Galas *et al.* [13], word statistics are used to detect over-represented motifs, and a definition of a general concept of "word neighborhood" is given similar to the ball definition we give here. However, the analysis there is restricted to over-representations at specific positions with respect to a common point of reference across all sequence, deeming it mostly appropriate for prokaryotic transcription or translation promoter region elucidation.

The general outline of our approach is similar to that of Wolferstetter *et al.* [27] and Vilo *et al.* [26]. Both search for over-represented words and try to extend them. Vilo *et al.* examine ℓ-mers of varying sizes that are identified by building a suffix tree for the promoter regions. Then, they use a binomial formula for evaluating significance. For the clustering they constructed, this resulted in a very large pool of sequences (over 1500). They use multiple alignment-like procedure for combining these ℓ-mers into longer consensus regions. Thus, to learn longer binding sites with variable position, they require overlapping subsequences to be present in the data. This is in contrast to our approach that uses PSSMs to extend the observed patterns, and so is more robust to highly variable positions that flank the conserved region.

Van Helden *et al.* [24] also use binomial approach. They try to take into consideration the presence of multiple copies of a motif in the same sequence, but suffer from resulting inaccuracies with respect to auto-correlating patterns. Our work can be seen as generalizing this approach in several respects, including the use of a hyper-geometric null model, the discussion of general distance functions and event space coarsening, and the iterative PSSM improvement phase.

There are several directions in which we can extend our approach, some of them embedding ideas from previous works into our context.

First, in order to estimate the sensitivity of our model it will be interesting to examine it on smaller, and known, gene families, as well as on synthetic data sets, as those advocated in [19]. Extending our empirical work beyond yeast should also provide new insights and challenges.

Our method treats the complete promoter region as a uniform whole. However, biological evidence suggests that the occurrence of binding sites can depend on the position within the promoter sequence [22]. We can easily augment our method by defining events on sub-regions within the promoter sequence. This will facilitate the discovery of subsequences specific to certain positions. Another biological insight already mentioned is the phenomena of two conserved patterns separated by a relatively long spacer. In the case of homeodimers we can easily expand our scope to handle events that require two appearances of the subsequence within the promoter region. Otherwise, we can try to extend our PSSMs further to flank the seed while weighting each column such as to allow for longer spacers between meaningful sub-patterns.

So far we have looked for contiguous conserved patterns within the binding site. More complex extensions involve defining new distance measures that incorporate preferences for more conserved positions in specific positions in the pattern, and random projection techniques, akin to [5], which will allow us to easily handle longer ℓ-mers. We can also further generalize our model by allowing ourselves to express our ℓ-mer centroids over the IUPAC alphabet. This allows both for a reduction of the event space and the natural incorporation of biological insight, as outlined above. Our current method for diluting the set of "covering" δ-balls is highly heuristic. Interesting theoretical issues include the formal criteria we should optimize in selecting this approximating set of δ-balls and how to efficiently optimize with respect to such a criterion. Finally, we intend to combine the putative sites we discover with learning methods that learn dependencies between different sites and between sites and other attributes such as expression levels and functional annotations [2].

Acknowledgments

The authors thank Zohar Yakhini for arousing our interest in this problem and for many useful discussions relating to this work, and the referees for pointing out further relevant literature. This work was supported in part by Israeli Ministry of Science grant 2008-1-99 and an Israel Science Foundation infrastructure grant. GB is also supported by a grant from the Ministry of Science, Israel. NF is also supported by an Alon Fellowship.

References

1. T.L. Bailey and Elkan C. Fitting a mixture model by expectation maximization to discover motifs in biopolymers. In *Proc. Int. Conf. Intell. Syst. Mol. Biol.*, volume 2, pages 28–36. 1994.
2. Y. Barash and N. Friedman. Context-specific Bayesian clustering for gene expression data. In *Proc. Ann. Int. Conf. Comput. Mol. Biol.*, volume 5, pages 12–21. 2001.
3. Y. Benjamini and Y. Hochberg. Controlling the False Discovery Rate: a practical and powerful approach to multiple testing. *J. Royal Statistical Society B*, 57:289–300, 1995.
4. A. Brazma, I. Jonassen, J. Vilo, and E. Ukkonen. Predicting gene regulatory elements in silico on a genomic scale. *Genome Res.*, 8:1202–15, 1998.
5. J. Buhler and M. Tompa. finding motifs using random projections. In *Proc. Ann. Int. Conf. Comput. Mol. Biol.*, volume 5, pages 69–76. 2001.
6. H. J. Bussemaker, H. Li, and E. D. Siggia. building a dictionary for genomes: identification of presumptive regulatory sites by statistical analysis. *PNAS*, 97(18):10096–100, 2000.
7. H. J. Bussemaker, H. Li, and E. D. Siggia. Regulatory element detection using a probabilistic segmentation model. In *Proc. Int. Conf. Intell. Syst. Mol. Biol.*, volume 8, pages 67–74. 2000.
8. S. Chu, J. DeRisi, M. Eisen, J. Mullholland, D. Botstein, P. Brown, and I. Herskowitz. The transcriptional program of sporulation in budding yeast. *Science*, 282:699–705, 1998.
9. J. DeRisi., V. Iyer, and P. Brown. Exploring the metabolic and genetic control of gene expression on a genomic scale. *Science*, 282:699–705, 1997.
10. R. Durbin, S. Eddy, A. Krogh, and G. Mitchison. *Biological Sequence Analysis : Probabilistic Models of Proteins and Nucleic Acids*. Cambridge University Press, 1998.
11. R. Durrett. *Probablity Theory and Examples*. Wadsworth and Brooks, Cole, California, 1991.
12. M.B. Eisen, P.T. Spellman, P.O. Brown, and D. Botstein. Cluster analysis and display of genome-wide expression patterns. *PNAS*, 95:14863–14868, 1998.
13. D. J. Galas, M. Eggert, and M. S. Waterman. Rigorous pattern-recognition methods for dna sequences: analysis of promoter sequences from *Escherichia coli*. *J. Mol. Biol.*, 186:117–28, 1985.
14. A. P. Gasch, P. T. Spellman, C. M. Kao, O. Carmel-Harel, M. B. Eisen, G. Storz, D. Botstein, and P. O. Brown. Genomic expression program in the response of yeast cells to environmental changes. *Mol. Bio. Cell*, 11:4241–4257, 2000.

15. T. R. Hughes, M. J. Marton, A. R. Jones, C. J. Roberts, R. Stoughton, C. D. Armour, H. A. Bennett, E. Coffey, H. Dai, Y. D. He, M. J. Kidd, A. M. King, M. R. Meyer, D. Slade, P. Y. Lum, S. B. Stepaniants, D. D. Shoemaker, D. Gachotte, K. Chakraburtty, J. Simon, M. Bard, and S. H. Friend. Functional discovery via a compendium of expression profiles. *Cell*, 102(1):109–26, 2000.

16. V. R. Iyer, C. E. Horak, C. S. Scafe, D. Botstein, M. Snyder, and P. O. Brown. Genomic binding sites of the yeast cell-cycle transcription factors sbf and mbf. *Nature*, 409:533 – 538, 2001.

17. L. J. Jensen and S. Knudsen. Automatic discovery of regulatory patterns in promoter regions based on whole cell expression data and functional annotation. *Bioinformatics*, 16:326–333, 2000.

18. C. E. Lawrence, S. F. Altschul, M. S. Boguski, J. S. Liu, R. F. Neuwald, and J. C. Wooton. Detecting subtle sequence signals: a Gibbs sampling strategy for multiple alignment. *Science*, 262:208–214, 1993.

19. P.A. Pevzner and S.H. Sze. Combinatorial approaches to finding subtle signals in dna sequences. In *Proc. Int. Conf. Intell. Syst. Mol. Biol.*, volume 8, pages 269–78. 2000.

20. F.P. Roth, P.W. Hughes, J.D. Estep, and G.M. Church. Finding DNA regulatory motifs within unaligned noncoding sequences clustered by whole-genome mRNA quantitation. *Nat. Biotechnol.*, 16:939–945, 1998.

21. S. Sinha and M. Tompa. A statistical method for finding transcription factor binding sites. In *Proc. Int. Conf. Intell. Syst. Mol. Biol.*, volume 8, pages 344–54. 2000.

22. P. T. Spellman, G. Sherlock, M. Q. Zhang, V. R. Iyer, K. Anders, M. B. Eisen, P. O. Brown, D. Botstein, and B. Futcher. Comprehensive identification of cell cycle-regulated genes of the yeast saccharomyces cerevisiae by microarray hybridization. *Mol. Biol. Cell*, 9(12):3273–97, 1998.

23. S. Tavazoie, J. D. Hughes, M. J. Campbell, R. J. Cho, and G. M. Church. Systematic determination of genetic network architecture. *Nat Genet*, 22(3):281–5, 1999. Comment in: Nat Genet 1999 Jul;22(3):213-5.

24. J. van Helden, B. Andre, and J. Collado-Vides. Extracting regulatory sites from the upstream region of yeast genes by computational analysis of oligonucleotide frequencies. *J. Mol. Biol.*, 281(5):827–42, 1998.

25. J. van Helden, A. F. Rios, and J. Collado-Vides. discovering regulatory elements in noncoding sequences by analysis of spaced dyads. *Nucl. Acids Res.*, 28(8):1808–18, 2000.

26. J. Vilo, A. Brazma, I. Jonassen, A. Robinson, and E. Ukkonen. Mining for putative regulatory elements in the yeast genome using gene expression data. In *Proc. Int. Conf. Intell. Syst. Mol. Biol.*, volume 8, pages 384–94. 2000.

27. F. Wolfertstetter, K. Frech, G. Herrmann, and T. Werner. Identification of functional elements in unaligned nucleic acid sequences by a novel tuple search algorithm. *Comput. Appl. Biosci.*, 12(1):71–80, 1996.

Comparing Assemblies Using Fragments and Mate-Pairs

Daniel H. Huson, Aaron L. Halpern, Zhongwu Lai, Eugene W. Myers,
Knut Reinert, and Granger G. Sutton

Informatics Research, Celera Genomics Corp.
45 W Gude Drive, Rockville, MD 20850, USA
Phone: +1-240-453-3356, Fax: +1-240-453-3324
Daniel.Huson@Celera.com

Abstract. Using current technology, large consecutive stretches of DNA (such as whole chromosomes) are usually assembled from short fragments obtained by shotgun sequencing, or from fragments and mate-pairs, if a "double-barreled" shotgun strategy is employed. The positioning of the fragments (and mate-pairs, if available) in an assembled sequence can be used to evaluate the quality of the assembly and also to compare two different assemblies of the same chromosome, even if they are obtained from two different sequencing projects. This paper describes some simple and fast methods of this type that were developed to evaluate and compare different assemblies of the human genome. Additional applications are in "feature-tracking" from one version of an assembly to the next, comparisons of different chromosomes within the same genome and comparisons between similar chromosomes from different species.

1 Introduction

Although current technology for DNA sequencing is highly automated and can determine large numbers of base pairs very quickly, only about (on average) 550 *consecutive* base pairs (bp) can be reliably determined in a single read [6]. Thus, a large consecutive stretch of source DNA can only be determined by "assembling" it from short fragments obtained using a *shotgun sequencing* strategy [5]. In a modification of this approach called *double-barreled* shotgun sequencing [1], larger clones of DNA are sequenced from both ends, thus producing *mate-pairs* of sequenced fragments with known relative orientation and approximate separation (typically, employing a mixture of 2kb, 5kb, 10kb, 50kb and 150kb clones). So, usually a sequencing project produces a collection of fragments that are randomly sampled from the source sequence. The average number x of fragments that cover any given position in the source sequence is known as the *fragment x-coverage*.

Given two different assemblies of the same chromosome-sized source sequence, possibly obtained from two different sequencing projects, how can one evaluate and compare them? The aim of this paper is to present some fast and simple methods addressing this problem that are based on fragment and mate-pair data obtained in a sequencing project for the source sequence. Additional applications are in tracking

O. Gascuel and B.M.E. Moret (Eds.): WABI 2001, LNCS 2149, pp. 294–306, 2001.

forward "features" from one version of an assembly to the next, comparison of different chromosomes from the same genome and of similar chromosomes from different species. Although each method on its own is just an implementation of a simple idea or heuristic, our experience is that the integration of these methods gives rise to a powerful tool. We originally developed this tool to compare different assemblies of the human genome, see Figures 6 and 7 in [7].

In Section 2 we discuss assembly evaluation and comparison techniques based on fragments. In particular, we introduce the concept of "segment discrepancy" that measures by how much the positioning of a segment of conserved sequence differs between two assemblies. Then we present some mate-pair based methods in Section 3, including a useful breakpoint detection heuristic. Finally, we demonstrate the utility of these methods in Section 4.

2 Fragment-Based Analysis and Comparison Methods

Several useful methods for evaluating a single assembly or comparing two assemblies—such as sequencing coverage, dot-plots, or line-plots—can be implemented in terms of the positions in an assembly to which fragments are assigned.

For our purposes, a *contig* is simply a finite string $A = a_1 a_2 \ldots$ of characters $a_i \in \{A, C, G, T, N\}$ representing a stretch of contiguous DNA, where A, C, G and T correspond to the four bases and N stands for "unknown". An *assembly* is a contig A that was obtained from the fragments of some sequencing project using some assembly algorithm, without elaborating on the details. A run of consecutive N's represents an undetermined sequence part, and the number of N's in the run is sometimes used to represent its estimated length.

A fragment is a string $F = f_1 f_2 \ldots$ of characters $f_i \in \{A, C, G, T\}$, of length $len(F)$ usually less than 900. We say that a fragment F *hits* (or is *recruited by*) an assembly A if F globally aligns to A with high identity (e.g. 94% or more). In this case, we use $s(F, A)$ and $t(F, A)$ to denote the position in A to which the first character and last character of F align to, respectively. In particular, a fragment aligns in the forward direction if $s(F, A) < t(F, A)$, whereas the alignment is against the reverse-complement of F if $s(F, A) > t(F, A)$. For simplicity, we will assume that all s values are distinct, i.e., $s(F) \neq s(G)$ for any two different fragments that hit A. (In practice, fragment coordinates do sometimes agree, but our experience is that one can simply ignore such fragments without a substantial loss of coverage.)

Given a set of fragments $\mathcal{F}$ and an assembly A, we use $\mathcal{F}(A)$ to denote the set of all fragments in $\mathcal{F}$ that hit A. If an assembly A was obtained by assembling fragments from a set $\mathcal{F}$, then the set $\mathcal{F}(A)$, and the values of $s(F, A)$ and $t(F, A)$ for all $F \in \mathcal{F}(A)$, are known. If an assembly A of a chromosome is obtained from one sequencing project, and the set of fragments $\mathcal{F}$ available was obtained from a different sequencing project studying the same chromosome, then a fast high-fidelity alignment program [4] can be used to compute $\mathcal{F}(A)$.

2.1   Fragment-Coverage Plot

Let A be an assembly and $\mathcal{F}(A)$ a set of fragments that hit A. For each fragment $F \in \mathcal{F}(A)$ define a *begin-event* $(\min\{s(F, A), t(F, A)\}, +1)$ and an *end-event* $(\max\{s(F, A), t(F, A)\}, -1)$. To obtain a *fragment-coverage plot* for A, consider all events (x, e) in order of their first coordinate x and for each begin-event, plot the number of fragments that span x, given by the number of begin-events minus the number of end-events seen so far, see Figure 1.

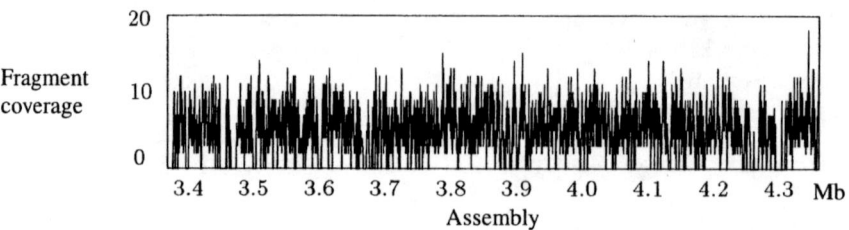

Fig. 1. Fragment-Coverage Plot for a 1 Mb Region of Chromosome 2 of Human [7]. The assembly A is represented by a line segment $[1, \mathrm{len}(A)]$ along the x-axis. The number of fragments uniquely hitting A is plotted as a function of their position.

A fragment-coverage plot is useful because poorly assembled regions often have low fragment-coverage, whereas regions of repetitive sequence can be identified as those stretches of sequence that are hit by unusually high numbers of fragments.

In practice, one can easily accomodate for fragments hitting multiple times. However, for ease of exposition, throughout this paper we will assume that $\mathcal{F}(A)$ is the set of all fragments that *uniquely* hit A.

2.2   Dot-Plot and Line-Plot

Consider two different assemblies A and B of the same chromosome, and assume that a set $\mathcal{F}$ of fragments obtained from a shotgun sequencing project for the chromosome is given. Once we have determined $\mathcal{F}(A)$ and $\mathcal{F}(B)$, how can we visualize this data?

Let $\mathcal{F}(A, B) := \mathcal{F}(A) \cap \mathcal{F}(B)$ denote the set of fragments that hit both assemblies. A simple dot-plot can be produced by plotting (x, y) with $x := s(F, A)$ and $y := s(F, B)$ for all $F \in \mathcal{F}(A, B)$, see Figure 2; at higher resolution, plot a line from $(s(F, A), s(G, B))$ to $(t(F, A), t(G, B))$. Alternatively, represent assembly A and B by a line segment from $(1, 0)$ to $(\mathrm{len}(A), 0)$ and from $(1, 1)$ to $(\mathrm{len}(B), 1)$, respectively. A simple line-plot showing matching regions of the two assemblies is obtained by drawing a line segment between $(s(F, A), 0)$ and $(s(F, B), 1)$ for all $F \in \mathcal{F}(A, B)$, see Figure 3.

If $\mathcal{F}(A)$ is given, but $\mathcal{F}(B)$ is unknown, then a short-cut to recruiting fragments to B is to compute $\mathcal{F}_A(B) := \{F \in \mathcal{F}(A) \mid F \text{ hits } B\}$ instead of $\mathcal{F}(B)$, at the price of obtaining a less comprehensive analysis. Alternatively, one could first compare the

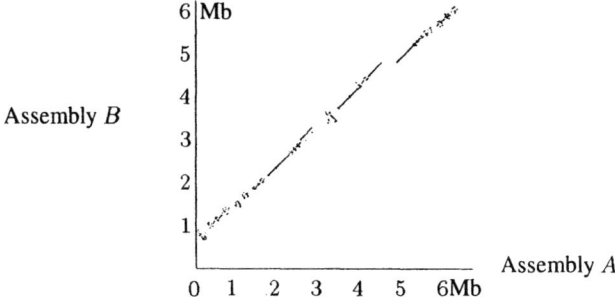

Fig. 2. Fragment based dot-plot comparison of two different assemblies of a 6Mb region of chromosome 2 in human. Each point represents a fragment that hits both assemblies.

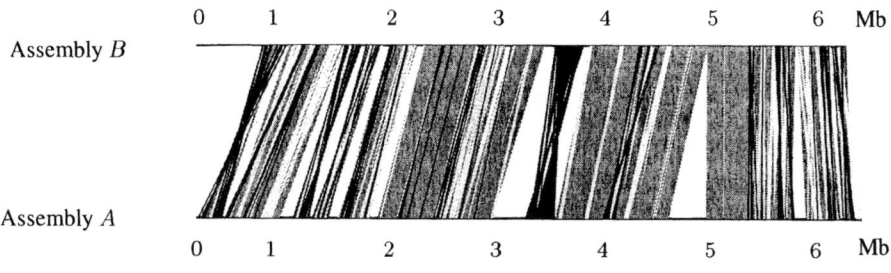

Fig. 3. Fragment Based Line-Plot Comparison. Each line segment represents a fragment that hits both assemblies. Medium grey lines represent fragments contained in the heaviest common subsequence (HCS) of consistently ordered and oriented segments, light grey lines represent consistently oriented segments that are not contained in the HCS, and dark grey lines represent fragments (or segments) that have opposite orientation in the two assemblies.

consensus sequence of assembly B directly against that of assembly A and then project fragments from A onto B wherever compatible with the segments of local alignment between A and B.

2.3 Fragment Segmentation

For analysis purposes and also to speed up visualization significantly, it is useful to segment the fragment matches by determining the maximal consistent and consecutive runs of them.

Consider a fragment $F \in \mathcal{F}(A, B)$. We say that F has *preserved orientation*, if and only if F has the same orientation in A and B, i.e., if either both $s(F, A) < t(F, A)$ and $s(F, B) < t(F, B)$, or both $s(F, A) > t(F, A)$ and $s(F, B) > t(F, B)$ hold. Let $\mathcal{F}^+(A, B)$ denote the set of all fragments that have preserved orientation and set $\mathcal{F}^-(A, B) := \mathcal{F}(A, B) \setminus \mathcal{F}^+(A, B)$.

For any two fragments $F, G \in \mathcal{F}(A, B)$, define $F <_A G$, if $s(F, A) < s(G, A)$, and define $F <_B G$, if $s(F, B) < s(G, B)$. Because we assume that all s values are distinct, these are both total orderings and we use $\text{pred}_A(F)$ and $\text{succ}_A(F)$ to denote the $<_A$-predecessor and $<_A$-successor of F, respectively.

A sequence $S = (F_1, F_2, \ldots, F_k)$ of fragments is called a *matched segment*, in either of the two following cases:

1. $\{F_1, F_2, \ldots, F_k\} \subseteq \mathcal{F}^+(A, B)$ and $\text{succ}_A(F_i) = \text{succ}_B(F_i)$ for all $i = 1, 2, \ldots, k - 1$, or
2. $\{F_1, F_2, \ldots, F_k\} \subseteq \mathcal{F}^-(A, B)$ and $\text{succ}_A(F_i) = \text{pred}_B(F_i)$ for all $i = 1, 2, \ldots, k - 1$.

A matched segment is called *maximal*, if it can't be extended.

Let $S := \mathcal{S}(\mathcal{F}(A, B)) = \{S_1, S_2, \ldots, S_n\}$ denote the set of all maximal matched segments of $\mathcal{F}(A, B)$, and let $\mathcal{S}^+$ and $\mathcal{S}^-$ denote the subset of such segments in cases 1 and 2, respectively. Both $\mathcal{S}^+$ and $\mathcal{S}^-$ can be computed in a simple loop that considers each fragment in $<_A$ order and decides whether it extends the current segment or defines the start of a new one.

The *A-support* of a matched segment $S = (F_1, F_2, \ldots, F_k)$ is defined as the interval $[s(S, A), t(S, A)]$, with $s(S, A) := \min_{F \in S}(s(F, A), t(F, A))$ and $t(S, A) := \max_{F \in S}(s(F, A), t(F, A))$. The *B-support* is defined similarly. Let $\text{len}(S)$ denote the minimum length of the A- and B-supports of S.

2.4   Heaviest Common Subsequence

Given two orderings O_1 and O_2 of the set of numbers $\{1, 2, \ldots, n\}$ (for some fixed number n) and a weight function $w : \{1, 2, \ldots, n\} \to \mathbb{N}^{\geq 0}$. A subsequence $H := H(O_1, O_2, w)$ of both orderings is called a *heaviest common subsequence*, if it has maximal weight $w(H) := \sum_{h \in H} w(h)$. The heaviest common subsequence can be computed in $O(n \log n)$ time and space, see [3].

For $S = (S_1, S_2, \ldots, S_n)$, let O_1 and O_2 denote the ordering of the indices $1, 2, \ldots, n$ induced by the orderings of S defined by $s(\cdot, A)$ and $s(\cdot, B)$, respectively. With weight function $w(i) := \text{len}(S_i)$, compute the heaviest common subsequence H of O_1 and O_2.

We call $\mathcal{H} := \{S_i \in S \mid i \in H\}$ the *heaviest common subsequence of matched segments*. We can distinguish between four categories of matched segments:

1. $\mathcal{S}^+ \cap \mathcal{H}$ is the set of segments that have the same ordering and orientation in both assemblies,
2. $\mathcal{S}^- \cap \mathcal{H}$ is the set of segments that have the same position in both assemblies, but are inverted with respect to each other,
3. $\mathcal{S}^+ \setminus \mathcal{H}$ is the set of segments that have transposed positions, and
4. $\mathcal{S}^- \setminus \mathcal{H}$ is the set of segments that appear both transposed and inverted.

The amount of sequence contained in each of these four categories is a good measure of how similar two assemblies are. In visualization, using different colors for each of them significantly enhances the dot-plot and line-plot representation described above, see Figure 3.

2.5 Segment Displacement

Consider two segments $S = (F_1, F_2, \ldots)$ and $T = (G_1, G_2, \ldots)$. We say that S and T are *parallel* if either both $s(F_1, A) < s(G_1, A)$ and $s(F_1, B) < s(G_1, B)$, or both $s(F_1, A) > s(G_1, A)$ and $s(F_1, B) > s(G_1, B)$ hold.

It seems reasonable to "trust" those portions of the two assemblies that are covered by segments from the heaviest common subsequence $\mathcal{H}$. Thus, we propose to measure the amount by which the positioning of a segment S not in $\mathcal{S}^+ \cap \mathcal{H}$ differs in the two assemblies as follows: We define the *displacement* $D(S)$ associated with S as the sum of lengths of all segments in $\mathcal{H}$ that are not parallel to S. In Figure 4 we plot segment length vs. segment displacement.

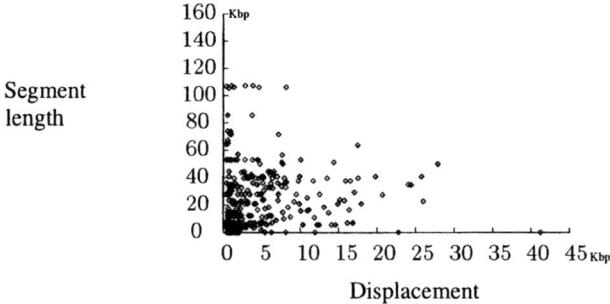

Fig. 4. Scatter-Plot Comparison of Two Assemblies: a dot (x, y) represents a sequence segment S of length $\text{len}(S) = x$ whose displacement $D(S)$ is y. In other words, the placement of S in the two assemblies differs by at least $D(S)$ bp. Note that points along the x-axis correspond to in-place inversions.

3 Mate-Pair-Based Evaluation Methods

Let A and B be two assemblies of a chromosome and let $\mathcal{F}$ be a set of associated fragments. Assume now that the fragments in $\mathcal{F}$ were generated using a "double-barreled" shotgun protocol in which *mate-pairs* of fragments are obtained by reading both ends of longer clones. For purposes of this paper, a *mate-pair library* $M = (L, \mu, \sigma)$ consists of a list L of pairs of *mated* fragments, together with a mean estimate μ and standard deviation σ for the length of the clones from which the mate-pairs were obtained, see Figure 5.

Typical clone sizes used to produce mate-pair libraries used in Celera's human genome sequencing were 2kb, 10kb, 50kb, and 150kb. The quality of shorter mate-pairs can be very good with a standard deviation of about 10% of the mean length, whereas the standard deviation can reach 20% for long clones. Also, because both ends of clones are read in separate sequencing reactions, there is a potential for mis-associating mates.

Source sequence

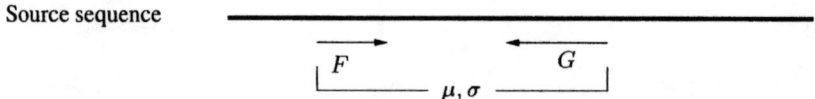

Fig. 5. Two fragments F and G that form a mate-pair with known mean distance μ and standard deviation σ. Note their relative orientation in the source sequence.

However, a high level of automation and electronic sample tracking can reduce the occurrences of this problem to below 1%. By construction, any fragment will occur in at most one mate-pair.

Given an assembly A with fragments $\mathcal{F}(A)$ and a collection of mate-pair libraries $\mathcal{M} = \{M_1, M_2, \ldots\}$, let $m = \{F, G\} \subset \mathcal{F}(A)$ be a mate-pair occurring in some library $M_i = (L, \mu, \sigma)$. Then m is called *happy* if the positioning of F and G in A is reasonable, i.e., if F and G are oriented towards each other (as in Figure 5) and $\mid |s(F, A) - s(G, A)| - \mu \mid \le 3\sigma$, say. An unhappy mate-pair m is called *mis-oriented* if the former condition is not satisfied, and *mis-separated* if only the latter condition fails.

3.1 Clone-Middle Plot

We obtain a *clone-middle plot* for A as follows: For each pair of fragments $F, G \in \mathcal{F}(A)$ that occurs in a mate-pair library M, draw a line segment from $(t(F, A), y)$ to $(t(G, A), y)$, where $y \in [0, 1]$ is a randomly chosen height. Lines can be shown in different colors depending on whether the corresponding mate-pair is happy, mis-separated or mis-oriented, see Figure 6, and also Figure 6 in [7]. The interval $[t(F, A), t(G, A)]$ (assuming w.l.o.g. $t(F, A) < t(G, A)$) is called the *clone-middle* (in A) associated with the pair F, G.

One draw-back of this visualization for large assemblies is that substantially misplaced pairs give rise to very long lines in the plot and obscure the view of local regions. To address this, we introduce the *localized* clone-middle plot (see Figure 7): Let $\{F, G\}$ be a mis-separated or mis-oriented mate from some library $M = (L, \mu, \sigma)$. Assume w.l.o.g. that $s(F, A) < s(G, A)$. Represent the mate-pair by a line that indicates the range in which F expects to see G, i.e., by drawing a line segment from $t(F, A)$ of length $\mu + 3\sigma - (\text{len}(F) + \text{len}(G))$ towards the right, if $s(F, A) < t(F, A)$, and to the left, otherwise. As above, define the *clone-middle* accordingly.

Mis-separated and mis-oriented mate-pairs indicate discrepancies between a given assembly and the original source sequence or chromosome, as follows.

3.2 Breakpoint Detection

Loosely speaking, a *breakpoint* of an assembly A is a position p in A such that the sequence immediately to the left and right of p in A comes from two separate regions of the source sequence.

Let $m = \{F, G\}$ be a mis-oriented mate-pair such that $s(F, A) < s(G, A)$. We distinguish between three different cases: *normal*-oriented: both fragments are oriented to the right; *anti*-oriented: both are oriented to the left; and *outtie*-oriented: F is oriented

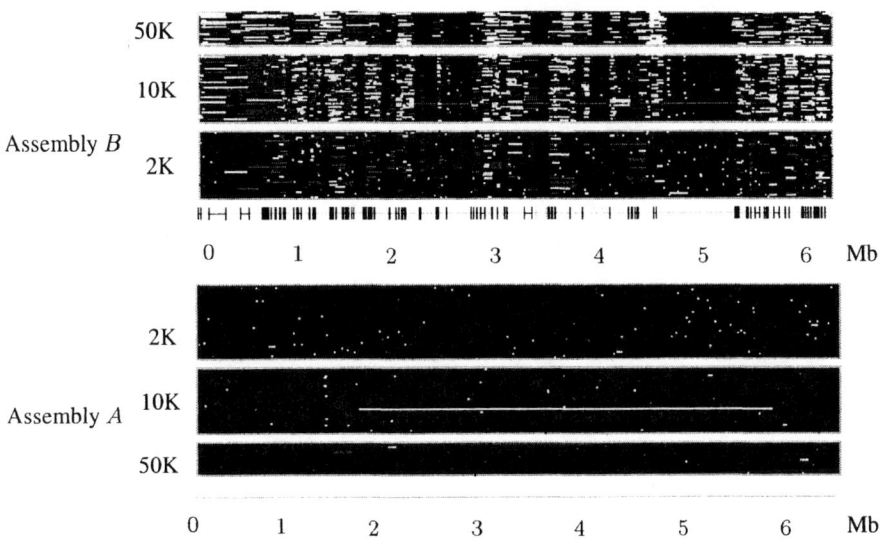

Fig. 6. Clone-Middle Diagram for Assemblies A and B. Each mate-pair m is represented by a horizontal line segment joining its two fragments, if m is mis-separated (shown in light grey) or mis-oriented (shown in dark grey). Happy mates are not shown. Mate-pairs are grouped by "library", labeled $2K$, $10K$ and $50K$. Ticks along the axis indicate putative breakpoints, as inferred from the mis-oriented mates.

to the left and G is oriented to the right. (Happy and mis-separated mates are *innie*-oriented).

We now describe a simple but effective heuristic for detecting breakpoints. Choose a threshold $T > 0$, depending on details of the sequencing project. (All figures in this paper were produced using $T = 5$.) An *event* is a three-tuple (x, t, a) consisting of a coordinate $x \in \{1, \ldots, \text{len}(A)\}$, a type $t \in \{\text{normal}, \text{anti}, \text{outtie}, \text{mis-separated}\}$, and an "action" $a \in \{+1, -1\}$, where $+1$ or -1 indicates the beginning or end of a clone-middle, respectively. We maintain the number of currently *alive* mates $V(t)$ of type t. For each event $e = (x, t, a)$ in ascending order of coordinate x: If $a = +1$, then increment $V(t)$ by 1. In the other case $(a = -1)$, if $V(t) \geq T$, then report a breakpoint at position x and set $V(t) = 0$, else decrease $V(t)$ by 1. (For a better estimation of the true position of the breakpoint, report the interval $[x', x]$, where x' is the coordinate of the most recent alive $+1$-event of type t.) Breakpoints estimated in this way are shown in Figure 7.

A useful variant of the breakpoint estimator is obtained by taking the current number of alive happy mates into account: Scanning from left to right, a breakpoint is said to be present at position x if there exists an event $e = (x, t, -1)$ such that the number of alive unhappy mates of type t exceeds the number of alive happy mates of type t.

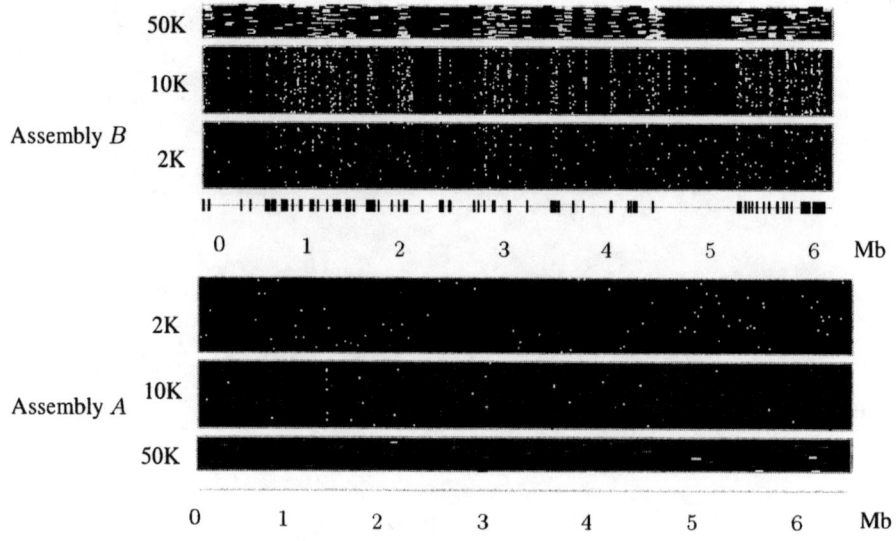

Fig. 7. A Localized Clone-Middle Diagram for Assemblies A and B. Here, each mis-separated or mis-oriented mate-pair is represented by a line that indicates the expected range of placement of the right mate with respect to the left one. Ticks along the axis indicate putative breakpoints, as inferred from the mis-oriented mates.

3.3   Clone-Coverage Plot

Similar to the fragment-coverage plot discussed in Section 2, one can use the clone-coverage events to compute a *clone-coverage* plot for each of the types of mate-pairs, see Figure 8.

Note that the simultaneous occurrence of both high happy and high mis-separated coverage may indicate the presence of a polymorphism in the fragment data.

3.4   Synthesis

Combining all the described methods into one view gives rise to a tool that is very helpful deciding by how much two different assemblies differ and, more, which one is more compatible with the given fragment and mate-pair data; see Figure 9. This latter capability is an especially powerful aspect of analysis in terms of fragments and mate-pairs.

4   Some Applications

The techniques described in this paper have a number of different applications in comparative genomics. Originally, our goal was to design a tool for comparing the similarities and differences of assemblies of human chromosomes produced at Celera with

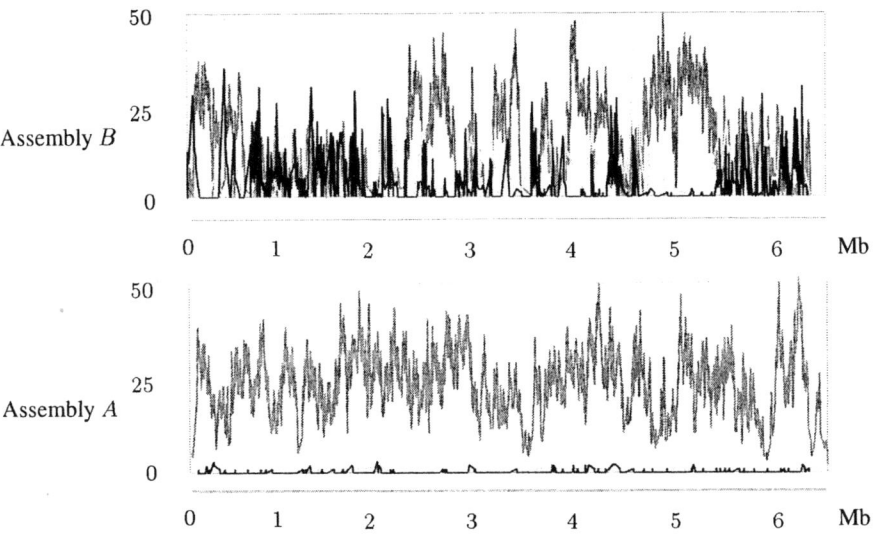

Fig. 8. Clone-coverage plot for assemblies A and B, showing the number of of happy mate-pairs (medium grey), mis-separated pairs (light grey) and mis-oriented ones (dark grey).

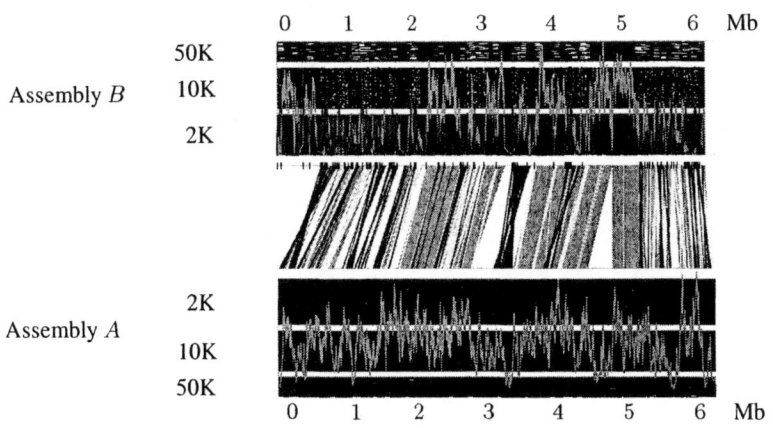

Fig. 9. A combined line-plot, clone-middle, clone-coverage and breakpoint view of the two assemblies A and B indicates that assembly A is significantly more compatible with the given fragment and mate-pair data than assembly B is.

those produced by the publicly funded Human Genome Project (PFP). A detailed comparison based on our methods is shown in Figures 6 and 7 of [7]. As an example, we show the comparison for chromosome 2 in Figure 10. For clarity, only segments of length 50kb or more are shown.

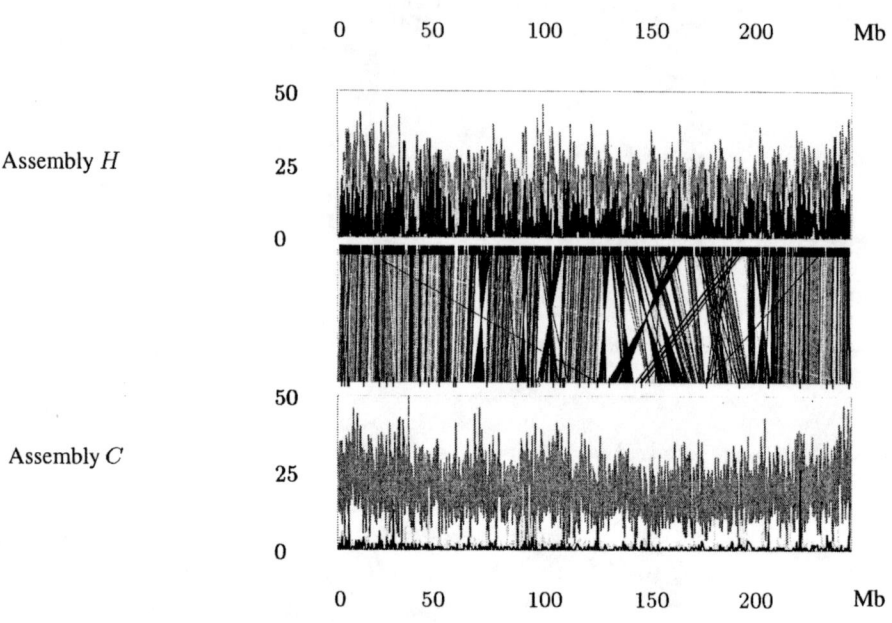

Fig. 10. Line-plot and breakpoint comparison of two different assemblies of chromosome 2 of human. Assembly C was produced at Celera [7] and assembly H was produced in the context of the publicly funded Human Genome Project and was released on September 5, 2000 [2]. The number of detected breakpoints (indicated as ticks along the chromosome axes) is 73 for C and 3592 for H.

4.1   Feature-Tracking

A second application is in tracking forward features from one version of an assembly to the next. To illustrate this, we consider two assemblies of chromosome 19 produced in the context of the PFP from publicly available data. Assembly H_1 was released on September 5, 2000 and assembly H_2 was released on January 9, 2001 [2].

How much did the assembly change and did it improve? The line-plot comparison of H_1 and H_2 in Figure 11 indicates that many local changes have taken place. A detailed analysis (not reported here) shows that many changes are due to a change of orientation of so-called "supercontigs" in the assembly. The number of detected breakpoints dropped from 723 to 488.

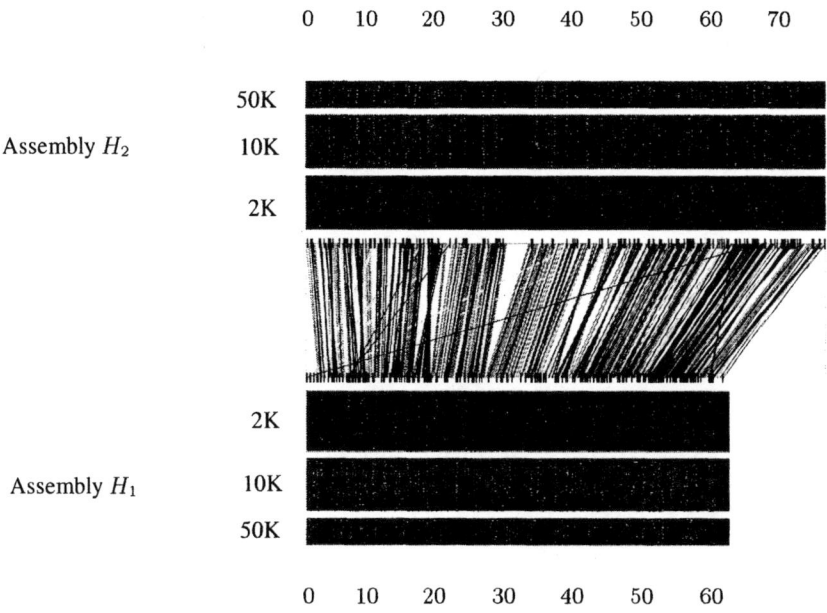

Fig. 11. Line-plot, clone-middle and breakpoint comparison of the PFP assembly H_1 of chromosome 19 as of September 5, 2000, and the a more recent PFP assembly H_2 dating January 9, 2001.

4.2 Comparison of Different Chromosomes

Additionally, our algorithms can be used to compare different chromosomes of the same species e.g. in search of duplication events, but also to compare different chromosomes from different species, in the latter case using a lower stringency alignment method to define fragment hits.

We illustrate this by a comparison of chromosome X and Y of human, as described in [7]. In this analysis we use only uniquely hitting fragments. In summary, we see approximately 1.3Mb of sequence in conserved segments, of which 164kb are contained in the heaviest common subsequence (relative to the standard orientation of X and Y), 82kb are contained in other segments of the same orientation and 1.05Mb in oppositely oriented segments, see Figure 12. We observe orientation preserving similarity at both ends of the chromosomes and a large inverted conserved segment in the interior of X.

References

1. A. Edwards and C.T. Caskey. Closure strategies for random DNA sequencing. *METHODS: A Companion to Methods in Enzymology*, 3(1):41–47, 1991.
2. D. Haussler. Human Genome Project Working Draft. http://genome.cse.ucsc.edu.
3. G. Jacobson and K.-P. Vo. Heaviest increasing/common subsequence problems. In *Proceedings 3rd Annual Symposium on Combinatorial pattern matching (CPM)*, pages 52–66, 1992.

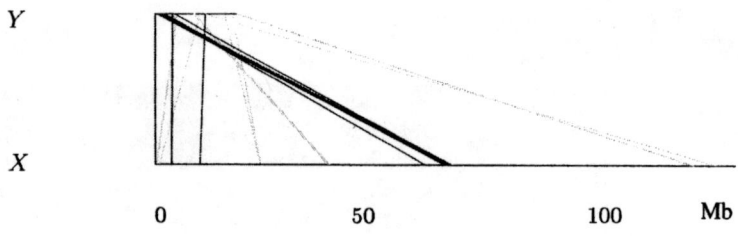

Fig. 12. A line-plot comparison of chromosome X vs. Y of human showing segments of highly conserved non-repetive sequence.

4. E. W. Myers, G. G. Sutton, A. L. Delcher, I. M. Dew, D. P. Fasulo, M. J. Flanigan, S. A. Kravitz, C. M. Mobarry, K. H. J. Reinert, K. A. Remington, E. L. Anson, R. A. Bolanos, H-H. Chou, C. M. Jordan, A. L. Halpern, S. Lonardi, E. M. Beasley, R. C. Brandon, L. Chen, P. J. Dunn, Z. Lai, Y. Liang, D. R. Nusskern, M. Zhan, Q. Zhang, X. Zheng, G. M. Rubin, M. D. Adams, and J. C. Venter. A whole-genome assembly of Drosophila. *Science*, 287:2196–2204, 2000.
5. F. Sanger, A. R. Coulson, G. F. Hong, D. F. Hill, and G. B. Petersen. Nucleotide sequence of bacteriophage λ DNA. *J. Mol. Bio.*, 162(4):729–73, 1992.
6. F. Sanger, S. Nicklen, and A. R. Coulson. DNA sequencing with chain-terminating inhibitors. *Proceedings of the National Academy of Sciences*, 74(12):5463–5467, 1977.
7. J. C. Venter, M. D. Adams, E. W. Myers, et al. The sequence of the human genome. *Science*, 291:1145–1434, 2001.

Author Index

Lecture Notes in Computer Science

For information about Vols. 1–2048
please contact your bookseller or Springer-Verlag